D0225834

Energetic Aspects of Muscle Contraction

MONOGRAPHS OF THE PHYSIOLOGICAL SOCIETY

A full list of titles in this series is given at the back of this book.

Monographs of the Physiological Society No. 41

Energetic Aspects of Muscle Contraction

ROGER C. WOLEDGE
Department of Physiology
University College London
London, England

NANCY A. CURTIN
Department of Physiology
Charing Cross and Westminster Medical School
London, England

EARL HOMSHER
Department of Physiology and
The Jerry Lewis Neuromuscular Research Center
University of California at Los Angeles
School of Medicine
Los Angeles, California
USA

1985

ACADEMIC PRESS
(Harcourt Brace Jovanovich, Publishers)
LONDON ORLANDO SAN DIEGO NEW YORK
TORONTO MONTREAL SYDNEY TOKYO

ACADEMIC PRESS INC. (LONDON) LTD.
24–28 Oval Road
LONDON NW1 7DX

United States Edition published by
ACADEMIC PRESS, INC.
Orlando, Florida 32887

British Library Cataloguing in Publication Data

Woledge, Roger C.
Energetic aspects of muscle contraction.----
(Monographs of the Physiological Society; no. 41)
1. Muscle contraction
I. Title II. Curtin, Nancy A.
III. Homsher, Earl IV. Series
612'.741 QP321

ISBN 0-12-761580-6

Woledge, Roger C.
Energetic aspects of muscle contraction.

Includes index.
1. Muscle contraction. 2. Energy metabolism.
I. Curtin, Nancy A. II. Homsher, Earl. III. Title.
[DNLM: 1. Muscle Contraction.
W1 MO569QW v.41 / WE 500 W852e]
QP321.W58 1985 599'.01852 84-20430
ISBN 0-12-761580-6 (alk. paper)

PRINTED IN THE UNITED STATES OF AMERICA

85 86 87 88 9 8 7 6 5 4 3 2 1

Contents

Preface

The physiological function of muscle is to convert energy from chemical reactions into mechanical work and heat. Therefore an understanding of muscle contraction requires consideration of this energy conversion, and this book is intended to present recent experiments relevant to this point of view. After the first chapter, which summarizes basic information for readers new to the study of muscle, we give in the next three chapters an account of the mechanical aspects of contraction, the chemical reactions that drive contraction, and the production of heat and work. We also describe experiments in muscle proteins *in vitro* and discuss how these can be compared with observations of muscle cells.

In these three chapters we have selected the experimental evidence for views that seem interesting to us, rather than either giving an exhaustive review of all possibly relevant work or presenting a summary of conclusions without the evidence on which they are based. Finally, some theories of contraction are described and considered in the light of the experiments on muscle outlined earlier in the book.

We would like to express our thanks to Professor D. K. Hill, F. R. S., for suggesting the theme of this book.

ROGER C. WOLEDGE
NANCY A. CURTIN
EARL HOMSHER

Introduction

A complete description of muscle contraction must account for the following aspects of contraction:

1. The mechanical properties, including the tension transients that accompany and follow changes in muscle fibre length
2. The biochemical behaviour of myosin and actin in muscle cells, including both steady state and transient rates of interaction of these proteins
3. The energy changes during contraction, including the reactions which supply the energy that is converted in muscle into heat and mechanical work
4. Control of contraction by calcium, particularly during the transition from the resting to the contracting state and during relaxation

Detailed theories of contraction, based on the crossbridge idea, have been proposed which attempt to explain (with varying degrees of complexity and success) some of the experimental facts about contraction. A few of these theories will be discussed later in the book; none of them is really complete, even in their explanation of the operation of crossbridges. Many questions about crossbridges and other aspects of contraction remain. How might these questions be answered? A single experimental approach will never be sufficient because the nature of the problem requires mechanical, biochemical, biophysical, and structural information. All of the following types of experiments seem necessary, as each has its own particular advantages, as well as limitations:

1. Studies of the mechanical properties of contracting fibres give the most direct information about force generation and work production by the crossbridges. During the past 20 years the use of techniques having finer time resolution has produced much new information about the mechanical behaviour of intact, isolated fibres. These techniques add primarily to the description of the force and movement, which are the "results" of coupling of chemical reactions to mechanical events, rather than the coupling process itself.
2. Experiments on skinned fibres offer in some respects a more direct

approach to the coupling process. For example, the concentration of ATP, ADP, and/or P_i in the solution surrounding the contractile proteins can be controlled. In this way the "driving process" can be manipulated and the response of the "driven process" (tension development or shortening) can be observed. However, this approach would be strenghened considerably if the extent of ATP splitting could be observed; recent studies with caged ATP seem very promising in this respect (Goldman *et al.*, 1982; Ferenczi *et al.*, 1983).

3. Experiments on isolated proteins in solution permit the study of single reactions under known and controlled chemical conditions. The results provide the most unambiguous evidence about the existence of certain intermediates, the rates of specific steps, and the energy changes accompanying them. They thus provide the information that forms the basis of detailed kinetic models for the actomyosin ATPase mechanism. The main limitation of such experiments is that the contractile proteins are free in solution and their interaction and reaction with ATP can neither convert chemical energy into mechanical work nor be constrained by mechanical forces. Thus, these studies cannot give much information about the coupling between chemical and mechanical processes which is central to contraction.

4. Measurements of energy changes during contraction of muscle have the advantage that they give information about both sides of the mechanochemical coupling process: chemical reactions that supply energy, and the resulting heat and mechanical work. The chemical reaction that has been for many years the main focus of attention in energetic studies is ATP hydrolysis. Observations of the extent of the reaction are useful because, after allowance has been made for nonactomyosin ATPase reactions, they give a measure of the number of completed crossbridge cycles. However, it now seems that the step in that cycle that splits ATP is different from the step during which work is done. The relation of the cycle to the amount of ATP split is therefore not a simple one.

Observations of energy output (heat and work) also have the advantage that they reflect all the processes occurring in the muscle. In addition to reactions like ATP hydrolysis, there are contributions from changes in the crossbridges themselves as they go through the various steps of the cycle. The fact that the energy output reflects a large number of reactions, some very undefined, has meant that the interpretation of heat measurements is complex and the literature is in some respects difficult. The situation is improved when the heat results are not viewed in isolation, but are considered along with results and ideas from other types of experiments. Because the heat and work production represents the sum of all the processes during contraction, a rigorous test of the success of a "complete" description of contraction would be how well it could predict the energy production.

This, of course, is not a complete list of the useful experimental approaches being used now, and new methods undoubtedly are being developed. In this book we bring together the results of different kinds of experiments which complement each other and we try to show how they add to our understanding of contraction. We begin with a very elementary summary of the structure and function of muscle and the contractile proteins, and some essential points about mechanical and energetic aspects of contraction. Most of the references in Chapter 1 are to books or reviews rather than the original publications of experimental evidence; these references are intended to be useful starting points for the beginner.

The reader already working on muscle will probably want to skip Chapter 1 completely or consult more detailed publications on particular topics, such as "The Structural Basis of Muscular Contraction" by Squire (1980) on structure, and "Reflections on Muscle" by A. F. Huxley (1980) on the relation of facts (mostly structural and mechanical) to theory. An excellent introductory account of muscle mechanics is given by Simmons and Jewell (1974). For a historical perspective on research in muscle contraction we recommend Needham's book, "Machina Carnis" (1971).

Most of the rest of the book is concerned with biochemical and energetic studies because we want to emphasize the contributions these methods make to understanding contraction. A description of the techniques used and a summary of the most important results are given, which should provide a useful background to the ideas and interpretation of these and also further experiments. Chapter 2 is concerned with a number of mechanical properties that are particularly relevant to energetics and theories of crossbridge operation. In Chapter 3 we consider the biochemical experiments on isolated actin and various forms of myosin and the kinetics models based on these results. The energetic aspects of contraction, ATP splitting and the production of heat and work, are described in Chapter 4. We give a brief account of a few influential theories of contraction in Chapter 5. We attempt to assess them in terms of the various kinds of experiments listed above, and also we point out some predictions that can be tested experimentally.

ROGER C. WOLEDGE
NANCY A. CURTIN
EARL HOMSHER

1
Initial Facts and Ideas

1.1 Structure of Muscle Cells

As an introduction to the rest of the book this chapter outlines the most important facts about the structure, physiology, and biochemistry of muscle and its constituent proteins and discusses these facts in terms of contraction mechanism. We concentrate on vertebrate skeletal muscle, which contains fibres that are large multinucleate cells between 10 and 200 μm in diameter and between 2 and 100 (or even more) mm in length. The cells are cross-striated at a periodicity of about 2.0 μm (at the slack length). The important components of these cells include the plasma membrane, the T tubules, the sarcoplasmic reticulum, the mitochondria, and, of course, the myofibrils themselves, which are responsible for the mechanochemical events of muscle contraction (Fig. 1.1). Each of these structures is described with some comments about its function.

1.1.1 PLASMA MEMBRANE

Each muscle fibre is surrounded by a continuous plasma membrane and is electrically separate from the other fibres. There are no gap junctions or tight junctions between fibres as there are in many tissues. The plasma membrane contains ion pumps responsible for maintaining the ionic composition of the inside of the cell. The action of these pumps and the selective permeability of the membrane to ions set the resting potential. In most muscle fibres the membrane can conduct an action potential. These fibres are called twitch fibres; in response to a sufficient stimulus either via the nerve or directly applied to the plasma membrane, the fibre produces a brief contraction, a twitch. In each twitch fibre there are just one or a few end plates, at which an action potential can be set up by the release of transmitter from the nerve terminal. Other fibres, found particularly in amphibians and reptiles, but also in a very few mammalian muscles, do not

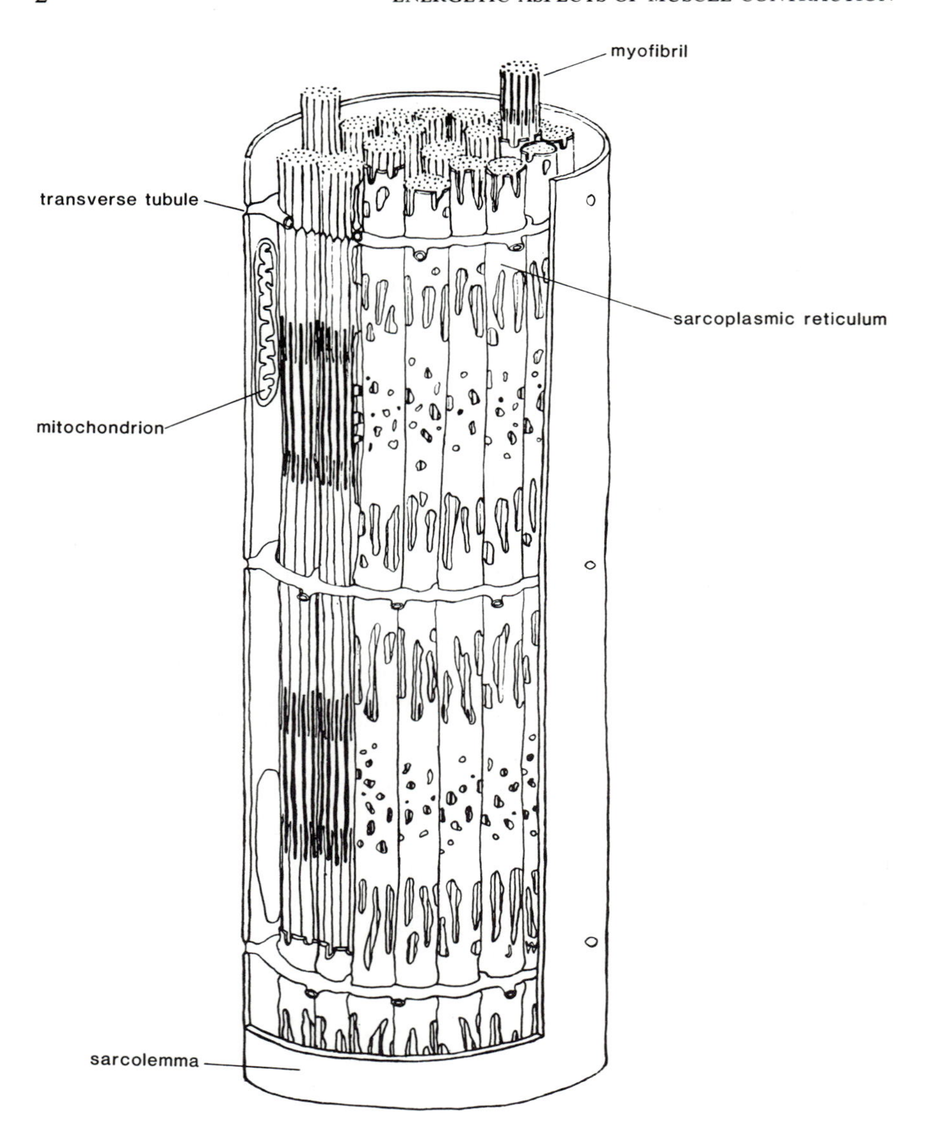

FIG. 1.1. A diagram of the relationship of organelles in a frog muscle fibre. (Partly based on the diagram by Peachey, 1965; drawing by J. Woledge.)

conduct action potentials (Kuffler and Vaughan Williams, 1953; Hess and Pilar, 1963; A. F. Huxley, 1964; Lannergren, 1975). In contrast to twitch fibres, these fibres have nerve terminals distributed all over their surface. At each nerve terminal, transmitter causes local depolarization. The depolarization resulting from a single nerve impulse is not sufficient to cause contraction, so these fibres do not twitch. In amphibians and reptiles these fibres are usually called slow fibres because their contractions are much slower than those of twitch fibres, but a better name would be non-twitch fibres; this would distinguish them from the slow fibres in mammalian muscle, which are mechanically slow but do conduct action potentials.

For more information about the electrical properties of the surface membrane the following short books should be consulted: "Nerve, Muscle and Synapse" by Katz (1966), and "Nerve and Muscle" by Keynes and Aidley (1981).

1.1.2 TRANSVERSE TUBULES

The transverse or T tubules are invaginations of the surface membrane which form a system of branched tubules about 0.04 μm in diameter running predominantly transversely into the fibres. As the T tubules are open at the surface of the cell their lumen can be considered part of the extracellular space (Fig. 1.2A). Large molecules such as ferritin or fluorescent dyes which do not readily cross the surface membrane enter the T tubules. The openings of these tubules are regularly arranged with a periodicity related to that of the myofibrils (Peachey, 1965). The position of the opening relative to the banding pattern depends on the muscle and species of animal (Smith, 1966). For example, in mammalian muscle the openings are at the level of the A–I junctions, whereas they are at the level of the Z line in frogs. A. F. Huxley (1971), in a review of this subject, describes how this difference was exploited in local stimulation experiments leading to discovery of the function of these tubules, which is to conduct the electrical changes causing activation into the fibre interior. This conduction is an active, regenerative mechanism (Costantin, 1975a,b); the tubules act like inside-out axons. The membrane capacitance of muscle is typically 5–10 $\mu F/cm^2$ surface area. This is greater than the value for a nerve fibre (1 $\mu F/cm^2$). This difference is probably due to the fact that the T tubules contribute to the observed capacity of muscle fibres (Falk and Fatt, 1964).

1.1.3 SARCOPLASMIC RETICULUM

The sarcoplasmic reticulum (SR) is a network of closed tubules of irregular shape surrounding each myofibril. The lumen is not continuous with the T tubules, or intracellular or extracellular space. Certain elements of the

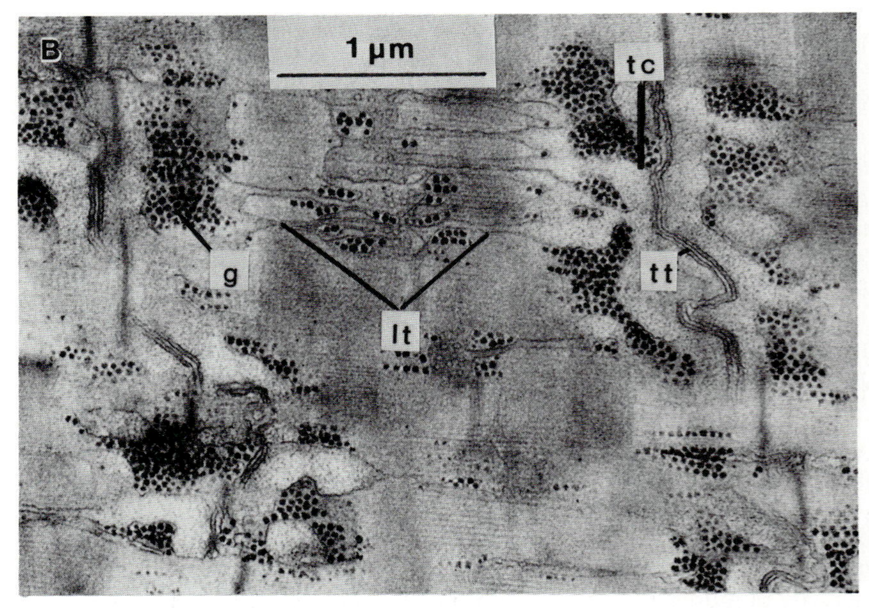

FIG. 1.2. (A) Electron micrograph of guinea pig psoas muscle prepared with lanthanum to show the T tubules (t) opening to the surface of the fibre. Sarcotubular elements (sr), filaments, Z lines, and glycogen granules are also visible. The distance between the Z lines is approximately 2.2 μm. (From Rayns *et al.*, 1968.) (B) Electron micrograph of a longitudinal section of frog muscle showing the T tubule (tt), the terminal cysternae (tc) closely apposed to it, and the longitudinal elements of the sarcoplasmic reticulum (lt). The dark particles are glycogen granules (g). (From Peachey, 1965.)

SR known as terminal cisternae are adjacent to the T system and form a structure called a triad (Fig. 1.2B). The SR can be isolated from muscle homogenates in the form of closed vesicles, and many biochemical studies have been done on these preparations. Under appropriate conditions these vesicles sequester calcium from the external solution by active transport. This process is one of the two *in vivo* functions of the SR (Endo, 1977). First, it reduces the calcium concentration around the myofibrils, and thus allows the muscle to relax. The energy for the active uptake of calcium comes from the hydrolysis of ATP. The SR splits ATP at a high rate, and its usage of ATP is an appreciable fraction of the total ATP used during contraction. The second function of the SR is to release calcium upon stimulation. This calcium, which activates the myofibrils, presumably comes from the terminal cisternae, the elements of the SR most closely associated with the T tubules. The release can be demonstrated by electron probe analysis of freeze-fixed tissues (Somlyo *et al.*, 1981) and by calcium indicators, such as aequorin and arsenazo III (Blinks *et al.*, 1976, 1982). The nature of the signal that is transmitted from the T tubule to the SR and that triggers this release is uncertain, but it is the subject of a number of intriguing lines of research. We are likely to learn much more about the electrophysiology of the SR in the next few years.

1.1.4 MITOCHONDRIA

Muscle mitochondria appear to be very similar in both structure and properties to those from other tissues. In most vertebrate skeletal muscles, mitochondria are not particularly numerous, compared, for instance, to insect flight muscle or cardiac muscle. However, there are significant differences in the mitochondrial content of fibres, and this difference is one of the characteristics used to distinguish different fibre types (Smith and Ovalle, 1973). The principal function of mitochondria is, of course, to provide a supply of energy to the muscle by means of oxidative phosphorylation of ADP. Their relative paucity is reflected in the fact that oxidative phosphorylation is not particularly rapid per gram of tissue. The muscle obtains a rapid supply of energy from anaerobic sources and then recharges these sources relatively slowly using oxidative phosphorylation. In some muscles a secondary role of mitochondria may be the uptake of calcium ions from the sarcoplasm. This is only likely to be important in muscles that have many mitochondria and a rather poorly developed SR (Gillis, 1972).

1.1.5 MYOFIBRILS

Most of the space within the muscle fibre is occupied by myofibrils, which cause the actual contractile behaviour of muscle. The myofibrils are about

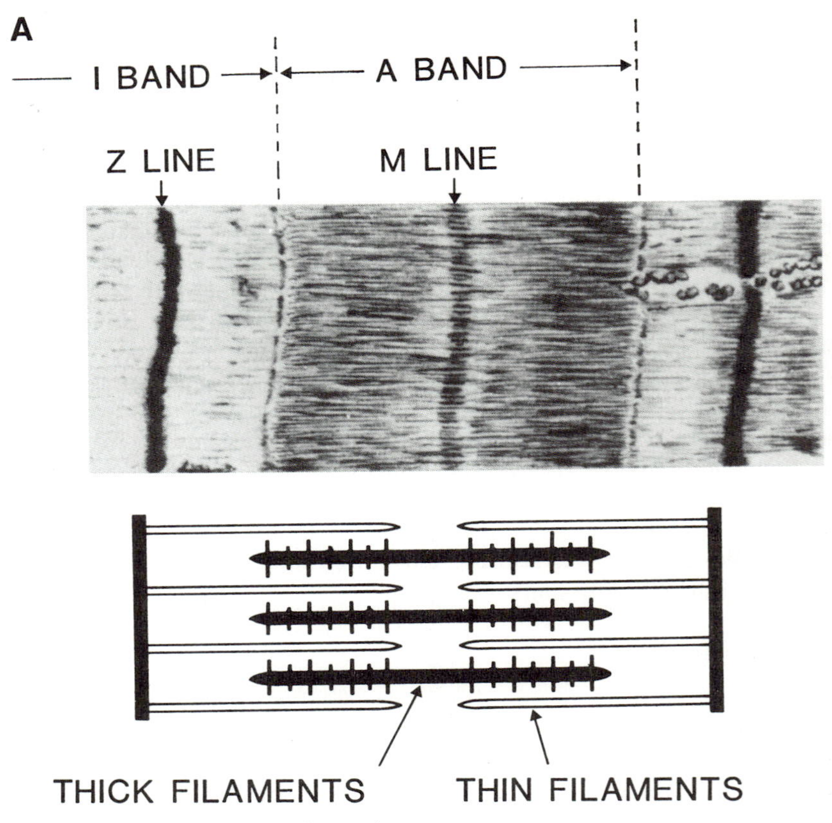

FIG. 1.3. (A) (see legend on opposite page).

1–2 μm in diameter (larger in nontwitch fibres) and run the length of the muscle fibre. The myofibrils are responsible for the striated appearance of the fibre because the striations of individual myofibrils are in register. These striations are due to the regular arrangement of two kinds of filaments within the myofibrils: the thin (or I) filaments and the thick (or A) filaments (Fig. 1.3A). The region containing only the thin filaments is called the I band because it is optically isotropic; i.e., it is not birefringent.[1] The region containing the thick filaments, as well as the part of the thin filaments that overlaps them, is called the A band (for anisotropic, as it is birefringent). The individual thin filaments are fixed to each other in the center of the I band at the Z line. The thick filaments are joined at the M line.

[1] In birefringent (or optically anisotropic) material the refractive index is dependent on the orientation of the material relative to the direction in which light is passing through it.

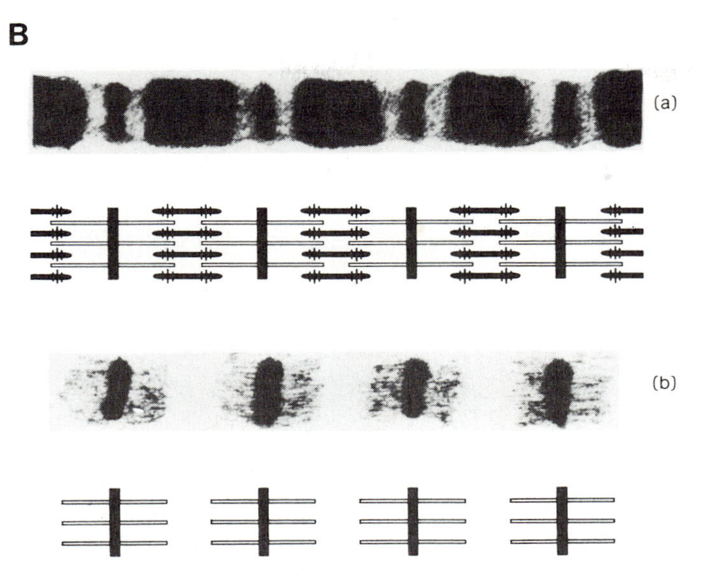

FIG. 1.3. (A) Diagram to show the arrangement of thick and thin filaments which produce the striated appearance of muscle. (From Carlson and Wilkie, 1974.) (B) Single myofibril before (a) and after (b) extraction of myosin, indicating that the A band contains myosin. (From Carlson and Wilkie, 1974.)

Unstimulated vertebrate muscle fibres are very extensible, as are those whole muscles that contain relatively little connective tissue. The large changes in length are made possible by the relative movements of the thick and thin filaments, which slide past each other without changing their length. This sliding also occurs when muscle actively shortens. The concept that length changes are accommodated in this way without change in the filament lengths is known as the sliding filament theory. The constancy of the filament length is crucial in narrowing the field of possible mechanisms of muscle contraction. It has been firmly established by three lines of evidence: (1) light microscopy of living fibres in which A-band length is shown to remain constant during length changes (A. F. Huxley and Niedergerke, 1954; H. E. Huxley and Hanson, 1954); (2) electron microscopy in which the length of the filaments is measured in fixed preparations (Page and Huxley, 1963); and (3) X-ray diffraction of living fibres in which various periodicities of the regular arrangement of proteins within thick and thin filaments are measured. There are no substantial changes in any of these spacings when the muscle is activated, or when it shortens (Wray and Holmes, 1981).

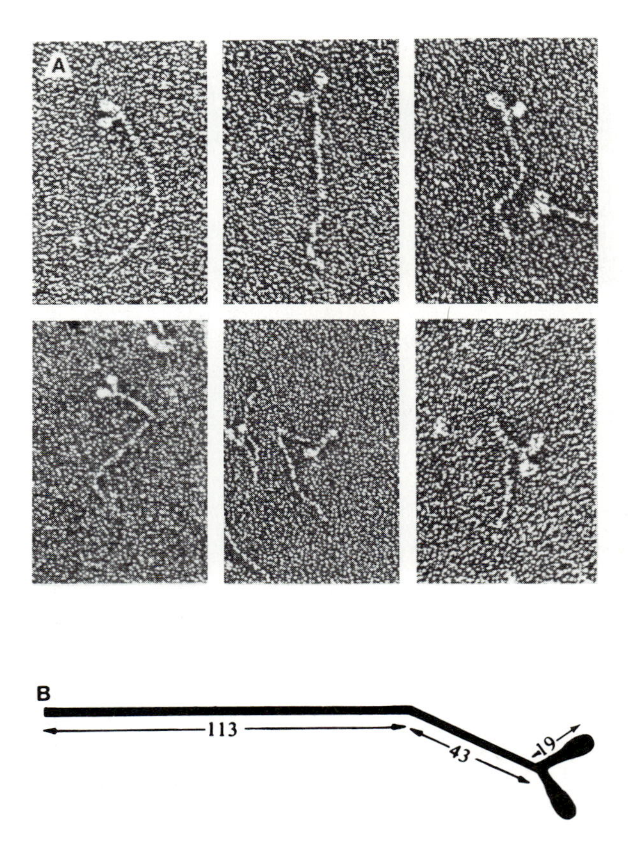

FIG. 1.4. Shape of the myosin molecule. (A) Electron micrographs of rabbit skeletal myosin, rotary shadowed platinum. Nominal magnification, ×180,000. (B) Dimensions (in nm) of the myosin molecule found from a large number of micrographs like those in (A). The diameter of the head is 64 nm at the widest part. (From Offer and Elliott, 1978.)

In some invertebrate muscles, however, there do seem to be changes in the length of the thick filaments on activation (Dewey *et al.*, 1979).

The main constituent of the thick filament is the protein myosin, as can be shown either by its selective removal using solutions of high ionic strength in which myosin is soluble (Fig. 1.3B), or by antibody staining. Myosin is a large molecule (MW 500,000) composed of two identical heads connected to a long tail (156 nm). There is a flexible hinge region in the tail 43 nm from the heads (Fig. 1.4). Each thick filament is about 1.6 μm long. The two halves of the filament on either side of the M line are of opposite polarity, with the tails of the myosin molecules pointing toward the center of the

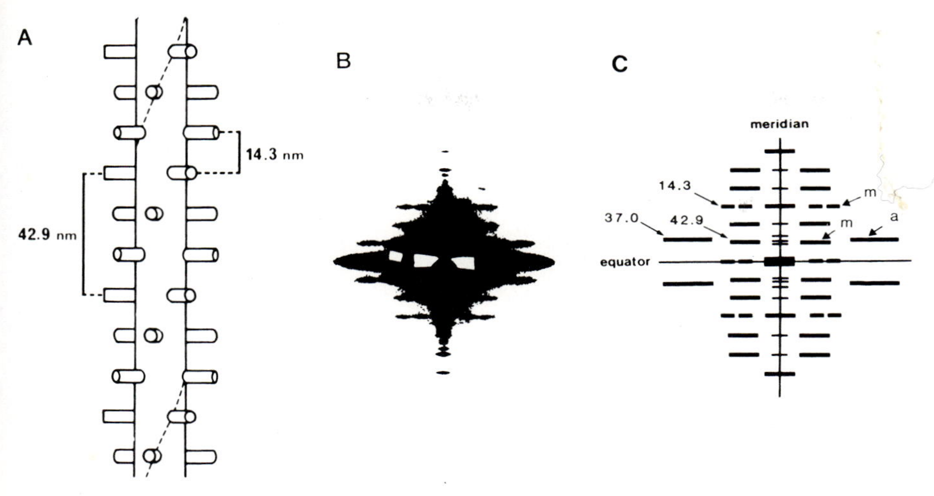

FIG. 1.5. (A) Diagram showing the 6/3 helical arrangement of myosin crossbridges along the thick filament. (From Offer, 1974.) The evidence for this arrangement is discussed by Squire (1980). (B) Low-angle X-ray diffraction pattern from frog sartorius muscle in the relaxed state showing the characteristic layer line patterns caused by the thick and thin filaments. Muscle axis is from top to bottom of page. The spacings (in nm) and probable origin of these layer lines (a, actin; m, myosin) are summarized in (C). (From Squire, 1981a.)

filament. The heads of the myosin molecules project from the thick filament everywhere except for a region 0.15 μm long in the center (the bare zone) where there are only overlapping myosin tails. The exact packing of myosin into the thick filament is uncertain. It seems likely that there are three strands (Squire, 1980) or subfilaments, in which case the thick filament contains about 300 molecules of myosin. The myosin heads projecting from the thick filament form a helical array with a periodicity of 42.9 nm and an axial interval of 14.3 nm (Fig. 1.5A). These periodicities contribute to the characteristic X-ray diffraction pattern of relaxed muscle and can be determined from it (Fig. 1.5B and C). In addition to myosin, the thick filaments contain much smaller amounts of other proteins whose role may be to control and define the structure of the filaments (Offer, 1974). Thick filaments are very uniform *in vivo*; in contrast, the artificial thick filaments that form spontaneously from purified myosin *in vitro* vary greatly in length and thickness.

The thin filament (Fig. 1.6A) contains three proteins of well-established function: actin, tropomyosin, and troponin. Each thin filament (the length varies in different species but is typically 1.0 μm) contains about 350 actin monomers and 50 molecules each of troponin and tropomyosin. The major

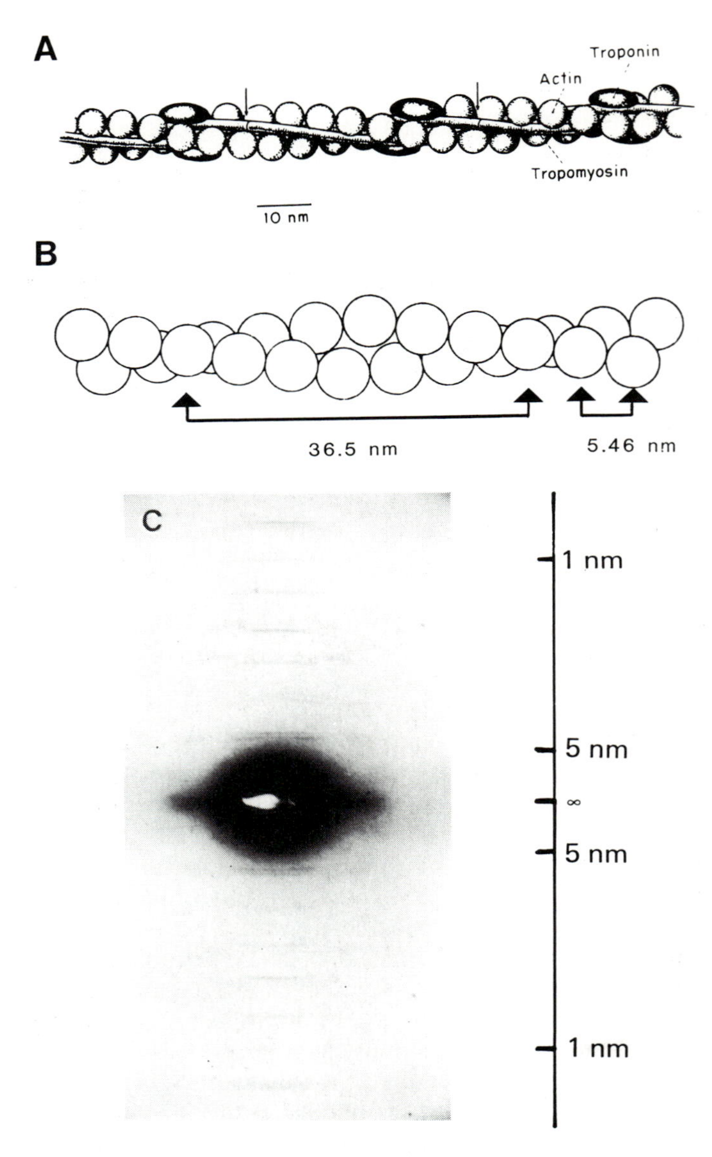

FIG. 1.6. (A) A model for the fine structure of thin filament. (From Ebashi, 1975.) (B) F-actin structure showing arrangement of G-actin monomers as deduced from X-ray diffraction and electron microscopy studies. (After Hanson and Lowy, 1963.) (C) Pattern of reflections produced by X-ray diffraction from frog sartorius muscle (fibres aligned vertically). The arrangement of actin in the thin filaments is responsible for the reflections in the range 0.5–5 μm. (Courtesy of H. E. Huxley.)

role of actin is to interact with myosin. Tropomyosin and troponin have a control function.

Actin monomers (G-actin, MW 41,700) are polymerized to form two helical strands of F-actin, which are wound round each other with a pitch of 73 nm. The axial separation of the G-actin monomers is 5.46 nm (Fig. 1.6B). These periodicities can be determined from characteristic X-ray reflections from living muscle and artificial thin filaments (Figs. 1.5B and 1.6C). For example, the cross-over points on the actin helix occur at one-half the pitch, 73/2 = 36.5, which produces the diffraction pattern shown in Fig. 1.5B. The actin filament is a polarized structure, as can be shown by the binding of myosin heads to it which results in the asymmetrical "arrowhead" appearance (Fig. 1.7). The sense of polarization is different in the filaments on either side of the Z line. This polarization of the thin filament, together with that of the thick filament, is responsible for the fact that the force acts in the direction toward the center of the sarcomere.

The tropomyosin molecules are rod shaped and polymerize end to end with a slight overlap. A filament of tropomyosin is bound to each actin strand and lies in the grooves between the actin strands. X-ray diffraction studies have suggested that the arrangement is asymmetrical so that each tropomyosin strand is nearer to one actin strand than to the other. The thin filament contains an equal number of tropomyosin molecules and troponin complexes. Each troponin complex consists of troponin I, troponin C, and troponin T. This complex binds to both the tropomyosin and to the actin. Troponin I together with tropomyosin inhibits the activation of myosin ATPase by actin, possibly by physically blocking the myosin-combining site on actin (Squire, 1981b). Each tropomyosin inhibits seven actin molecules, probably by each of the actins interacting with one of seven rather similar regions on the tropomyosin molecule. Because of its affinity for calcium, troponin sensitizes the thin filament to calcium (Ebashi and Endo, 1968; Ebashi *et al.*, 1969). When troponin C is combined with calcium, it is able to remove the inhibitory effects of tropomyosin and thus allows contraction.

An examination of an electron micrograph of a transverse section of striated muscle (Fig. 1.8A) shows that both thick and thin filaments are arranged in a regular array. The thick filaments are arranged in an hexagonal array and the thin filaments are at the trigonal points, each equidistant from three thick filaments (Fig. 1.8B). There are twice as many thin filaments as thick ones; each square micrometer of myofibril cross section has about 500 thick filaments and 1000 thin filaments. The spacing between the filaments can be deduced from the equatorial X-ray diffraction pattern, and experiments have shown that the filament lattice exhibits constant volume behaviour during changes in muscle length. When the sarcomeres are longer the spacing between the filaments is less.

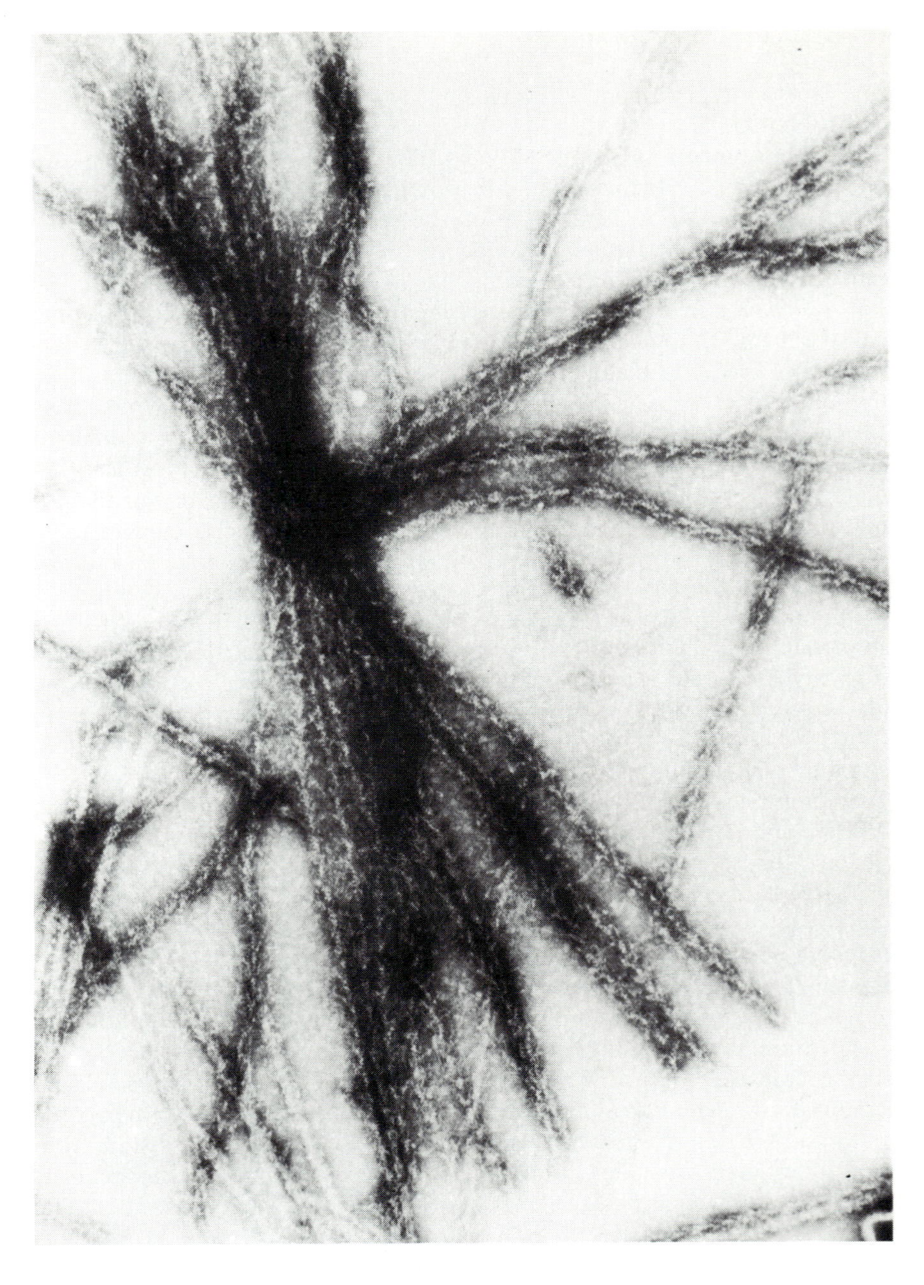

FIG. 1.8. (A) Electron micrograph of a cross section of myofibrils of frog sartorius. The meromyosin. The arrowhead structures result from binding of the heavy meromyosin to actin. Note that the heads point in opposite directions on the two sides of the Z line. (Courtesy of H. E. Huxley.)

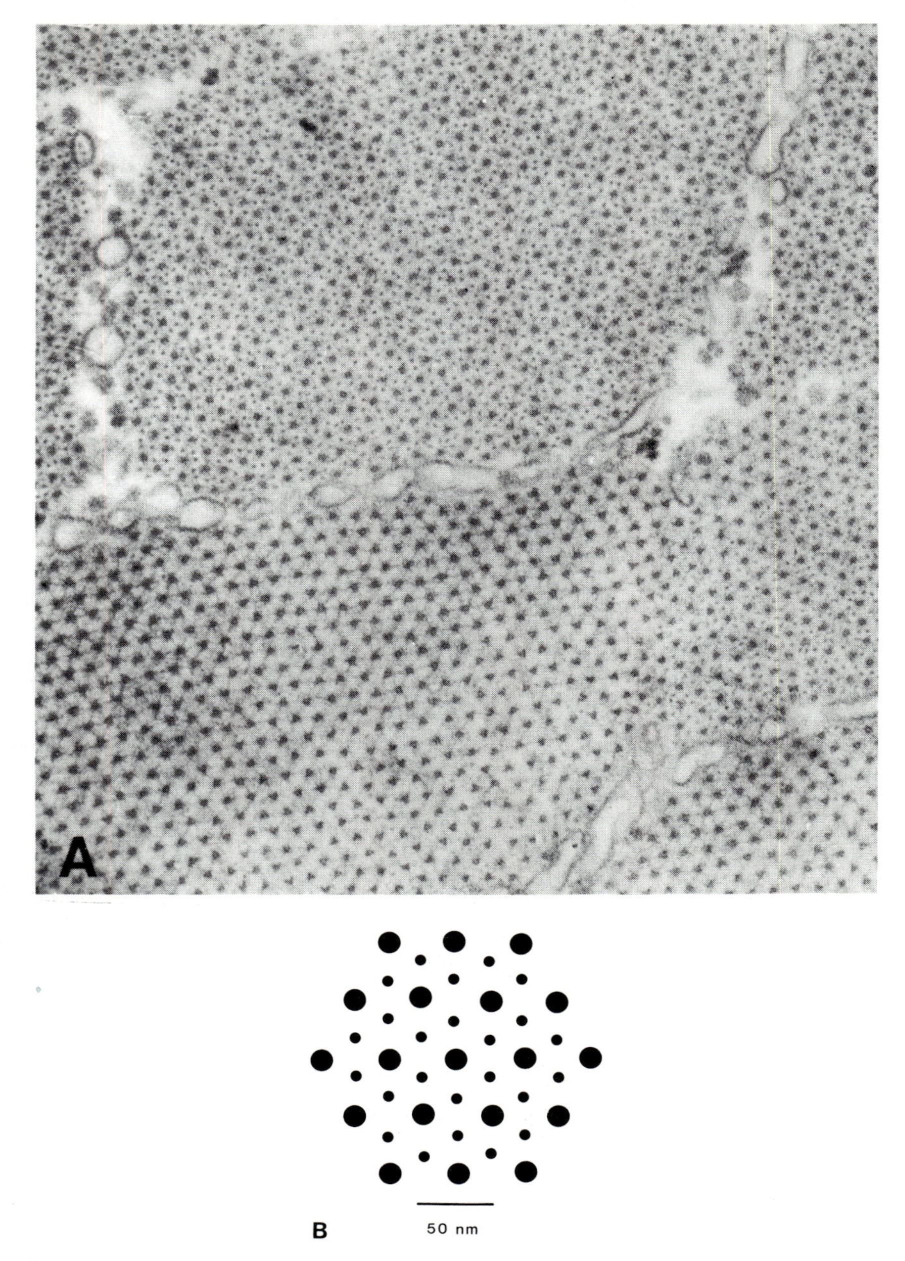

FIG. 1.8. (A) Electron micrograph of a cross section of a myofibrils of frog sartorius. The hexagonal array of thick filaments can be seen in one myofibril and the array of thick and thin filaments in another myofibril. Magnification, ×100,000. (S. G. Page, personal communication.) (B) Arrangement of filaments in a transverse section of the A band of a myofibril of a vertebrate muscle.

1.2 The Crossbridge Hypothesis

The development of force by myofibrils and their active shortening requires that there is some way of transmitting force from thick to thin filament along the fibril. Although some workers consider that electrostatic, hydraulic, or osmotic forces can provide this coupling between the filaments (Ullrick, 1967; Elliott *et al.*, 1970; Yu *et al.*, 1970; Morel *et al.*, 1976; Tirosh *et al.*, 1979; Iwazume, 1979), it is widely accepted that the link between them is some form of chemical bonding between the myosin head and actin. The variable distance between thick and thin filaments is accommodated by the myosin head moving away from the thick filament backbone by bending at the "tail hinge." The link is referred to as a crossbridge. It is further supposed that force is developed by a change in some portion of the actin–crossbridge complex. As the relative movement of thick and thin filaments that can occur (about 1.0 μm per half-sarcomere) is many times the length of a myosin molecule (about 0.15 μm), it is natural to suppose that during shortening the crossbridges break and reattach further along the thin filament. These cycles of attachment, pulling, detachment, and recovery enable mechanical work to be done, and thus must be driven by some spontaneous process. This process is considered to be ATP splitting.

The crucial elements of the crossbridge hypothesis are the following points. (1) The myosin head can bind to the thin filament to form a crossbridge. (2) Force can be generated in the crossbridge, and that force tends to move the thin filaments toward the center of the sarcomere. (3) The crossbridge can cyclically attach and detach from the thin filament; this allows movements of the filaments much greater than the length of an individual crossbridge. (4) ATP is split during the crossbridge cycle and provides the energy for the mechanical events of contraction.

Point (1) is supported by several fairly direct lines of evidence. First, it can be shown, for instance by *in vitro* studies, that purified myosin and actin combine to form a strongly bound complex. Second, the equatorial X-ray diffraction pattern from muscle changes on activation in a manner that suggests the movement of myosin heads away from the thick filament toward the thin filament. This movement is just what would be expected if myosin heads are able to bind to actin on activation. Third, in electron micrographs of muscles fixed during rigor, continuous structures can be seen linking thick to thin filaments (Fig. 1.9). These structures are seen particularly clearly in insect asynchronous flight muscles. In this tissue the angle the crossbridge makes to the thick filament has been shown to be different in preparations fixed at rest and in rigor (Reedy *et al.*, 1965). In the former case the angle is about 90°, and in the latter case it is about 45°. It is as though the crossbridges had rotated about a hinge on the thick filament

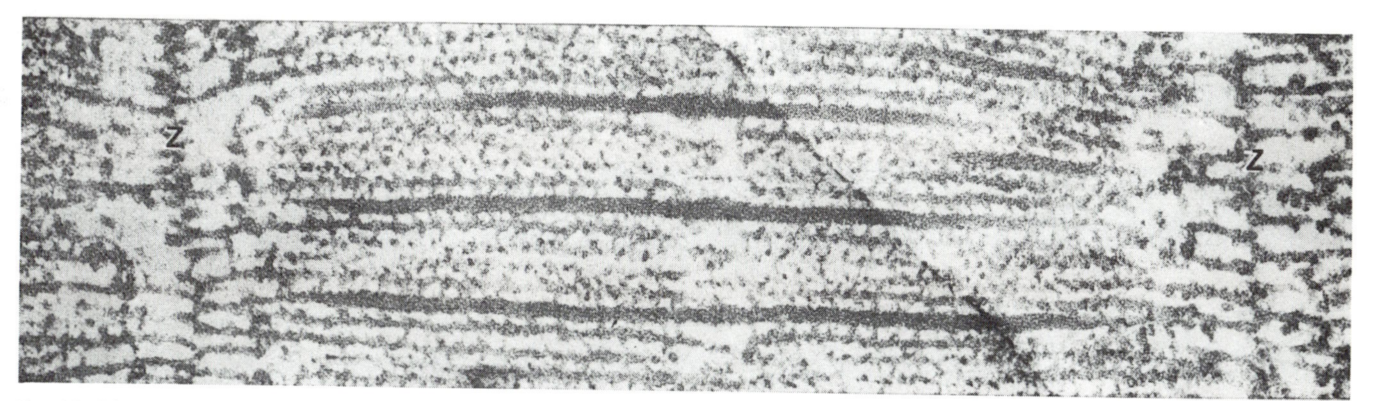

FIG. 1.9. Electron micrograph of a longitudinal section of frog skeletal muscle showing crossbridges between the thick and thin filaments. Z, Z line. (From C. Franzini-Armstrong, personal communication.)

as they pushed the actin toward the M line. The evidence on points (2), (3), and (4) is discussed in Section 1.3. The crossbridge hypothesis has found general acceptance because it reconciles so many seemingly unrelated structural, biochemical, and physiological observations. In the remainder of the chapter we illustrate the unifying nature of the crossbridge concept by considering various facts about muscle contraction and muscle proteins in the light of this hypothesis.

There are, of course, many unanswered questions about how crossbridges might work in muscle. It is perhaps appropriate to list some here, and others are suggested in the section that follows:

1. How many myosin heads constitute a crossbridge? If more than one, how do they interact?
2. How many actin molecules are involved in a crossbridge?
3. Do crossbridges interact with each other? If so, how is the interaction mediated?
4. What is the nature of the bonding between actin and myosin: covalent, electrostatic, hydrophobic, etc.?
5. How many crossbridges are attached at any one time under different conditions?
6. How many ATP molecules are split in a crossbridge cycle? Is the value constant? Is ATP splitting obligatory in each cycle?
7. Over what distance can the crossbridge remain attached? How does this compare to the distance over which it can develop force?
8. Where in the crossbridge–actin complex is force generated? What are the mechanical properties of the parts of the complex involved in transmitting this force?

1.3 The Mechanical and Energetic Properties of Muscle

In this section are briefly described the main facts about the mechanical behaviour of muscle and their possible interpretation. The resting muscle fibre is very extensible, but on stimulation it rapidly becomes much less extensible. This change occurs even before tension development or shortening starts. If the ends of the fibre are held (isometric contraction), tension develops. If the ends are allowed to move, active shortening occurs; if there is an opposing force, work is done.

How do the forces and the movements of individual crossbridges add together within the muscle fibre? The answer is that while movements add up along the length of the fibrils, forces add up across the cross section of the fibril. Consider first an individual thin filament and a half-thick filament that overlaps it (Fig. 1.10A). All the crossbridges between the two filaments

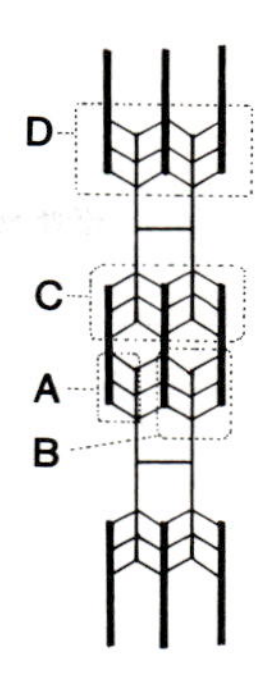

FIG. 1.10. Diagram to illustrate the discussion of how forces and movements add up in muscle. For further explanation see text.

are in parallel, and the forces they produce add up. The relative movement of these two filaments is the same, regardless of how many crossbridges connect them; thus the movements produced by these crossbridges do not add to each other. Also, in parallel with these crossbridges are all the bridges between the other filaments within the myofibril at the level of the same half-sarcomere (for example, B in Fig. 1.10). The crossbridges in the other half of the same sarcomere (C), however, are mechanically in series with those in the first half. The *force* they produce does not add to that produced by the first set, but *movements* generated in the second half of the sarcomere do add to those generated in the first as do movements in the next sarcomere (D). Thus, a set of n uniform half-sarcomeres in series produces only the same tension as one half-sarcomere, but n times as much movement. These facts have implications for the correct way of "normalizing" measurements of muscle force and movement (Table 1.I). The way in which movements and forces add up in a nonuniform series of sarcomeres is very complicated (Morgan *et al.*, 1982).

1.3.1 TENSION AND LENGTH

From these considerations it is apparent that the amount of tension produced in a uniform series of sarcomeres should be proportional to the number of sites at which crossbridges can form within each half-sarcomere. This number can be varied by varying the sarcomere length, and thus, the overlap of thick and thin filaments. The well-known experiments of Gordon *et al.* (1966b) on the dependence of active tension on sarcomere length are shown in Fig. 1.11. When the sarcomere is 3.65 μm in length, corresponding to the right end of the graph in Fig. 1.11A, the thick and thin filaments do not

TABLE 1.I

Normalized measurements of force, velocity, and power

When measurements of the mechanical performance of muscles are to be compared it is often desirable to eliminate variations due to (1) the size and shape of the muscle used, (2) the arrangement of the fibres within the muscle, and (3) sometimes also the length of the thick or thin filaments. This table gives the formulae required to do this in various simple cases.

The abbreviations used are

P	force
P_{norm}	normalized force
V	velocity
V_{norm}	normalized velocity
l_m	muscle length
l_f	fibre length. In all the cases considered below all the fibres within the muscle have the same length
vol	volume of muscle fibres (the intracellular space)
wt	wet weight of muscle (including extracellular space)
f_i	fraction of total volume that is intracellular
D	density of muscle
A	total cross-sectional area of muscle fibres
s	sarcomere length
N_s	number of sarcomeres in series in a fibre

The volume of a muscle, or muscle fibre, can be obtained by measuring the length and cross-sectional area, or it can be obtained from the weight of the whole preparation.

$$\text{vol} = \frac{f_i}{D} \cdot \text{wt}$$

The density of muscle is about 1.06 mg/mm^3 (Hill, 1931). In multifibred preparations f_i is about 0.8. It is usual, however, for the purposes of normalization, to take $D = 1$ mg/mm^3 and to take f_i as 1. That is, the approximation used is vol (in mm^3) = wt (in mg).

(A) Muscle consists of parallel fibres running the whole length of the muscle. Because shortening velocity adds up along the fibre, it will be proportional to muscle length. Normalized shortening velocity is, therefore, given by

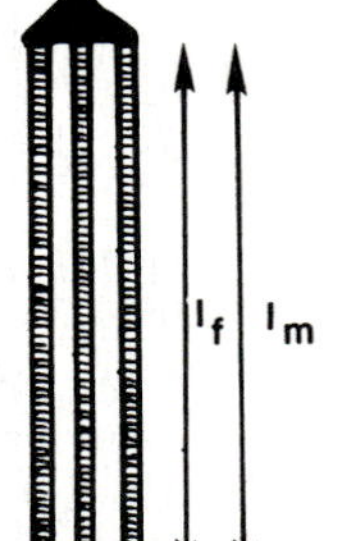

$$V_{norm} = \frac{V}{l_f} \approx \frac{V}{l_m}$$

As the fibres are in parallel force adds up across the cross section and is proportional to the total cross section of the fibres within the muscle. Normalized force is therefore given by

$$P_{norm} = \frac{P}{A} = \frac{Pl_f}{\text{vol}} \approx \frac{Pl_m}{\text{wt}}$$

The normalized power output (pow_{norm})

$$\text{pow}_{norm} = P_{norm} V_{norm} = \frac{PV}{\text{vol}} \approx \frac{PV}{\text{wt}}$$

TABLE 1.I *(Continued)*

(B) Muscle consists of parallel fibres shorter than the muscle length. Obviously in this case $l_m \neq l_f$ but otherwise the above formulae apply.

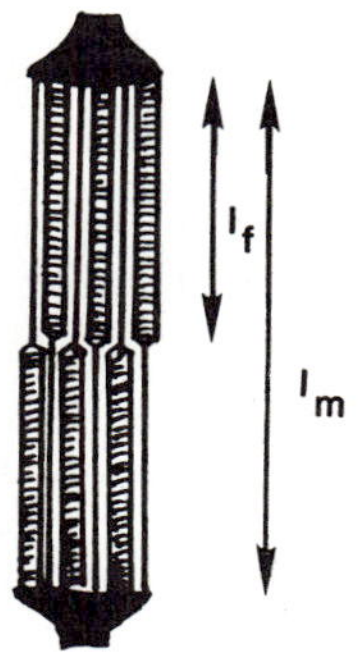

$$V_{norm} = \frac{V}{l_f}$$

$$P_{norm} = \frac{Pl_f}{vol} \approx \frac{Pl_f}{wt}$$

$$pow_{norm} \approx \frac{PV}{wt}$$

(C) Muscle is pennate, that is, the fibres lie at an angle θ to the axis of the muscle. Force and shortening are measured along the axis of the muscle.

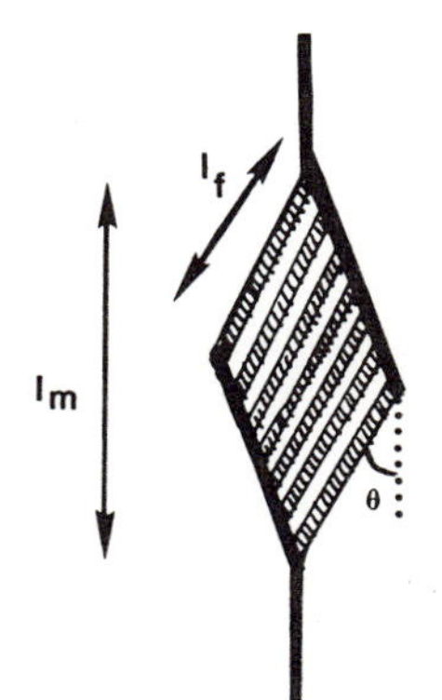

$$V_{norm} = \frac{V}{l_f} \cdot \cos\theta$$

$$P_{norm} = \frac{Pl_f}{vol} \cdot \frac{1}{\cos\theta} \approx \frac{Pl_f}{wt} \cdot \frac{1}{\cos\theta}$$

$$pow_{norm} = \frac{PV}{vol} \approx \frac{PV}{wt}$$

Note. It is assumed that, during shortening, there are no significant frictional forces as the fibres move past each other.

(D) To allow for the effect of different thin filament lengths as, for example, when comparing frog and mouse muscle, it is necessary to express the shortening velocity per sarcomere. Thus for (A) or (B) the formula is

$$V_{norm} = \frac{V}{N_s} = \frac{V}{l_f} s$$

Sometimes the velocity is expressed per half-sarcomere, because this gives the actual velocity at which thick and thin filaments are passing each other. The force exerted by a muscle is uninfluenced by the length of the thin filaments, if they are longer than the thick filaments.

The effects of variations in thick filament length are not considered here because such variations do not occur in vertebrate muscles.

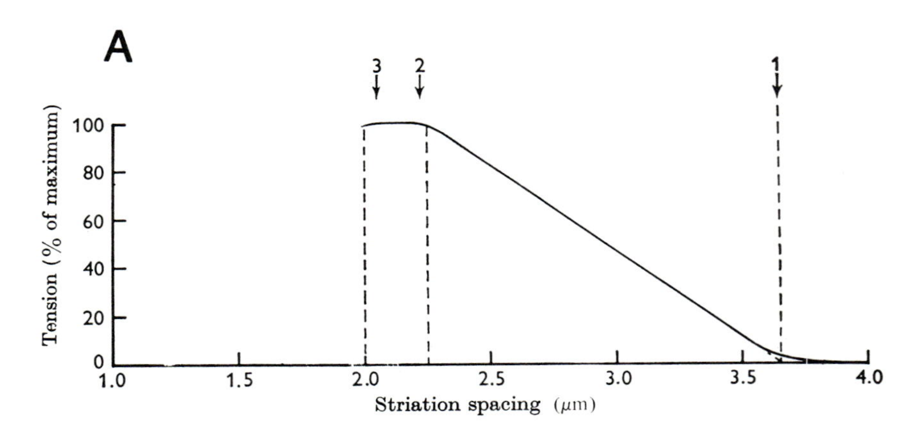

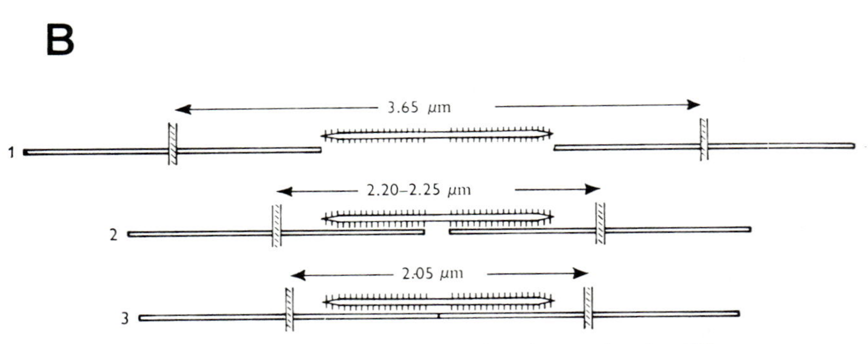

FIG. 1.11. (A) Schematic summary of observations of tension produced at different sarcomere lengths. (B) Diagrams showing the filament overlap at the three sarcomere lengths labelled in (A). (After Gordon *et al.*, 1966.)

overlap and no tension is developed. As the amount of overlap is increased by reducing sarcomere length, tension rises proportionally until the ends of the thin filaments reach the bare zone. Tension is the same for all sarcomere lengths between 2.25 and 2.00 μm. These results are compatible with the development of force being due to the interaction of myosin heads with actin.

The force that muscle produces in isometric contraction is about 0.3 N/mm^2 cross section. Assuming that there are 500 thick filaments/μm^2 and 150 myosin molecules in each half of the thick filament, the force per myosin head would be about 4 pN (2.4 GN/mol). However, as we shall describe, it is generally supposed that crossbridges are continually attaching and detaching. At any moment, only a proportion of them are attached. The actual force per attached crossbridge is thus greater than 4 pN.

1.3.2 STIFFNESS AND COMPLIANCE

The word *stiffness* means the instantaneous dependence of tension (*P*) on length (*l*). It is the slope, dP/dl, of the graph shown in Fig. 1.12. Compliance (dl/dP) is the reciprocal of stiffness. The stiffness of a muscle fibre can be measured by very rapidly changing its length while recording the tension. Unstimulated muscle fibres have a low stiffness, suggesting that thick and thin filaments can move past each other with little restraint. When the muscle fibre is stimulated it becomes very much stiffer. The stiffness rises with the same time course as the tension and the stiffness also varies with overlap in approximately the same way as tension. This result suggests that relative movements of the thick and thin filaments are restrained in the active muscle by the crossbridges between them, and, also, that the filaments themselves have a high stiffness. In active muscle a very quick release of 6 nm per half-sarcomere is sufficient to reduce the tension to zero. This distance is small compared to the size of the crossbridge and this result is compatible with crossbridge hypothesis. The recovery of tension after a quick release has a complex time course (Fig. 1.13A), with a rapid phase, and then a pause before a slower part resembling the rise of tension at the onset of contraction. The rapid recovery phase may be due to new cross-bridges attaching, or to further development of force within existing cross-bridges (Podolsky and Nolan, 1973; Huxley and Simmons, 1971b).

1.3.3 MAXIMUM SHORTENING VELOCITY

If active shortening occurs without an opposing force, the speed of shortening is at its maximum (V_{max}, Fig. 1.13B). V_{max} is found to be independent of sarcomere length (Edman, 1979). This observation can be understood from

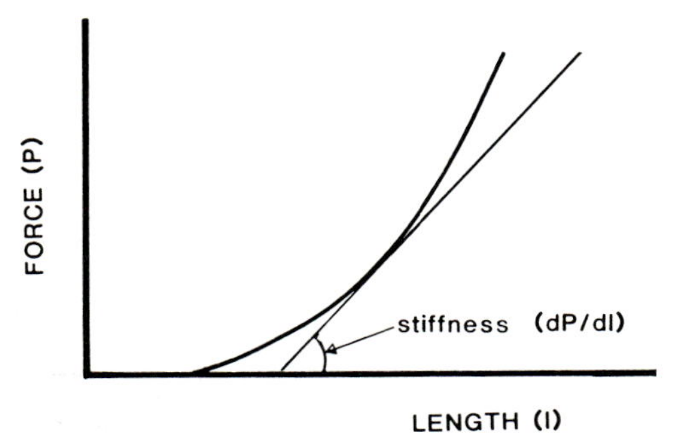

FIG. 1.12. Diagram of the relation between force and length in an elastic body. The slope of the tangent to the curve is the stiffness.

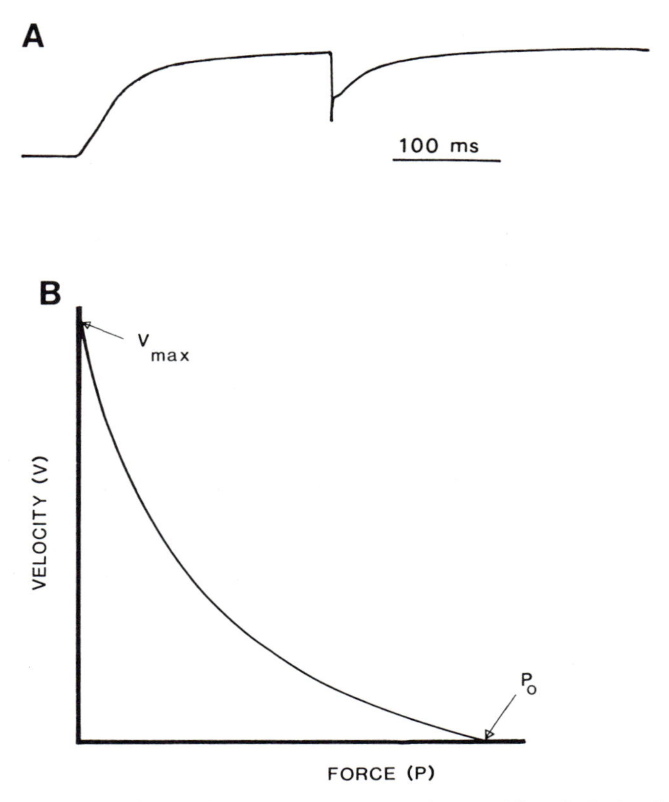

FIG. 1.13. (A) Tension changes in response to a step change of length. Isolated frog fibre, 4°C. Tracings of a record with a release of 8.8 nm per half-sarcomere. (From Huxley and Simmons, 1973.) (B) The relation between force and velocity of shortening in muscle.

the limited range of action of the crossbridge if we add the further idea that the crossbridge can "push," as well as "pull." Each crossbridge has a relation between the tension it exerts and its length. Its length depends on the position in which it formed, and on any relative movement of the thick and thin filaments that has taken place since then. When it is formed a crossbridge presumably exerts force in the forward direction (a pull), but, if shortening occurs while it remains attached, the force will fall. If this process continues, the force will fall to zero and then reverse. The bridge will then be "pushing." Thus, particularly during rapid shortening the filaments experience forces in both directions from the crossbridges. At V_{max}, where there is no external force opposing shortening, the forward and backward forces balance. Therefore, V_{max} should be independent of the number of attached crossbridges, and, consequently, independent of filament overlap. When the muscle is shortening at V_{max}, crossbridges can remain attached for less time than during isometric contraction (a quantitative example is given below in Section 1.3.5). Since presumably a finite time is required for the reattachment of a broken crossbridge, the number attached should be fewer during shortening. In agreement with this expectation, it is found that the stiffness of rapidly shortening muscle is less than that of isometrically contracting muscle (Julian and Sollins, 1975), and that the relative intensity of the 1:1 and 1:0 X-ray reflections is changed in a way that is compatible with there being fewer crossbridges attached (H. E. Huxley, 1979).

1.3.4 ATP AND MUSCLE

It is accepted that the splitting of ATP to ADP and inorganic phosphate provides the energy for muscle contraction. Some of the evidence for this is as follows:

1. Dead muscle is stiff (rigor mortis); it contains no ATP. When ATP (without any calcium) is added to muscle in rigor (after first removing the surface membrane chemically or mechanically, so that the ATP can reach the myofibrils), the muscle again becomes extensible. This "relaxation" of the muscle in the absence of calcium does not involve ATP being split. If calcium ions are present, as well as ATP, the muscle will actively shorten, develop tension, and split ATP.
2. Analogous processes can be seen using a solution of actin filaments and myosin. In the absence of ATP the solution is turbid, because the myosin is attached to the actin filaments and effectively scatters light. When ATP is added the solution clears because myosin is dissociated from the actin and the smaller particles scatter the light less effectively. If myosin is in the form of filaments, the network of actin and myosin filaments actively contracts, squeezing out water as it splits ATP; this process is called "super-precipitation."
3. In intact, living muscle ATP is split on stimulation. In the absence of metabolic inhibitors what is actually seen is the conversion of phosphorylcreatine (PCr) to creatine (Cr) and inorganic phosphate (P_i). This is because creatine kinase rapidly rephosphorylates the ADP that is formed. When creatine kinase is blocked by dinitrofluorobenzene (DNFB), net ATP splitting is seen (Davies, 1965a).
4. Even in the absence of actin, myosin is an ATPase, but a very slow one. Binding of ATP to myosin and its cleavage to ADP and P_i on myosin are rapid processes, but the dissociation of the products from myosin is slow. Actin greatly activates myosin ATPase by binding to the myosin–product

complex, after which the products can dissociate more rapidly. The actin–myosin complex (actomyosin) then binds ATP, and actin may dissociate from the complex before the ATP is split (these events are discussed in more detail in Chapter 3). This biochemical cycle for ATP splitting by the proteins in solution fits in neatly with the crossbridge hypothesis. In each complete cycle *in vitro*, one ATP is split and actin and myosin are dissociated and reassociated, which seems analogous to the breaking and reforming of a crossbridge (Fig. 1.14). The "working" stroke of the crossbridge might be tentatively identified with the release of products from the actomyosin–product complex (step 4 in Fig. 1.14) and the "recovery" stroke with the splitting step (step 2). It is worth noting that in this scheme ATP is split on myosin alone, not when actin and myosin are combined. Also, external work is not done at the time of ATP splitting; ATP splitting puts myosin into a state in which it can do work.

At this stage it is interesting to consider the amount of work that can be done by a single crossbridge and to compare it with the maximum work that can be done by splitting one molecule of ATP (the free energy change). As noted above, the force exerted by a single crossbridge in isometric contraction is at least 4 pN. The maximum force one crossbridge can exert is greater than this because only a portion of all possible crossbridges are formed at one time, and not all crossbridges are exerting the maximum force. The tension–length behaviour of one crossbridge can be determined in quick release experiments (see Chapter 2). The work that can be done by a crossbridge is the area under its tension–length curve and is at least 3.7×10^{-20} J per crossbridge or 22 kJ/mol of bridges. This is of the same order of magnitude as the free energy change for the splitting of ATP, which

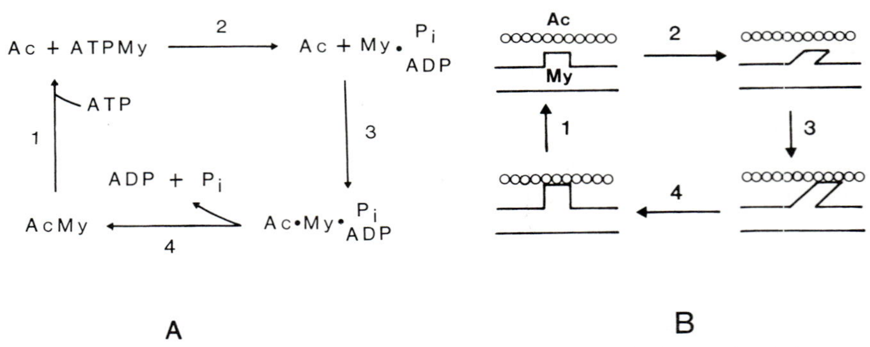

FIG. 1.14. A simple analogy between ATP splitting by actomyosin in solution (A) and a crossbridge cycle (B). Ac, Actin; My, myosin.

is about 50 kJ/mol (Kushmerick and Davies, 1969). Had these numbers been very different from each other it would have created difficulties for any simple crossbridge hypothesis. Had the work turned out to be greater than the free energy, the hypothesis would be thermodynamically impossible. If the work had turned out very much smaller than the free energy, the hypothesis would have difficulty in explaining the relatively high efficiency of muscle in converting the free energy of ATP splitting into work (up to 50% in frog muscle).

1.3.5 ATP SPLITTING IN ACTIVE MUSCLE

In intact muscle ATP is split during an isometric contraction by two ATPases which are relevant here: the actomyosin ATPase and the ATPase of the SR calcium pump. Both might be expected to become very active on stimulation. If muscle is stretched before stimulation so that most actin–myosin interaction is prevented, the quantity of ATP split is about a quarter of that at a length optimum for tension development. Assuming that the SR ATPase is length independent, this suggests that three-quarters of the ATP used at optimum length is split by actomyosin. The crossbridge interpretation of this ATP usage is that crossbridges break during isometric contraction and are replaced. This breakage might be either spontaneous in the absence of relative movement of thick and thin filaments or related to small movements of the filaments which may occur during isometric contraction. If we make two naive assumptions, we can estimate the rate of crossbridge turnover from the rate of the ATP hydrolysis. The assumptions are (1) that one ATP is split for each crossbridge cycle, and (2) that all possible crossbridges are active. For example, 1 mm^3 of frog muscle contains 100 pmol of crossbridges and splits ATP during isometric tetanus (at 0°C) at about 300 pmol/s; 75% of this is due to actomyosin ATPase. Thus, each crossbridge cycle lasts on average 450 ms. It is interesting to compare this value with a rough estimate of the duration of attachment of a bridge during shortening at V_{max}, which in frog muscle (at 0°C) is 1.5 μm/s per half-sarcomere. If bridges can stay attached for a relative movement of thick and thin filaments of 30 nm (twice the distance over which forward force can be exerted), their lifetime would be only 20 ms. Though these calculations are rough, they illustrate why, in the crossbridge hypothesis, it is necessary to suppose that crossbridge lifetime is less when shortening is occurring than when it is not. In other words, bridges break faster during shortening than during isometric contraction. One might wonder about the alternative assumption that in isometric contraction bridge lifetime is also less than 20 ms. This would mean that no more than 10% of the bridges were attached at any one time. The force exerted by each would then have to be >40 pN, and the work it could do would be clearly greater than that available from splitting one molecule of ATP.

1.3.6 FENN EFFECT

If stimulated muscle is allowed to shorten against an opposing force, thus doing an appreciable amount of work, the rate of ATP splitting is greatly increased (Kushmerick and Davies, 1969; see Chapter 4). The energy output (heat + work) is also increased, as was first observed by Fenn (1923, 1924). Both phenomena are now known as the Fenn effect. The crossbridge hypothesis provides an explanation of the Fenn effect. In the previous section we showed why it is natural to assume that shortening increases the rate at which crossbridges break. If the rate constant for the formation of bridges is not altered by shortening, then the increased rate of breaking will lead to an increased turnover of bridges, and, hence, an increased rate of ATP splitting and energy output. This type of explanation is discussed further in Chapter 5.

1.3.7 ENERGY PRODUCTION

Heat production measurements by A. V. Hill (see Hill, 1965, for references) and his collaborators allowed the energy output of contracting muscle to be determined at a time before methods had been developed for measurement of ATP splitting (indeed long before ATP was discovered). It has often been assumed (initially for want of information, and more recently in spite of it) that the observed energy output is all derived from simultaneous ATP splitting and the reactions that resynthesize ATP. This idea has been shown to be wrong (see Chapter 4 for evidence). In fact, although a large part of the energy output does come from the ATP splitting, there is also a considerable fraction that does not, both in isometric contraction and during shortening.

2
Mechanics of Contraction

2.1 Introduction

This chapter gives an account of the mechanical properties of muscle and is illustrated with experimental results. We comment on some important technical advances and the technical points that might limit the interpretation of the experiments. Particular attention is paid to results that are important for understanding the mechanism of contraction and therefore also for understanding muscle energetics. As appropriate, we outline the interpretation of these experiments within the general framework of the crossbridge idea of muscle action introduced in Chapter 1 and indicate where such interpretation is difficult.

2.1.1 EXPERIMENTAL PREPARATIONS

Experiments on mechanics can be made on a wide range of muscle preparations: at one extreme several kilograms of synergistic muscle in the body undergoing a voluntary contraction, and, at the other, a segment of a skinned fibre, weighing only a microgram. Experiments on different preparations should be complementary in developing an idea of how muscles perform their function. As the emphasis here is on the contractile event itself, we shall consider only experiments made on smaller isolated preparations. Therefore, we do not consider the way in which the nervous system selects and activates motor units appropriate to the required movement nor how their energetic requirements are met. Nevertheless, a wide range of experimental preparations is left for consideration, and so we indicate briefly some of the advantages and limitations of each.

2.1.1.1 *Whole Muscle Preparations*

The mechanical behaviour of whole muscle is described in this section for comparison with the results of energetic studies, most of which have been made on whole muscle, and because it is relevant to understanding the behaviour of muscles *in situ*. An advantage of whole muscles is the relative ease of experimenting with them; with only a little practice they can be dissected from the animal without damage and they are large enough that measurements of force and displacement can be made with simple, robust transducers.

Some of the problems that may have to be faced in work with whole muscles are listed below. Needless to say, they do not all apply at any one time (otherwise whole muscle experiments would be useless indeed).

1. *Differences in fibre type.* Many muscles are known to contain fibres of several types differing in their mechanical and biochemical properties. Measurements obtained in such muscles are thus not typical of any one type of fibre and may not even represent an average of all the fibres. Figure 2.1 shows the force–velocity relation calculated for a whole muscle consisting of fibres with different maximum velocities of shortening. Even in muscles which have previously been considered homogeneous (e.g., frog sartorius), there are fibres of very different size and of different histological appearance, and which may well have different energetic and mechanical properties.

2. *Stimulation.* In thick muscles, such as gastrocnemius, stimulation sufficient to activate all the fibres may damage fibres near the electrodes. If nerve stimulation is used to avoid this problem, complications can arise from the partial blocking of the neuromuscular junctions (Kuffler, 1952); this blockage can be length dependent (Hutter and Trautwein, 1956). This problem can also arise with "direct" stimulation, which is often actually stimulation of nerve twigs unless a neuromuscular blocking agent, such as curare, is used.

3. *Diffusion limitations.* In thick muscles the diffusion of oxygen, CO_2, and ions may not be fast enough to maintain the muscle in a stable steady state (A. V. Hill, 1965). This problem is particularly acute in experiments at high temperature.

4. *Passive tension.* There are often substantial elastic structures in parallel with the fibres in whole muscles. The passive tension may therefore be high at long muscle lengths, and even near the *in situ* length in some muscles. In principle, it can be subtracted from the tension observed during contraction, but this is not always a straightforward procedure because (a) the passive tension changes with time and with the history of length changes imposed on the muscle and (b) the muscle fibres and the parallel elastic structures may share a series compliance, such as a tendon. A further consequence of the presence of parallel elastic structures is that, if they tear in one part of the muscle, the force in the intact parts can cause stretching and damage to the muscle fibres underlying the damaged connective tissue. An

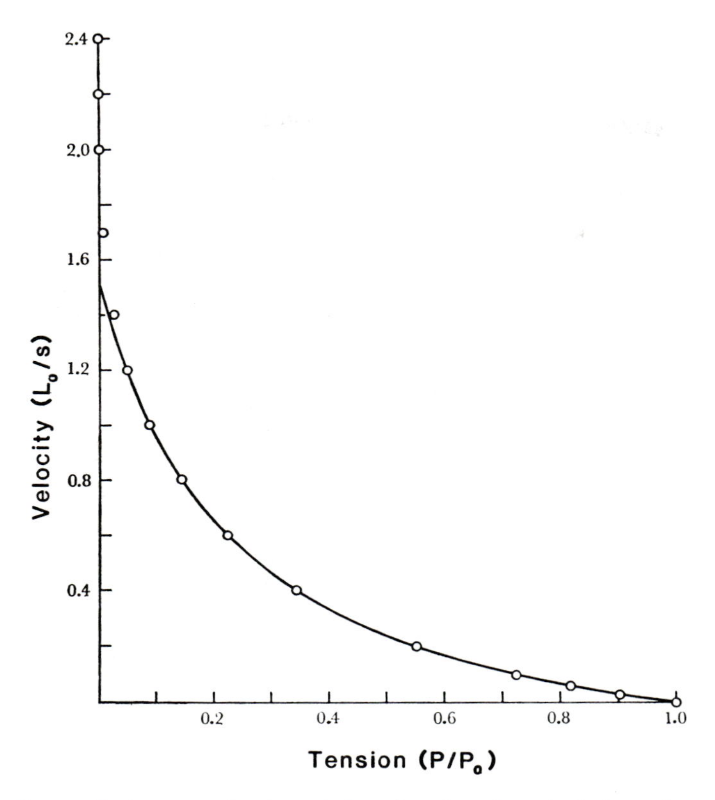

FIG. 2.1. Force–velocity curve showing relation given by a hypothetical muscle containing fibres with a range of maximum velocities of shortening (V_{max}). It is assumed (for simplicity) that the fibres all have the same length L_0, the same maximum tension P_0, and the same value of $a/P_0(=0.25)$; but the individual fibres have widely different values of the maximum velocity V_{max}. The hypothetical muscle contains 82 fibres which can be divided into 10 groups, each group having a different maximum speed, V_{max}, as follows:

v_{max} (L_0/s)	2.4	2.2	2.0	1.8	1.6	1.4	1.2	1.0	0.8	0.6
No. of fibres	1	3	7	13	17	17	13	7	3	1

When the muscle is shortening with velocity V, the fibres whose $V_{max} < V$ become slack and exert no force; but those whose $V_{max} > V$ are taut and contribute to the total force. The usual force–velocity equation was employed to calculate the force produced by each group of fibres.

$$P/P_0 = \frac{1.25V_{max}/4}{V + V_{max}/4} - 0.25$$

For a number of values of V, the force produced by each group of fibres was calculated and weighted according to number of fibres in that group. The sum of the forces produced by all the groups at each velocity is shown in the figure. The apparent V_{max} for the muscle is about 1.5 L_0/s, which is considerably less than that of many of the fibres. (From Hill, 1970.)

advantage of the parallel elastic structures is that they stabilize the uniformity of sarcomere length (see next section).

5. *Observations of sarcomere length.* Sarcomere length measurements in whole muscles are usually made on a smaller fraction of the sarcomere population than in experiments on single fibres. This is particularly the case in thicker muscles. In such muscles microscopic observation is restricted to surface fibres which may not be typical of those deeper in the muscle. Light diffraction methods can provide a value for the sarcomere length, but they give information only on regions where the sarcomeres are well ordered, and the results can be influenced by the tilt of the sarcomere with respect to the incident light and by changes in light scattering (Rudel and Zite-Ferenczy, 1979a,b; McCarter, 1981). Diffraction gives no information about the ends of the fibres, where light scattering by the tendons prevents the production of a useful diffraction pattern. These factors make investigation of sarcomere inhomogeneities in most whole muscles less reliable than in single fibres.

2.1.1.2 *Isolated Intact Fibres*

Most amphibian muscles contain cells that are larger (up to more than 150 μm in diameter) and more easily dissected than those of mammals, birds, and reptiles. These large cells, obtained from sartorius, iliofibularis, semitendinosus, anterior tibialis, and toe muscles, have been used in many mechanical studies. Experiments by Lannergren and colleagues (1978, 1982) have shown that in some of these muscles there are a number of distinct types of fibres differing in their mechanical properties. Regrettably many authors do not seem to be aware of this fact, and consciously or unconsciously select for their experiments a particular fibre type, which is not specified. It is not clear whether it would be practicable to do mechanical studies on much smaller isolated fibres (diameter <50 μm). All fibres in mammalian muscle and a large proportion of those in amphibian muscle are of this size, and remain to be investigated.

Using single fibres has these advantages:

1. Adequate stimulation of single muscle fibres, particularly the "twitch" fibre types usually used, is simple and easily monitored.
2. The mechanical, biochemical, and energetic behaviour of different fibre types can, in principle, be compared with each other.
3. The preparation is likely to be more homogeneous than a whole muscle, for example, in its myosin isoenzyme content (Gauthier *et al.*, 1978). There may still be important inhomogeneities, however, such as (a) inequality of sarcomere length along the length of the fibre, which is discussed below, and (b) inequalities across the width of the fibre due to submaximal activation at short fibre length (Gonzalez-Serratos, 1971). Edman and Reggiani (1982) have suggested that there may be (c) differences in the maximum speed of

shortening in different regions of the muscle fibre. These could be due to variations in the myosin isoenzyme composition.

4. Studies at long sarcomere length are easier because most of the external connective tissue has been removed and consequently the passive tension is smaller. However, the sarcomere length becomes nonuniform, and this leads to difficulties in interpreting the results of experiments made at long sarcomere lengths.

5. The distribution of sarcomere spacings within much of the preparation can be observed (by microscopy or light diffraction) for comparison with the mechanical behaviour of the fibre.

Some problems are inherent to the use of isolated single fibres:

1. Experiments are more difficult than with whole muscles because of the small size of the preparation.

2. It may not be possible to use the types of muscle fibres that have very small diameters.

3. Inhomogeneities in sarcomere length along the fibre length can be very troublesome. As first shown by Huxley and Peachey (1961), the sarcomeres at the end of an isolated muscle fibre are shorter than those in the centre (Fig. 2.2A). During contractions at sarcomere length beyond 2.2 μm these differences become progressively greater. The end sarcomeres can exert more force than those in the centre because they have a greater degree of filament overlap; they therefore shorten and stretch those in the centre (Huxley and Peachey, 1961; A. V. Hill, 1970; Julian and Morgan, 1979a,b; J. D. Altringham, personal communication). These effects are present also in whole muscle but seem to be less marked, probably because the forces exerted by connective tissue outside the muscle fibre have a stabilizing action (Curtin and Woledge, 1981).

2.1.1.3 *Skinned Muscle Fibres*

The use of "skinned" muscle fibres, from which the surface membrane has been removed either by dissection or by chemical treatment, allows the composition of the myofibrillar space to be altered. With these preparations many interesting experiments are possible. The early stages of preparation of the fibres are easier than preparation for experiments with intact fibres because damage to the surface membrane, which is ultimately removed, is of less importance and because only a short length of fibre is usually required. Small mammalian fibres can be used as well as large amphibian fibres. There are a number of special problems in these experiments:

1. An extra compliance may be introduced in the damaged regions of the fibres to which clips or hooks are attached. This compliance may be nonlinear and time dependent.

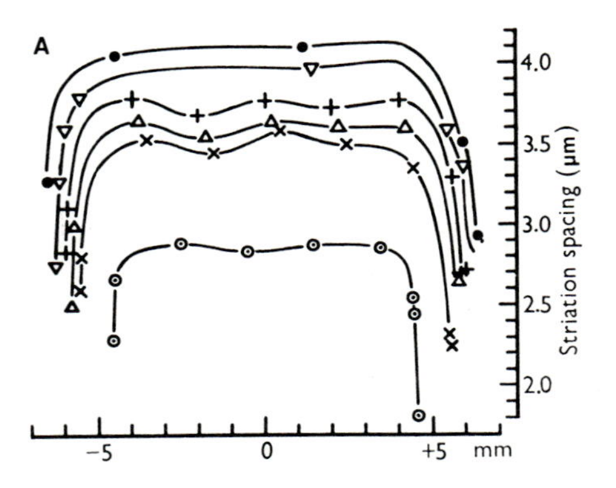

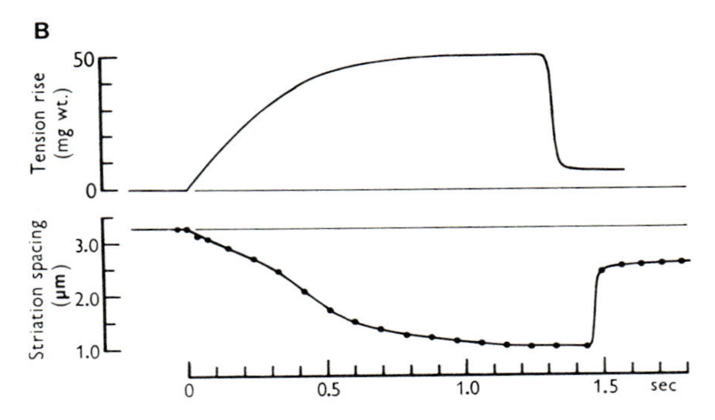

FIG. 2.2. (A) Variation of striation spacing along an isolated muscle fibre (from iliofibularis muscle) at rest. Abscissa, distance from centre of fibre; ordinate, striation spacing, measured on photomicrographs. The various curves were obtained at different degrees of elongation. Each point gives the mean value of the striation spacing found by measuring several (usually three or four) rows of 15–30 striations at different parts of the same negative, except in a few cases near the ends of the fibres where the striations were partly obscured by overlying fibrous tissue, and could only be measured in one or two small fields. (From Huxley and Peachey, 1961.) (B) Time course of shortening near one end (lower curve) and tension rise (upper curve), recorded during isometric tetani of single fibre from semitendinosus (frog). Shortening obtained from measurements on cine film (From Huxley and Peachey, 1961.)

2. The sarcomeres may become disorganized rather easily. During the activation of these preparations, which is achieved by increasing the calcium concentration in the bathing solution, activation may not occur uniformly and simultaneously over the entire cross section of the preparation. Evidently removal of the surface membrane destroys, or removes, some influence that tends to stabilize the organization of the myofibrils. It is important, therefore, in these experiments to check that orderly arrangement has not been

lost. Because full activation is associated with this disorganization, many experiments have had to be done with less than maximal activation.

3. The composition of the experimental solution must be carefully chosen so that the levels of substrates and products are properly controlled. When substrate must diffuse into the fibre, and product diffuse out, concentration gradients of these substances exist in the fibre and may influence its mechanical behaviour. To create an osmotic environment that maintains a filament lattice spacing similar to that *in vivo*, the solution must contain high molecular weight polymers which are excluded from the filament lattice (Godt and Maughan, 1977; Goldman and Simmons, 1978).

2.1.2 TECHNIQUES

The study of muscle mechanics depends on the measurement of the force (tension) and the length of the whole, or specific parts, of the muscle. Either variable may be controlled by the experimenter or made to follow a predetermined sequence of changes, while the other variable is observed. Such control, sometimes called *clamping*, may be achieved by a variety of different techniques: isotonic lever (Jewell and Wilkie, 1958), mechanical ergometer (Levin and Wyman, 1927), electromechanical ergometer (Edman and Hwang, 1977; Brady, 1966), or "spot follower" (Gordon *et al.*, 1966a). Controlling tension can be difficult if rapid length changes are required, and controlling length is usually preferred.

In the simplest experiments the entire muscle length is held constant during contraction; this is an *isometric* or *fixed-end* contraction. The term *isotonic* is used to refer to a contraction during which (or during part of which) tension is controlled while the changes in muscle length are observed. *Isovelocity* refers to change in length at a controlled steady velocity.

Tension recordings. Tension transducers (that is, instruments for converting the force into an electrical signal) are usually either silicon strain gauges or variable capacitance transducers (Fig. 2.3). For some studies a very high frequency response is necessary. A variable capacitance transducer with a frequency response (50 kHz) and sensitivity suitable for use with single fibres is described by A. F. Huxley and Lombardi (1980).

For most purposes the tension everywhere along a muscle fibre can be regarded as equal to that measured by the transducer at one end of it. This is because the force produced by the muscle is large compared to the forces that accelerate the mass of the muscle fibre, overcome viscous resistance, and cause the fibre to move through the solution around it. However, these nonmuscle forces may not be negligible during very rapid movements, and it may be necessary to make allowance for them (e.g., Ford *et al.*, 1981). These effects are smaller if short muscle fibres are used, and, for this reason, shorter

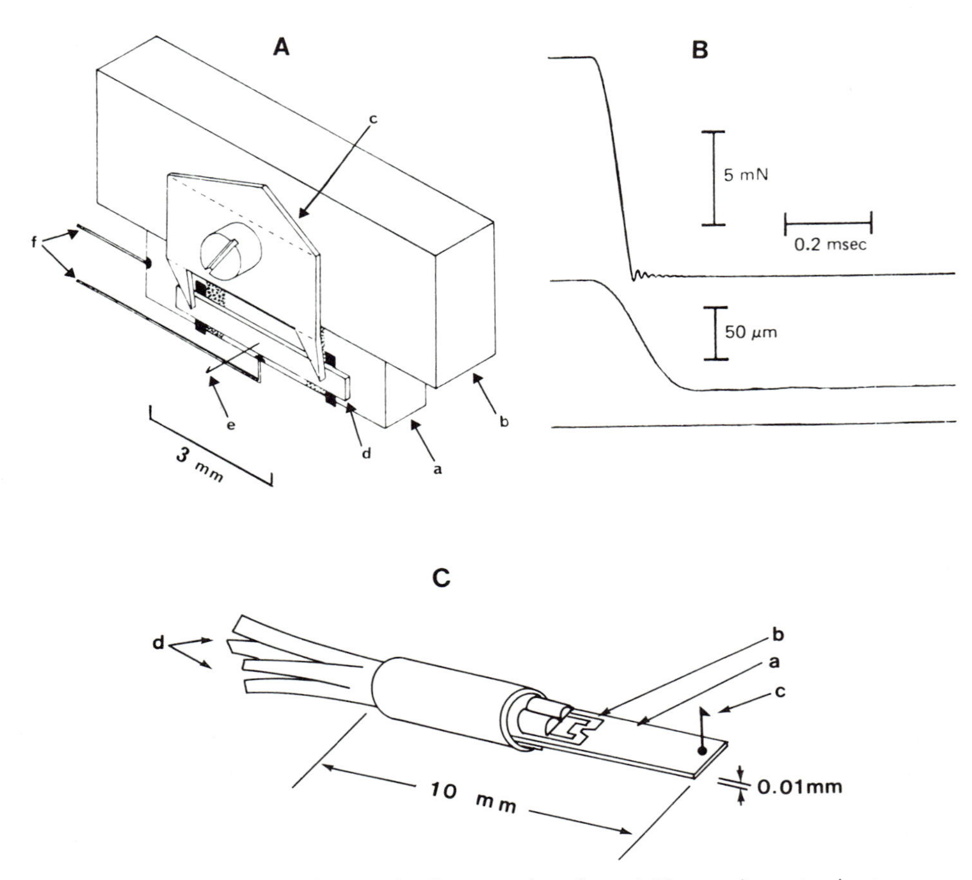

FIG. 2.3. (A) Diagram to show mode of construction of a variable capacitance tension transducer. The fixed quartz plate (a) is stuck to the lower edge of a brass block (b), which carries a three-point clamping piece (c). This holds the flexible strip (d) in place; the separation between this flexible strip (d) and the fixed quartz plate (a) is governed by the gold-leaf strips shown in black. Stippling indicates metallization of front of fixed plate removed. (e) The hook to which fibre is attached. (f) Leads. (B) Response of the transducer. Upper trace shows output signal when the transducer is rapidly unloaded; a polyester loop connects the transducer pin to the arm of a motor (Ford *et al.*, 1977) whose displacement is shown in the lower trace. (From Huxley and Lombardi, 1980.) (C) Diagram of a silicon strain gauge tension semiconductor transducer (AE800 series, Aksjeselskapet Mikro-Electronik, 3191 Horten, Norway). The silicon beam (a) has one diffused resistor (b) on each surface of the beam. Bending of the beam causes a change in the value of these resistors. The fibre is attached to hook (c). (d) Leads.

fibres are preferred in experiments on rapid tension transients (for example, fibres about 6 mm long from anterior tibialis rather than 12 mm fibres from semitendinosus).

Length recordings. It is a fairly simple matter to observe and record the distance between the two hooks to which a muscle fibre is attached. This includes the length of the muscle fibre(s) and also the tendons. The compliance of the tendons is usually not negligible. It can be measured and allowed for in the experiments or, in the case of single fibres, it can be largely eliminated by gripping the tendons in metal clips very close to the ends of the fibre (Ford *et al.*, 1977). While the distance between the ends of the tendons is proportional to the average sarcomere length of all the sarcomeres in the fibre, there can be large systematic variations in the sarcomere along the length of a fibre. Furthermore, these variations may increase in a systematic way during contraction, as is illustrated and explained in Fig. 2.2B. This effect can be reduced by monitoring the length of the central portion of a muscle fibre and holding it constant, instead of holding the overall length of the preparation constant. The experiment is then being performed only on this central portion of fibre, within which the sarcomere length is relatively uniform, while the behaviour of the ends is no longer of concern. This technique is referred to as a *length clamp*. Gordon *et al.* (1966a) describe apparatus for this purpose which measures the distance between two metal foil markers attached to the fibre surface. An alternative method, used by Goldman and Simmons (1979) with short segments of skinned fibres, is to measure and control the sarcomere length deduced from the diffraction pattern.

In all these schemes it is necessary to move one end of the muscle fibre to achieve the control of either length or tension. These movements are made by specially designed motors which move very rapidly (in some cases in as little as 50 μs). One such motor is described by Ford *et al.* (1977).

2.1.3 A. V. HILL'S TWO-COMPONENT MODEL

A. V. Hill (1938) suggested a model that has had a profound influence on the design and interpretation of many experiments. Although this model has now been shown to be wrong by the experiments described below, its influence persists.

The essence of Hill's idea can be appreciated by considering what happens to the tension in a muscle during and after a rapid decrease in muscle length (a *release*). As illustrated by the record in Fig. 2.4A, tension drops quickly during the release, and after the release is over, returns more slowly to its original level. Hill suggested that the muscle should be regarded as containing two components (Fig. 2.4B): (1) a spring referred to as the series elastic

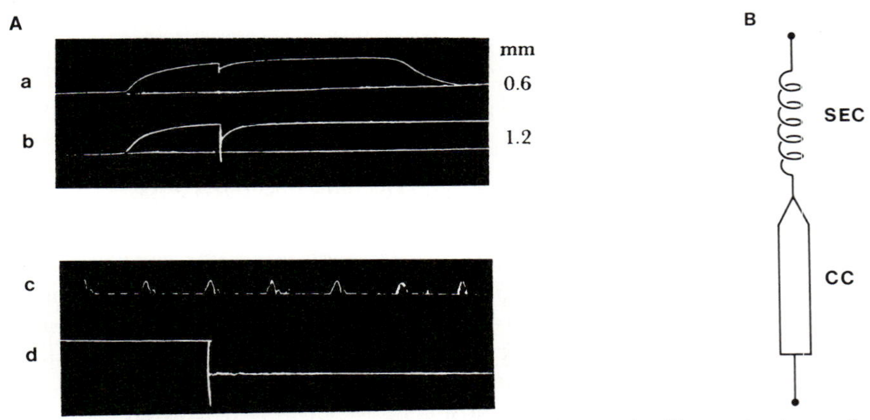

FIG. 2.4. (A) Effect of quick release on the tension recorded in a maximal isometric tetanus of a frog sartorius muscle. Top two records (a, b) are tension during an isometric tetanus with a quick release. The distance of the release is shown at the right of these traces. Trace c shows time marks at 0.2-s intervals. Trace d shows tension for a release of resting muscle under tension. All records were made at room temperature. (From Gasser and Hill, 1924.) (B) Diagram of Hill's two-component model to show the relation of the series elastic component (SEC) and the contractile component (CC).

component (SEC), whose length depends only on the force in the muscle—because the spring is undamped the SEC can change its length rapidly; and (2) a contractile component (CC), which can shorten but not so quickly as the SEC. The velocity of shortening by the CC depends only on the force. (Note: We are here considering the behaviour of a muscle at a length near the plateau of the length–tension curve.)

During a sufficiently rapid release the length of the CC does not change appreciably; all the change in muscle length occurs within the SEC. The tension therefore drops along the stress–strain curve of this spring. After the release is over the CC shortens, the SEC is restretched, and the tension rises again. The rate of tension redevelopment (dP/dt) at each moment is fixed by the speed at which the CC can shorten (dx/dt) and the stiffness of the SEC (dP/dx).

$$dP/dt = (dP/dx)(dx/dt)$$

The force–velocity relation of the muscle is determined only by the CC. It was further assumed that the properties of the CC and the SEC were the same throughout the contraction. Consequently all the mechanical properties of the muscle (at the optimum length) were completely specified by the force–velocity curve of the CC and the stress–strain curve of the SEC; both relations are assumed to be independent of time, once the muscle had become active. According to this model the process of activation at the start of stimulation consists in abruptly switching on the contractile component. Since it then develops tension by the same processes supposed to occur after a quick release, the time course of tension rise would also be the same, on the basis of this model.

This idea was reinforced by a number of experiments, of which perhaps the most striking (Fig. 2.5A) is the observation that when muscle is stretched shortly after the start of stimulation, by just the right amount, the tension rises to P_0 during the stretch and then remains steady. According to the two-component model this result is obtained because the SEC has been extended by just the amount required to raise the tension to P_0. The CC therefore neither shortens nor lengthens and tension remains constant. In this experiment the stretch was regarded as a procedure that probed the muscle's preexisting "capacity to exert force" or "active state." The experiments were extended to times when the muscle was not fully active, and sudden length changes were used to assess the time course of "active state," for instance during twitches.

The two-component model has been abandoned for the following reasons:

1. The stiffness of the muscle, which according to the two-component model is that of the SEC, is not constant. For example, muscle stiffness increases as the force increases during tension development (Hill, 1970; Bressler and Clinch, 1974; Huxley and Simmons, 1971b) and is less during shortening than in isometric contraction (Julian and Sollins, 1975).
2. The time course of tension development after a quick release, and more particularly at the beginning of a contraction, is slower than that predicted by the two-component model (Fig. 2.5B) (Jewell and Wilkie, 1958).
3. The velocity during isotonic shortening is found not to be an instantaneous function of force. It approaches its steady value with a complex time course after the imposition of a step change in force (Fig. 2.26) (Podolsky, 1960).
4. Most of the compliance of muscle is found to be in the crossbridges themselves (Ford *et al.*, 1981). There is therefore no shortening, in the sense of filament sliding, required to extend this compliance during tension development.
5. In experiments designed to measure "active state" in relaxing muscle the results are found to depend on the size of the length changes imposed on the muscle (Jewell and Wilkie, 1960). This makes it seem unlikely that these experiments really measure a "capacity to exert tension" existing in the unperturbed muscle.

One more aspect of the two-component model deserves comment. In this model, mechanical energy is stored in the SEC of the muscle while it is

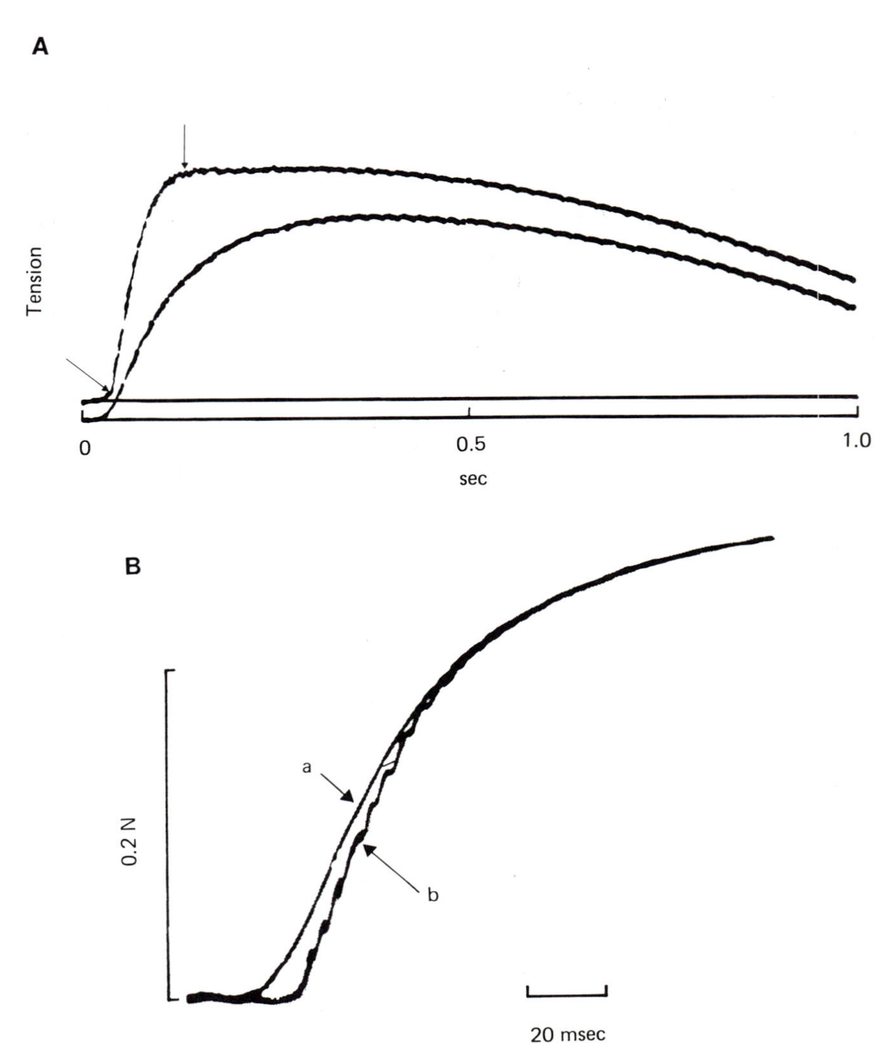

FIG. 2.5. (A) Records of tension in an isometric twitch (lower record) and in a twitch with stretch (upper record). Sartorius muscle from toad (*Bufo bufo*) at 0°C. The muscle length was 27.5 mm and it was stretched by 1 mm starting 38 ms after the stimulus, marked by the first arrow. The stretch followed an approximately exponential time course with a time constant of 32 ms. It was 95% complete at the second arrow. (From Hill, 1970.) (B) Experimental records of (a) the initial rise of tension in a 1-s tetanus (30 shocks/s) and (b) the redevelopment of tension after a quick release at 0.7 s. Frog sartorius muscle at 0°C. The tension rise before the release is not shown; the vibrations are produced by the sudden tightening of the connecting link. After the first 60 ms, during which the initial rise occurs more slowly than the redevelopment, the two traces coincide. (From Jewell and Wilkie, 1958.)

exerting tension (and only in the SEC). The amount of stored energy could easily be found (as a function of tension) from the stress–strain curve of the SEC. In an isometric contraction it was supposed that energy was put into the SEC during tension development and that this energy was dissipated as heat during relaxation. With the demise of this model it is necessary to reexamine the question of storage of mechanical energy in a muscle that exerts tension. What is the site of any stored energy? When is it released? How much is released?

2.2 Intact Muscles and Fibres

2.2.1 FORCE AND LENGTH

In Chapter 1 the relation between muscle length and the force produced in an isometric tetanus was described briefly. The interpretation of this relation in terms of sliding filaments was also outlined. In this chapter the experimental results are considered in more detail, with particular attention to quantitative comparison with predictions based on the sliding filament theory.

The results of a number of experiments on the relation between muscle or fibre length and the force exerted in fixed-end tetanic contractions are shown in Fig. 2.6. Length is expressed as a percentage of L_0, which is defined as the length in the centre of the plateau on the length–tension curve. (The symbol L_0 has been defined differently by other authors. For example, A. V. Hill used it to denote the length of a muscle *in situ*, which may not be the same as the length optimum for force development.) For all these experiments the extra force developed on stimulation, in excess of the passive tension, is shown by the points.

Figure 2.6D is from experiments by Close (1972b) on whole sartorius muscle from frog. His success in obtaining repeatable results with this preparation at very long muscle lengths is probably due to the particular care he took during dissection to keep the connective tissue at the ends of the muscle attached to the tendons. The relation of tetanic force to length has a broad maximum and falls to zero at 170% L_0 (as shown by the linear extrapolation in the figure).

Figure 2.6A is from the work of Ramsey and Street (1940) on single fibres dissected from frog semitendinosus muscle. The results differ from those for whole muscle in that force at long length falls to zero at about 200% L_0. Figure 2.6B and C are also for isolated single fibres from frog muscle and are in general similar to the results of Ramsey and Street except that the lengths at which force would be zero (indicated by the extrapolations on the

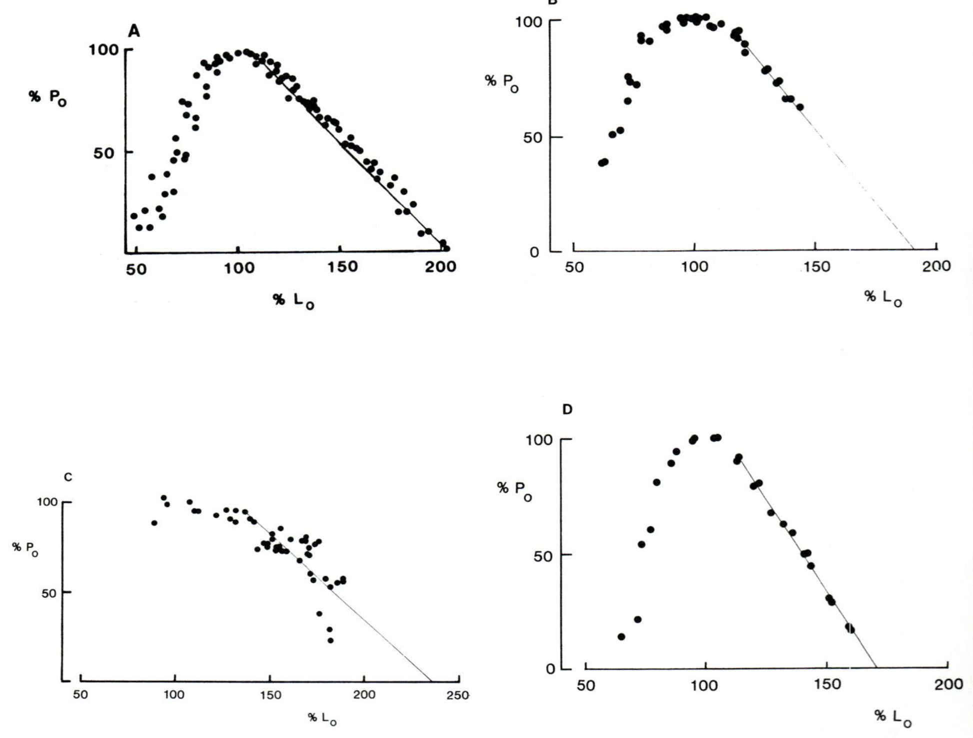

FIG. 2.6. Relation between length and force for various muscle preparations. In each case force is plotted as a percentage of the maximum value and length is shown as a percentage of L_0 (where L_0 is the length at the middle of plateau region where force is constant). The lines drawn on the figures are extrapolations of the linear fall of force as length is increased beyond L_0. (A) Single fibres from frog semitendinosus muscle (*Rana pipiens*). (From Ramsey and Street, 1940.) (B) Single fibres from frog semitendinosus muscle (*Rana temporaria*). (From Edman, 1966.) (C) Single fibres from frog anterior tibialis muscle (*R. pipiens*). (From J. D. Altringham, personal communication.) (D) Frog sartorius muscle (*R. temporaria*). (From Close, 1972b.)

figures) is in one case less and in the other greater than 200% L_0 (Edman, 1966; J. D. Altringham, personal communication).

Figure 2.7A from Gordon *et al.* (1966a,b) is obtained from experiments with a length-clamped segment in the centre of an isolated fibre from frog semitendinosus; the length of this segment was used to determine the abscissa in the figure. The region used for the length clamp was chosen for its uniformity of sarcomere length. Tension is found to be independent of length over a narrow range of lengths near L_0. There is a change in the slope of the curve at 80% L_0. Below this length, tension falls to zero at

65% L_0. In these experiments two different measurements of force were made from the records obtained at long length (Fig. 2.7A, inset): the maximum force in a 1-s tetanus and the force reached in the rapid phase of tension development, obtained by the extrapolation shown. The rapid phase declines linearly with segment length, and extrapolates to zero at a length of 172% L_0, although a very small amount of tension is observed at longer lengths than this. However, the maximum tension in a 1-s tetanus declines with length more slowly than the rapid phase does and extrapolates to zero at a length of about 180% L_0.

Figure 2.7C shows results obtained with mammalian muscle (Stephenson and Williams, 1982.) It can be seen that the tension declines from its optimum to zero at 143% L_0. This set of results was obtained with skinned fibres and is shown here because it is very complete. In experiments with intact fibre bundles from rat extensor digitorum longus (EDL) muscle, ter Keurs *et al.* (1981) found that the tension was zero at 147% L_0.

An understanding of these experimental results requires information about how the force produced by one sarcomere would be expected to vary with its length and also information about how the lengths of the sarcomeres within each experimental preparation are related to its overall length. Assuming the crossbridges act as independent force generators, the relation between the length of a single sarcomere and the force it produces can be predicted. This requires information about the length of the filaments and the distribution along them of sites at which crossbridges can form. In vertebrate muscle the length of the thick filament is the same in all species that have been investigated (Page and Huxley, 1963). The best estimate of this length is 1.56 μm (Craig and Offer, 1976). The length of the thin filament, however, is different in different species, varying from 1.98 (frog) to 2.38 μm (rat). The relationships between sarcomere length and isometric force predicted for these filament lengths is shown in Fig. 2.8A, B, and C for frog and for mammalian muscle.

Force is zero at a sarcomere length of 3.54 μm for frog (and 3.94 μm for mammalian muscle) because filament overlap is zero. At sarcomere lengths less than 3.54 (3.94) μm, isometric force should rise linearly as overlap increases until the overlap between the thin filaments and the crossbridge-bearing region of the thick filaments is maximal at a sarcomere length of 2.13 (2.54) μm. From 2.13 to 1.98 (2.54 to 2.39) μm, isometric force should not change because the number of force-generating sites is not changing. At lengths less than 1.98 (2.39) μm, it is likely that filaments will interfere with one another and quantitative predictions would require additional assumptions, for example about the compressibility of the filaments. In qualitative terms, changes in the slope of the force–length curve are expected at the following sarcomere lengths: at 1.98 (2.39) μm, below which the thin

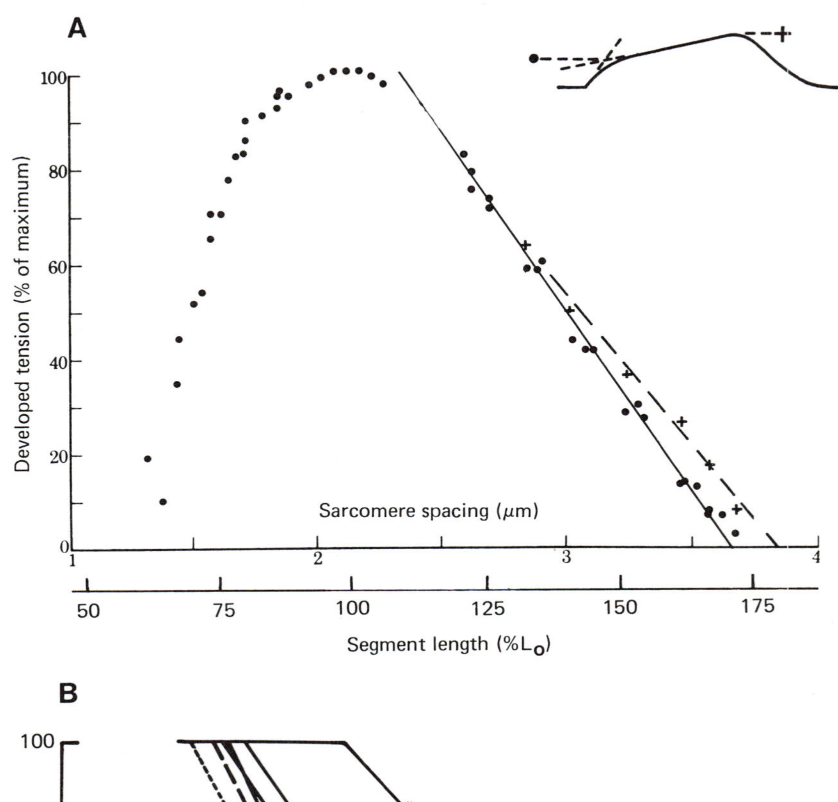
A
Developed tension (% of maximum)
100
80
60
40
20
0
Sarcomere spacing (μm)
1
2
3
4
50
75
100
125
150
175
Segment length (%Lo)

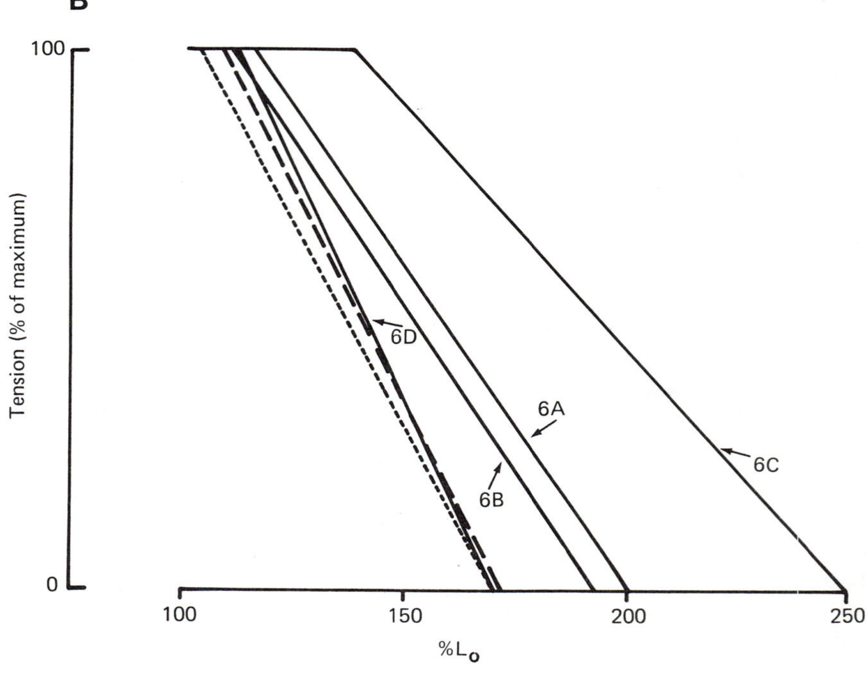
B
Tension (% of maximum)
100
0
6D
6A
6B
6C
100
150
200
250
%Lo

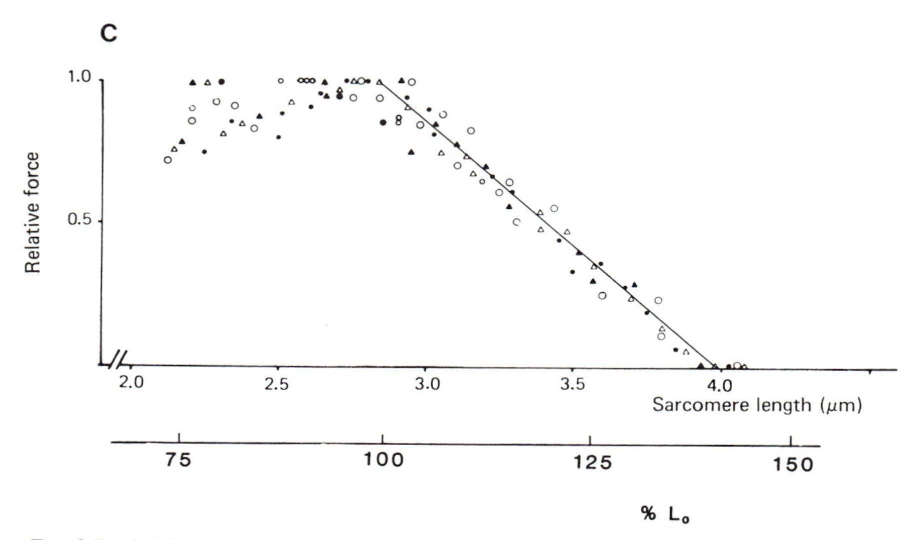

FIG. 2.7. (A) The relation between force and sarcomere length for length-clamped segments of single fibres from frog (*R. temporaria*) semitendinosus muscles. At long sarcomere lengths two measurements were made of tension as illustrated in the inset (which is a diagram of a force record for a 1-s tetanus); the points (●) were obtained by the extrapolation shown; the points (+) are the maximum force recorded. (B) A comparison of the lines from Fig. 6A, B, C, and D (solid lines) and Fig. 2.7A (dashed line). The dotted line is the predicted tension–length curve. See Fig. 2.8. (C) Sarcomere length–tension relation of soleus ($n = 9$) and extensor digitorum longus ($n = 7$), obtained using skinned muscle fibres from rat at low (●, ▲) and room temperature (○, △). Soleus fibres (●, ○); EDL (▲, △). The solid line is drawn through the points by eye. (From Stephenson and Williams, 1982.)

filaments cross over each other, and at 1.62 (1.62) μm, below which the thick filaments are compressed by the Z lines.

If each muscle fibre consisted only of a set of uniform sarcomeres, sarcomere length and fibre length would be proportional and the relation between fibre length and force would be as shown in Fig. 2.8D; force would be constant between 1.98 and 2.13 μm, it would decline linearly to zero between 2.13 and 3.54 μm, and there would be a change in the slope of the force–length relation at 1.62 μm. In Fig. 2.7B, the tension–length relation predicted from filament lengths is shown as a dotted line with summary lines from experiments superimposed for comparison. The predicted curve is very similar to the force–length curve (dashed line) obtained for the rapid phase of tension development in length-clamped segments of isolated fibres, in which the sarcomere length is known to be very uniform at the start of the contraction. In particular, Fig. 2.7A shows that force is constant between 2.00 and 2.25 μm, it declines to zero at 3.65 μm, and the force–length

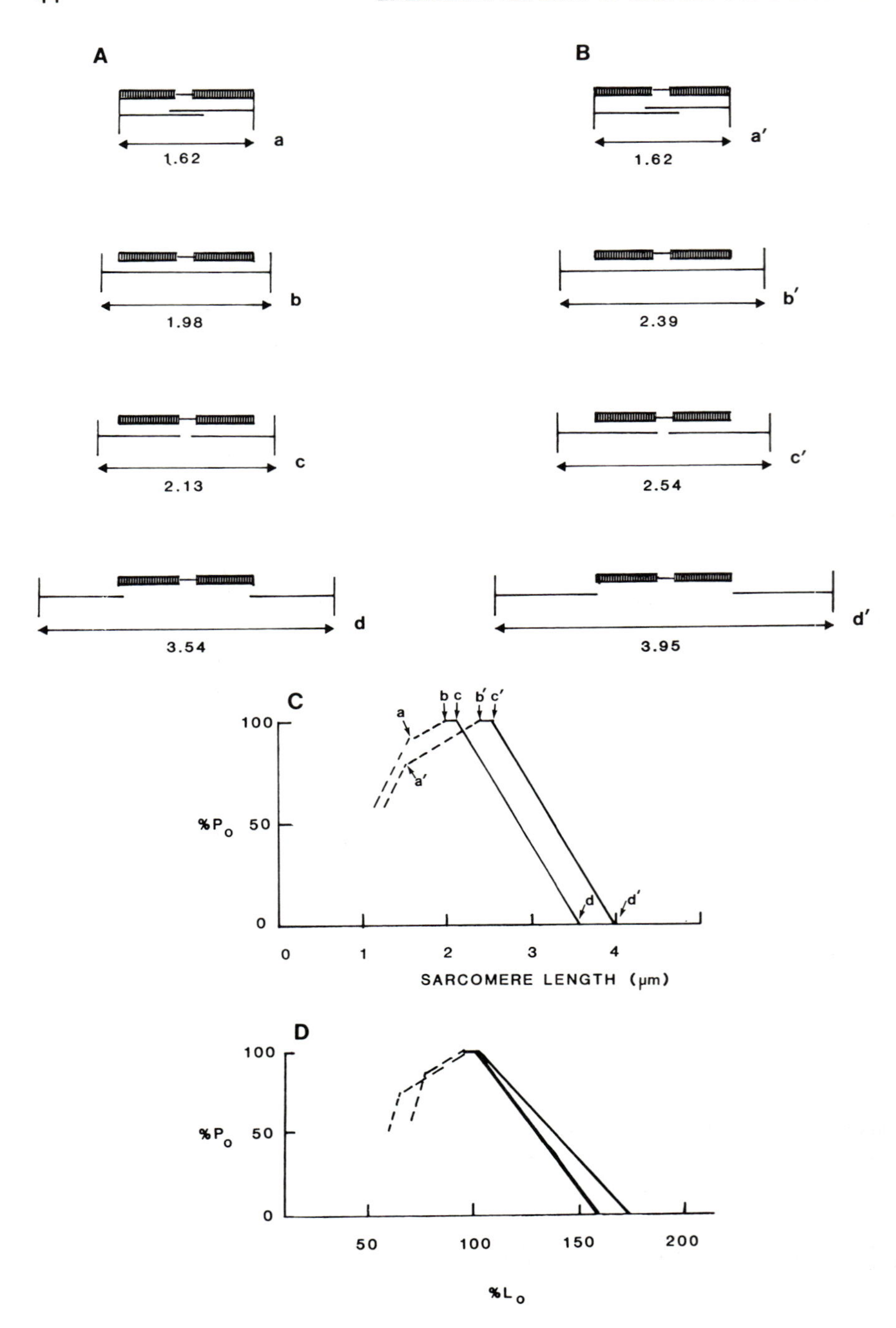
A
B
a
1.62
a′
1.62
b
1.98
b′
2.39
c
2.13
c′
2.54
d
3.54
d′
3.95
C
100
%P$_o$ 50
0
0 1 2 3 4
SARCOMERE LENGTH (μm)
D
100
%P$_o$ 50
0
50 100 150 200
%L$_o$

relation changes slope at 1.65 μm. The small discrepancies between the sarcomere lengths that are predicted and those observed are probably unimportant. Very similar results from experiments in which very small segments of a single fibre were length clamped have been reported recently by Edman and Reggiani (1983).

The result with whole frog muscle (Fig. 2.7B, line 6D) is very similar to that for length-clamped segments. This agreement shows that muscle length and sarcomere length are proportional in sartorius muscle, as they are in a length-clamped segment of a fibre. The results with mammalian muscle (Fig. 2.7C) agree with predictions in that tension declines to zero at a longer sarcomere length (4.0 μm) than in frog muscle, as expected from the longer thin filaments.

The other results with isolated fibres (Fig. 2.7B, lines 6A, B, and C) are different in that tension reaches zero at longer lengths and, at the right-hand end of the plateau, the data points lie below the corner formed by the extrapolated lines. These differences are due to the fact that the relation between fibre length and sarcomere length is more complex in these preparations. First, sarcomere length was probably not uniform along the length of the fibre. Huxley and Peachey (1961) have shown that when an unstimulated fibre is longer than its slack length, the last few sarcomeres at the ends are shorter than those in the rest of the fibre. This was also seen by J. D. Altringham (personal communication) in the fibres used for the experiments in Fig. 2.6C and was presumably also present in the experiments of Ramsey and Street (1940) and of Edman (1966). Second, at the level of the sarcomere, the conditions were probably not genuinely isometric during the fixed-end tetani in these experiments. The studies of Huxley and Peachey (1961) and of Gordon *et al.* (1966a) showed that when isolated fibres at long lengths are stimulated, the end sarcomeres, which have more filament overlap and therefore develop more force, shorten, and elongate the rest of the fibre. If there is filament overlap in the central sarcomeres, they are being stretched and are exerting more force than they would under isometric conditions (see next section). Alternatively if the fibre is so long that there is no filament overlap except at the ends, then parallel elastic elements provide

FIG. 2.8. (A) Diagram for frog muscle of the arrangements of the thick and thin filaments at different sarcomere lengths at which the slope of the tension–length curve would be expected to change. (B) As in (A) but for rat muscle, which has longer thin filaments. (C) Predicted tension–length relation of frog and rat muscle. The small letters refer to (A) and (B) above. (D) As in (C) but force is plotted against length, as a percentage of L_0, instead of sarcomere length.

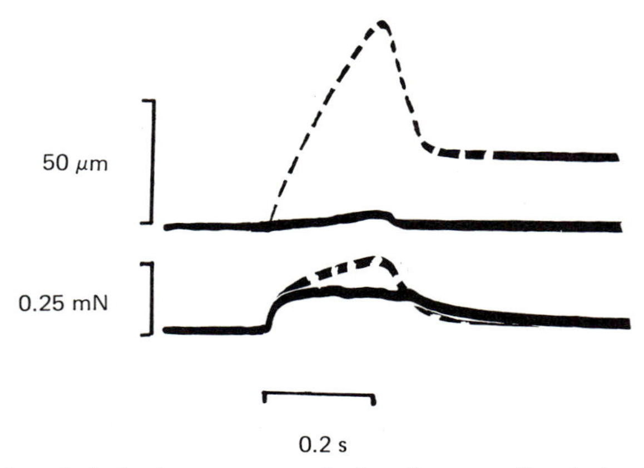

FIG. 2.9. Effect of eliminating servo control of marker separation during tetanic contractions of stretched muscle fibre, from an experiment by Julian *et al.* (1978) on a segment between two markers of a single fibre from frog semitendinosus muscle. The top traces are the output of the spot follower indicating change in marker separation (lengthening up). The bottom traces are the force output recorded at one fibre end. The initial marker separation was 2.79 mm, and this corresponded to an average sarcomere length of 3.23 μm. Temperature, 20°C. At full overlap P_0 was about 300 mN/mm^2. In the records shown by the full lines servo control was used to maintain a constant marker separation. In the next contraction, shown by the broken lines, the tendons of the fibre were fixed while the spot follower was used only to monitor changes in marker separation. The force response still contains an initial rapid upstroke but this is followed by a distinct creep phase. The creep phase is reflected in the upward deflection in the marker separation trace, which means that the ends of the fibre are stretching the middle.

continuity and are stretched. These ideas can explain the high forces observed at long fibre lengths. J. D. Altringham (personal communication) observed striation spacing by direct microscopy and found that shortening of the ends of the fibres did in fact occur in the experiments shown in Fig. 2.6C.

The experiments of Julian *et al.* (1978) show that these phenomena are also responsible for the "creep," that is the second, slower phase of increasing force during fixed-end tetani of fibres at long sarcomere lengths (Fig. 2.9). In contrast there was no creep at shorter lengths. They observed that, when the length clamp was not used, the gold markers they had stuck to the surface of the fibres moved. During the creep phase, these movements showed that the ends of the fibre shortened and the centre section was elongating. Length clamping the centre region reduces, but does not entirely eliminate, the creep phase of tension development. It seems that the residual creep of tension is due to nonuniformity of sarcomere length within the clamped region, or to nonuniformity of filament arrangement within the sarcomeres (A. F. Huxley, 1981), but direct evidence on this point is lacking.

2.2.2 FORCE AND VELOCITY

2.2.2.1 *Introduction*

When a muscle is stimulated, the amount of force it produces depends on whether, and how, its ends are restrained. If the ends are completely free to move, no force is produced, and the muscle shortens at its maximum velocity, V_{max}. If one end of the muscle is fixed and a small force is applied to the other end, the muscle shortens at a steady velocity V, less than V_{max}; the larger the force the slower the velocity of shortening. A sufficient force, P_0, will prevent shortening; this is an isometric contraction. Forces greater than P_0 will produce lengthening of the muscle, that is, V will be negative.

The force–velocity relation, that is, the absolute values of P_0 and V_{max} and the shape of the curve, summarizes quantitative information about the mechanical behaviour of muscle, which can be taken as the behaviour of a large population of crossbridges operating in a steady state. Therefore, an understanding of the crossbridge cycle should provide an explanation of the force–velocity curve and any serious theory must be designed to produce a force–velocity relation of the correct form.

In this section the relation between force and velocity of shortening is considered first. It has received more attention, perhaps because mechanical work is done by the muscle during shortening. The relation for lengthening is then described.

2.2.2.2 *The Force–Velocity Curve for Shortening*

In his 1938 paper, A. V. Hill described the relation between force and the velocity of shortening as part of a rectangular hyperbola:

$$(P + a)(V + b) = (P_0 + a)b$$

where P is the force during shortening at velocity V, P_0 is the force during an isometric tetanus, and a and b are constants. The curvature of the hyperbola (Fig. 2.10A) is specified by the value of a/P_0 (which is equal to b/V_{max}, Table 2.I). Hill's "characteristic equation" has been the most popular such equation, but others describe the relation of force and velocity just as well (for example, Aubert, 1956). Hill originally thought that a and b could be obtained from heat measurements as well as from mechanical measurements, and therefore, the equation was believed to be of fundamental significance. It is now clear that this is not so (see Chapter 4, Section 4.2.2).

Experiments by Edman *et al.* (1976) show that, in fact, the observations for frog twitch fibres systematically deviate from a hyperbola. This deviation occurs only at the high-force ($P > 0.8P_0$), low-velocity end of the curve (Fig. 2.10C). If a hyperbola is fit to the data points for forces less than $0.8P_0$,

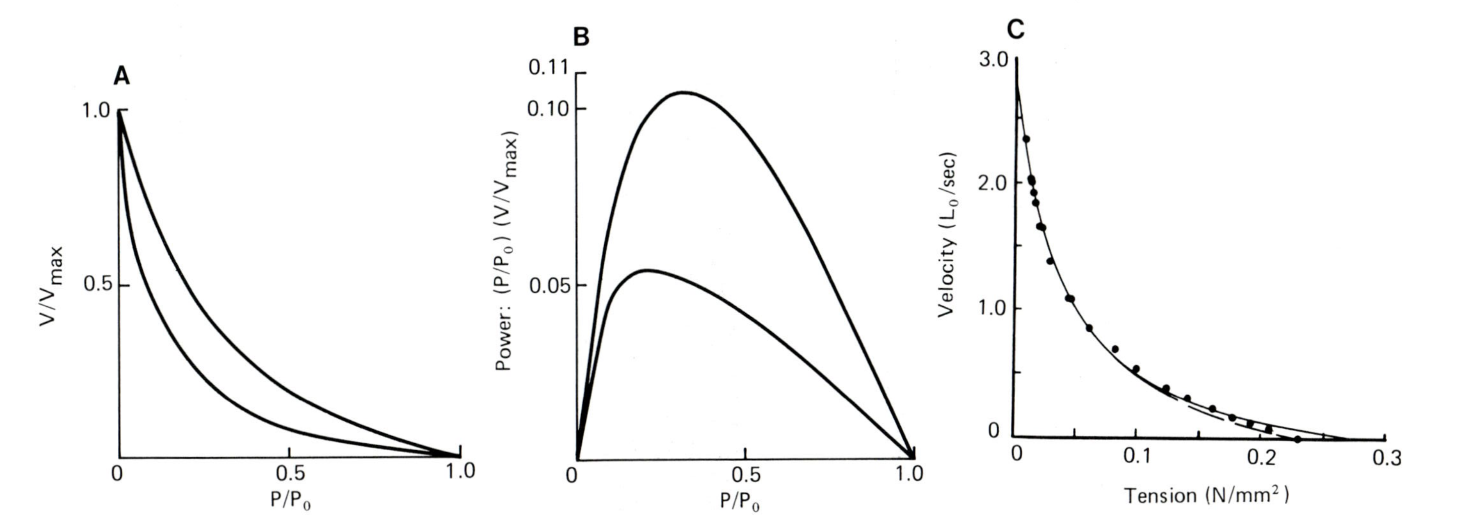

FIG. 2.10. (A) Force–velocity curves calculated from the normalized form of the Hill equation (see Table 2.I)

$$\frac{V}{V_{max}} = \frac{(1 - P/P_0)}{1 + (P/P_0)(P_0/a)}$$

where V_{max} and P_0 are the maximum velocity of shortening and isometric force, respectively, P is the force produced during shortening at velocity V, and a/P_0 is either 0.3 (upper line) or 0.1 (lower line). (B) Power–force curve calculated from the information in (A) using power = force × velocity. Upper line, $a/P_0 = 0.3$; lower line, $a/P_0 = 0.1$. Note that a more curved force–velocity curve, which has a smaller value of a/P_0, also has a lower maximum power output and is produced at a lower force. (C) Force–velocity curve determined from load-clamp data (●) from a single fibre from frog (*R. temporaria*). All data refer to the plateau of a tetanus at approximately 2.1 μm sarcomere length. Fibre length and cross-sectional area at 2.24 μm sarcomere length, 11.59 mm and 12.2×10^{-3} mm^2. Temperature, 2.7°C. The continuous line is a least-squares fitting of a hyperbola to velocity data obtained at loads less than or equal to 80% of the tetanic force. The dashed line is a hyperbola fitted to the data but constrained to pass through the maximum force value. (From Edman, 1979.)

TABLE 2.I

Some properties of Hill's equation

Hill's equation, giving the relation between force P and velocity V, is generally written as

$$(P + a)(V + b) = (P_0 + a)b \qquad \text{(T-1)}$$

where P_0 is the force exerted when $V = 0$ and a and b are constants having the dimensions of force and velocity, respectively. It can be seen that the equation is a rectangular hyperbola with asymptotes at $-a$ and $-b$.

The velocity at $P = 0$ is termed V_{max}. Making these substitutions in Eq. (T-1):

$$a(V_{max} + b) = (P_0 + a)b$$

$$aV_{max} + ab = P_0 b + ab$$

$$aV_{max} = P_0 b$$

$$V_{max}/b = P_0/a \qquad \text{(T-2)}$$

This ratio is a dimensionless constant. It is sometimes convenient to write Hill's equation in a form in which all terms are dimensionless, that is, as the relation between P/P_0 and V/V_{max}. This is achieved by dividing Eq. (T-1) by $P_0 V_{max}$, which gives

$$(P/P_0 + a/P_0)(V/V_{max} + b/V_{max}) = (1 + a/P_0)b/V_{max}$$

Writing P' for P/P_0, V' for V/V_{max}, and G for P_0/a [$= V_{max}/b$, see Eq. (T-2) above]

$$(P' + 1/G)(V' + 1/G) = (1 + 1/G)(1/G)$$

$$P'V' + P'/G + V'/G = 1/G$$

$$P'(V' + 1/G) = (1 - V')(1/G)$$

$$P' = (1 - V')/(1 + V'G) \qquad \text{(T-3)}$$

$$V' = (1 - P')/(1 + P'G) \qquad \text{(T-4)}$$

In this dimensionless form the equation is symmetrical, as is shown by comparison of Eqs. (T-3) and (T-4). Both asymptotes are equal to $-1/G$. The maximum power output (force $\times$ velocity) occurs when $P' = V'$ and is, in dimensionless terms, $P'V'$. Writing M for the value of P' at which force is equal to V'

$$M = (1 - M)/(1 + MG)$$

$$GM^2 + 2M - 1 = 0$$

Solving for M gives

$$M = [\sqrt{(1 + G)} - 1]/G \qquad \text{(T-5)}$$

and the normalized maximum power output (M^2) is

$$M^2 = (2 + G - 2\sqrt{(1 + G)})/G^2 \qquad \text{(T-6)}$$

The power output is of course $M^2 \cdot P_0 \cdot V_{max}$, which would be in milliwatts if P_0 is in newtons and V_{max} is in millimeters per second.

From Eqs. (T-5) and (T-6) it is clear that the value of the maximum power output (M^2) and the force (and velocity, M) at which maximum power is produced both depend on the curvature ($G = P_0/a$) of the force–velocity curve.

the observed forces above $0.8P_0$ fall below this hyperbola and the curve extrapolates to a P_0 value approximately 20% higher than that observed (solid line in Fig. 10C). However, if the fitted hyperbola is constrained to pass through the observed P_0 (as is often done), then the observations for force between about $0.5P_0$ and P_0 lie above the fitted curve (dashed line in Fig. 2.10C). The same result is obtained with afterloading and with quick-release isotonic shortening, and with bundles of fibres as well as single fibres (Edman *et al.*, 1976). Twitch fibres from *Xenopus* have a similarly nonhyperbolic force–velocity curve (Lannergren, 1978; Lannergren *et al.*, 1982). The results for slow fibres from *Xenopus* deviate at the high-velocity end of the curve as well as at the high-force end (Lannergren, 1978; Lannergren *et al.*, 1982). Thus it seems that under a range of conditions the force–velocity curve has a nonhyperbolic shape. Since Hill's "characteristic equation" is descriptive only, accepting this modification does not require that any fundamental ideas be revised.

It seems surprising that this behaviour was not detected earlier, considering that the "characteristic equation" was so well known. The explanation seems to be that earlier force–velocity curves have few, if any, points at forces between $0.8P_0$ and P_0 (see, for example, Hill, 1938; Katz, 1939; Jewell and Wilkie, 1958; A. V. Hill, 1970).

Table 2.II is a collection of results about the force–velocity properties of various muscles and muscle fibres as described by Hill's equation. Table 2.III summarizes published information about the effect of temperature on the parameters of the force–velocity curve.

There is considerable variation between muscles in V_{max} and to a lesser extent in P_0 (not shown in Table 2.II) and in a/P_0. The variations in V_{max} can be attributed to these factors:

1. The different temperatures at which the muscles have been studied. V_{max} has a Q_{10} of about 2 and is more sensitive to temperature change than are the other parameters of the force–velocity curve.
2. Variations between different types of fibres in the same animal.
3. The different sizes of the animals used. As pointed out by Hill (1950b, 1956) and by Close (1972a), one would expect on dimensional arguments that the speed of shortening of muscle would be inversely proportional to the cube root of the weight of the animal.
4. Other factors. For example, the muscles of the toad are slower than those of the frog, an animal of comparable size (Hill, 1950b). Similarly the muscle of the sloth are slower than those of the cat (Marechal *et al.*, 1963).

P_0 is usually expressed relative to the cross-sectional area of the whole muscle (see Table 1.I) and would be expected to be influenced by the density of filament packing and the amount of extracellular space in a muscle. Some of the variations in normalized P_0 values doubtless due to these factors. Variations are also due to the different temperatures used. It has been suggested by Luff (1981) and Close and Luff (1974) that some variations between different muscles in the force exerted per unit cross section are not due to these factors. This conclusion is based on the low values of P_0 they observed in rat and mouse inferior rectus, an eye muscle, compared to the limb muscles from the same animals.

The value of a/P_0 is less temperature dependent than either V_{max} or P_0. The variations in a/P_0 appear to be correlated with variations in V_{max}. This is illustrated in Fig. 2.11 which shows that for (1) the same preparation at different temperatures, (2) different muscles or muscle fibres from the same animal, and (3) analogous muscles from different species, larger values of V_{max} are in most cases accompanied by larger values of a/P_0.

a. Interpretations of the Force–Velocity Curve. As mentioned above, theories must take account of the force–velocity relation. The values of P_0, V_{max}, and a/P_0, which actually specify the curve, are of more general interest because on the basis of ideas about crossbridges, there are simple counterparts to these parameters.

1. P_0. The value of P_0 depends on the number of crossbridges that are attached, assuming that the force per attached crossbridge is a constant under isometric conditions.

2. V_{max}. The value of V_{max} reflects the maximum rate of crossbridge turnover, but is independent of the number of bridges that are operating.

3. a/P_0. A simple counterpart of the constant a/P_0, which is a measure of the curvature of the force–velocity relation, is not so obvious. It can, however, be understood in terms of the hypothesis that there are two processes responsible for the decline in force as velocity rises (Chapter 5, Section 5.1): decreased number of bridges exerting positive force and increased number exerting negative force. At high forces the first of these reasons would be more important; it determines the slope of the force–velocity curve near P_0. At high velocity the second process is more important and determines the slope near V_{max}. The curvature of the force–velocity curve (a/P_0) is determined by the ratio of these two slopes, and thus its value indicates the relative contribution of these two processes.

The value of a/P_0 is related to the shortening velocity (V/V_{max}) at which mechanical power output is maximal (see Table 2.I). Woledge (1968) has proposed that a/P_0 is also related to the thermodynamic efficiency of the muscle.

4. *Other terms.* As described above, the hyperbolic curve does not describe the complete force–velocity relation. Other terms are required to fit the

TABLE 2.II

Force–velocity properties of various muscles and single fibres

V_{max} (μm/s per half-sarcomere)[a]	V_{max} (L_0/s)	a/P_0	b (L_0/s)	n (muscles used)	T (°C)	Muscle and species	Reference
1.29	1.23	0.22	0.27	1		Sartorius, *R. temporaria*	Hill (1938)
1.38	1.31	0.26 (0.18–0.33)	0.34 (0.24–0.42)	12	0	Sartorius, *R. temporaria*	Katz (1939)
3.10	2.95	0.38 (0.32–0.43)	1.12 (1.1–1.15)	3	10.9		Katz (1939)
0.73	0.58	0.11	0.032	6	16.2	Retractor penis, tortoise (*Testudo* sp.)	Katz (1939)
1.95	1.86	0.28	0.52	1	0	Sartorius, *R. temporaria*	Ritchie and Wilkie (1958)
8.75	7.0 (3.0–11.1)	0.27 (0.14–0.43)	1.89	10	37	Tenuissimus, cat	McCrory *et al.* (1966)[b]
9.1 ±1.2	7.3	0.17 (0.15–0.20)	2.48	≥4	35	Soleus, rat	Close (1964)
21.3 ±2.3	17.0	0.25 (0.19–0.32)	6.84	≥4	35	EDL, rat	Close (1964)

15.8	12.7	0.279	7.09	2	35	Soleus, mouse	Close (1965)
30.3	24.2	0.258	12.5	2	35	EDL, mouse	Close (1965)
0.28	0.23 ±0.03	0.072 ±0.008	0.016	8	0	Rectus femoris, tortoise (*Testudo graeca* or *Testudo hermanii*)	Woledge (1968)
1.94	1.85	0.39	0.72	6	0	Twitch fibre, anterior tibialis, *R. pipiens*	Julian and Sollins (1972)
14.9	11.9	Results not fitted to Hill's equation		8	35	Inferior rectus, rat	Close and Luff (1974)
11.8	9.4			4	35	EDL, rat	
1.24	1.18	0.197	0.235	1	0	Semitendinosus, *R. esculenta*	Cavagna and Citterio (1974)[b]
2.35	2.24	0.25	0.56	3	0	Sartorius, *R. pipiens*	Curtin and Davies (1975)
1.89	1.80 ±0.060	0.290 ±0.023	0.501 ±0.047	16	1–2.5	Twitch fibre, semitendinosus, *R. temporaria*	Edman *et al.* (1976)
1.56	1.49 (1.31–1.66)	0.35 (0.30–0.45)	0.52 (0.42–0.63)	4	5	Twitch fibre, iliofibularis, *X. laevis*	Lannergren (1978)
2.68	2.55 (2.38–2.81)	0.35 (0.31–0.42)	0.88 (0.79–1.03)	4	10		Lannergren (1978)
5.46	5.20 (4.64–5.89)	0.38 (0.27–0.47)	1.97 (1.53–2.24)	6	20		Lannergren (1978)
1.16	1.1	0.1	0.11	7	21–24	Slow fibre, iliofibularis, *X. laevis*	Lannergren (1978)

(*Continued*)

Table 2.II-*continued*

Force–velocity properties of various muscles and single fibres

V_{max} (μm/s per half-sarcomere)[a]	V_{max} (L_0/s)	a/P_0	b (L_0/s)	n (muscles used)	T (°C)	Muscle and species	Reference
2.49	2.37 ±0.18	0.34 ±0.05	0.77 ±0.08	6	4 ± 0.3	Twitch fibre, semitendinosus, *R. esculenta*	Cecchi *et al.* (1978)
9.03	8.60 ±0.29	0.44 ±0.03	3.77 ±0.25	15	19.8 ± 0.2		
10.07	9.59	0.30	2.48	8	20	Twitch fibre, semitendinosus, *R. esculenta*	Cecchi *et al.* (1981)
25.8 ±0.8	20.6	0.329	6.51	6	35	Inferior rectus, mouse	Luff (1981)
24.7 ±0.3	19.8	0.367	6.97	6	35	EDL, mouse	Luff (1981)
19.8 ±0.8	15.8	0.265	4.05	6	35	Diaphragm, mouse	Luff (1981)
11.5 ±0.3	9.2	0.180	1.60	6	35	Soleus, mouse	Luff (1981)

5.04	4.8 ±0.15	0.35 ±0.14	1.43 ±0.28	8	10	Type 1 fibre, iliofibularis, *X. laevis*	Lannergren *et al.* (1982)
7.46	7.1 ±0.36	0.47 ±0.12	2.69 ±0.28	11	15		
9.66	9.2 ±0.36	0.48 ±0.09	3.51 ±0.23	13	20		
3.68	3.5 ±0.35	0.26 ±0.10	0.75 ±0.17	9	10	Type 2 fibre, iliofibularis, *X. laevis*	
5.46	5.2 ±0.31	0.27 ±0.11	1.15 ±0.22	8	15		
7.46	7.1 ±0.37	0.26 ±0.07	1.53 ±0.16	15	20		
8.1	6.5 ±0.1	0.212 ±0.01	0.99	19	35	Soleus, rat	Ranatunga (1982)[b]
6.5	5.2 ±0.4	0.18 ±0.02	0.68	5	30		
4.6	3.7 ±0.2	0.157 ±0.01	0.42	5	25		
2.8	1.6 ±0.1	0.137 ±0.01	0.22	6	20		
16.9	13.5 ±0.4	0.415 ±0.01	3.20	13	35	EDL, rat	
12.6	10.1 ±0.4	0.429 ±0.02	2.47	5	30		
10.1	8.1 ±0.5	0.385 ±0.03	1.77	5	25		
7.5	6.0 ±0.3	0.283 ±0.01	0.96	5	20		

[a] Calculation of V_{max} per half-sarcomere assumed that the sarcomere length of amphibian muscle was 2.1 μm and that of other muscles was 2.5 μm.
[b] In these experiments the muscle fibre lengths were not determined and were probably less than L_0. V_{max} and b may, therefore, be underestimated.

TABLE 2.III

The effects of temperature on the parameters of the force–velocity curve

Q_{10} values						
P_0	V_{max}/L_0	a/P_0	b/L_0	T (°C)	Preparation	Reference
1.35	1.66	1.37	2.27	10–20	Fibre Type 1, from iliofibularis, *Xenopus*	Lannergren *et al.* (1982)
1.19	2.08	1.00	2.08	10–20	Type 2	
1.41	1.75	1.80	3.15	10–15	Type 1	
1.29	1.58	1.04	1.64	15–20		
1.20	2.36	1.08	2.55	10–15	Type 2	
1.18	1.83	0.93	1.70	15–20		
1.22	2.25	1.18	2.66	4–19.8	Fibre, semitendinosus, *R. esculenta*	Cecchi *et al.* (1978)
1.24	2.67	—	—	2–12	Fibre, semitendinosus or anterior tibialis, *R. temporaria*	Edman (1978)
1.30	—	—	—	0–10	Sartorius, *R. temporaria*	Bressler (1981)
1.22	—	—	—	10–20		
1.23	2.83	1.31	3.71	20–25	Soleus, rat	Ranatunga (1982)
1.09	1.98	1.31	2.59	25–30		
0.98	1.53	1.39	2.13	30–35		
1.14	1.84	1.85	3.40	20–25	EDL, rat	
1.05	1.56	1.24	1.93	25–30		
0.99	1.80	0.94	1.69	30–35		

results for high forces. Additional studies of this part of the relation under other experimental conditions are needed. The results might give insight into how this feature is involved in crossbridge function.

b. Effect of Sarcomere Length on the Force–Velocity Curve. The relation between P_0 and sarcomere length is described in Section 2.2.1. How does V_{max}, which should be independent of the number of crossbridges, depend on sarcomere length? Are the results consistent with those for P_0? Edman (1979) has investigated these questions using single fibres from frog muscle and a new method of determining V_{max} (Fig. 2.12). This method is a development of a technique described by A. V. Hill (1970). The fibre is released rapidly during a tetanus so that the tension drops to zero, and a measurement is made of the time to the start of tension redevelopment. This is measured for different distances of release, and the *slope* of the relation between the size of the release and the measured time gives V_{max}. As the slope is not

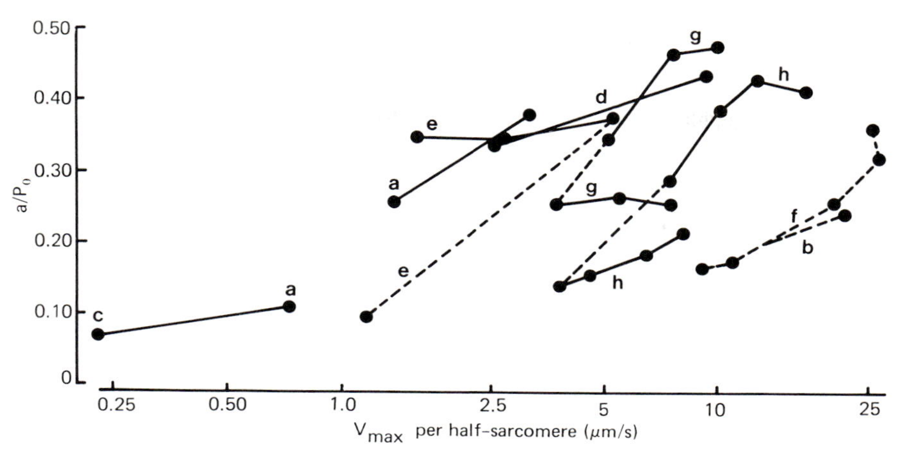

FIG. 2.11. The relation between a/P_0 and V_{max}. The points represent mean values from Table 2.II. The solid lines join points representing the same type of preparation studied at different temperatures. The dashed lines join points representing different types of muscle, or fibres, from the same species. Most of the lines have a positive slope. References: (a) Katz (1939), (b) Close (1964), (c) Woledge (1968), (d) Cecchi *et al.* (1978), (e) Lannergren (1978), (f) Luff (1981), (g) Lannergren *et al.* (1982), and (h) Ranatunga (1982).

influenced by elasticity in series with the fibres, the properties of such elastic elements need not be evaluated. This method was shown to give the same values of V_{max} as extrapolation of the force–velocity curve (Edman, 1979). His results on the relation of V_{max} and sarcomere length are shown in Fig. 2.13A and can be summarized as follows:

1. In the range of sarcomere lengths 1.65–2.7 μm, V_{max} was constant, whereas P varied as expected on the basis of filament overlap. The independence of V_{max} and P_0 between 2.0 and 2.3 μm agrees with expectations for crossbridge properties. The results for sarcomere lengths between 1.65 and 2.0 μm, however, do raise some questions when the isometric force at these lengths is considered. If the reason for there being less isometric force at lower sarcomere lengths is that fewer crossbridges are operating, then V_{max} would be expected to remain constant, as Edman observes that it does. On the other hand, it could be that isometric force is lower because there is a force within the fibre that is opposing that produced by the crossbridges. If this is the case, V_{max} would also be diminished. Recently, Morgan and Julian (1981a) have given a brief account of stiffness in isometric contraction, which shows that it was not proportional to tension at sarcomere lengths between 1.65 and 2.2 μm. If stiffness is taken as a measure of the number of attached bridges, this suggests that there is a force within the muscle opposing that produced by the crossbridges. Thus the simple interpretation of the stiffness

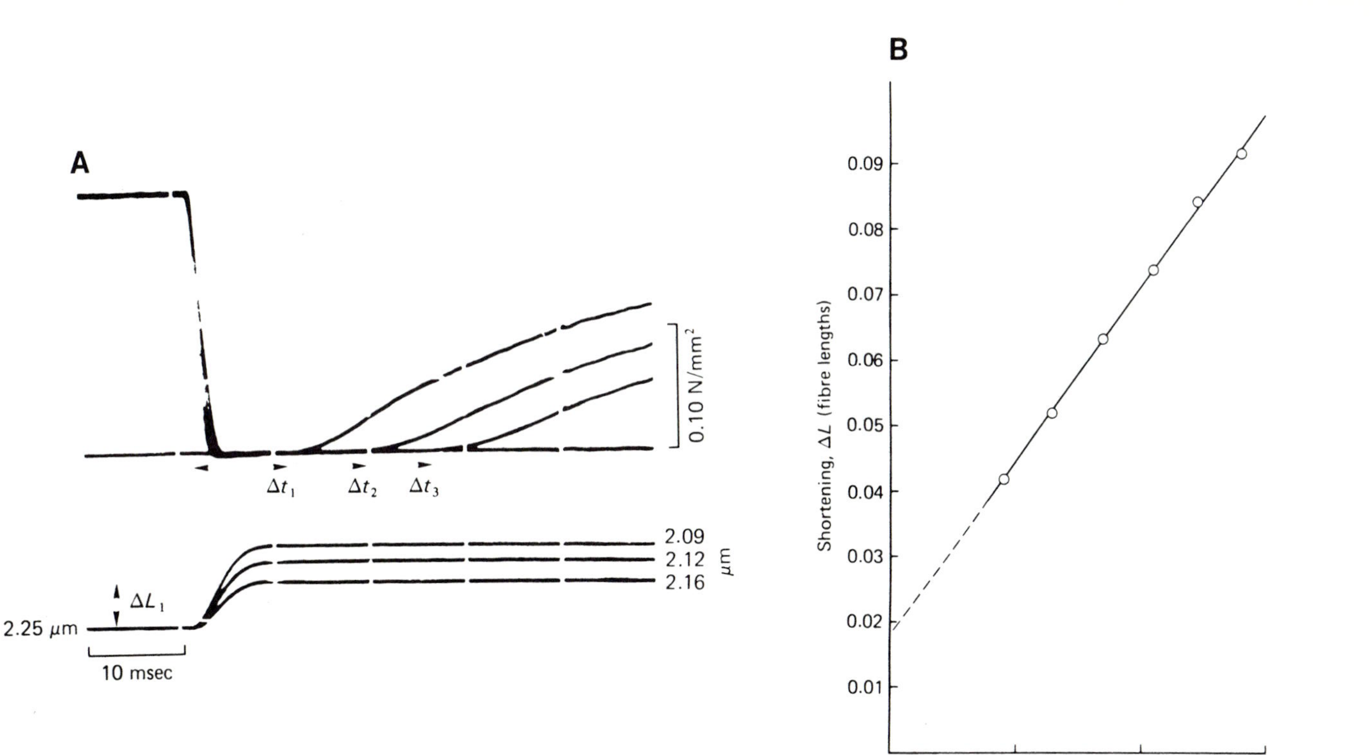

FIG. 2.12. (A) Examples of experimental records. Superimposed oscilloscope records of a single fibre from frog (*R. temporaria*) showing three quick releases of different amplitudes during plateau of fused tetanus. Upper traces, tension. Lower traces, release steps calibrated in micrometers sarcomere length. Fibre length and cross-sectional area at 2.25 μm sarcomere spacing, 9.00 mm and 0.012763 mm^2. Temperature, 0.87°C. (B) Graph of results from an experiment like that in (A), but from a different fibre, showing relationship between amplitude of shortening (ΔL) and time from onset of release to beginning of force redevelopment (Δt). Release performed from 2.25 μm sarcomere length during plateau of tetanus. Each data point is the mean of five release recordings. Straight line, least-squares regression of Δt upon ΔL (correlation coefficient = 0.988; n = 30). Slope of line, V_{max}. Intersection of line with ordinate, total series compliance. Fibre length and cross-sectional area, 12.5 mm and 0.0076 mm^2. Temperature, 2.4°C. (From Edman, 1979.)

3. At sarcomere lengths longer than 2.7 μm, V_{max} increases with increasing sarcomere length. Unstimulated fibres develop passive tension over this range of lengths and can shorten rapidly. The passive force seems to act as a compressive force (that is, directed toward the centre of the fibre) enhancing V_{max} above that observed at shorter sarcomere lengths, but not above that observed in unstimulated fibres at the same sarcomere length.

Edman (1979) uses this argument to extend the force–velocity curve beyond $P = 0$ in the following way. The value of the force in the unstimulated fibre muscle at each length was taken to be a negative force on the fibre ($P < 0$), and the corresponding value for the velocity of shortening of active muscle was measured. The results, expressed relative to the results for sarcomere length 2.1 μm, were added to the force–velocity curve for sarcomere length 2.1 μm (see Fig. 2.13B). The interesting feature is that the points for negative forces lie on a smooth continuation of the positive force–velocity curve. The fact that the force–velocity curve does not change its shape as it passes through the $P = 0$ point suggests that the properties of the bridges do not change here. This behaviour is in contrast to the other end of the force–velocity curve, where there may be discontinuity in the relation between force and velocity as velocity changes sign (see below).

c. Force–Velocity Curve at Early Times in a Tetanus. At the start of an isometric tetanus of frog muscle at 0°C, the force increases to its maximum value in about 200 ms (Jewell and Wilkie, 1958; Cecchi *et al.*, 1978). How rapidly does the muscle shorten if it is released during this time? Assuming that each crossbridge operates independently and that Ca^{2+} acts simply as a switch permitting the interaction of actin sites on the thin filament with crossbridges, V_{max} would be the same during this time as during the time when tension is maximum in an isometric tetanus. Results of experiments by Jewell and Wilkie (1958) indicate that the velocity of shortening is the same as these two stages of a tetanus. In their study, shortening was always isotonic, that is, the muscle lifted a load. In one set of experiments (after-loaded isotonics), the muscle started shortening as soon as it produced enough force to move the load. (The time interval between the start of stimulation and the beginning of shortening for each load is given in the legend to Fig. 2.14.) In the other sets of experiments (quick releases) the muscles started shortening much later in the tetanus. After the full tension had been developed, the muscle was suddenly subjected to a load smaller than the isometric force; it shortened rapidly for a few milliseconds, then continued shortening at a slower steady rate. It is this second phase of shortening that is of interest here. The results for these two types of experiments were extremely similar and a single force–velocity curve fitted the results very well (see Fig. 2.14). It seems from these results that, as soon after

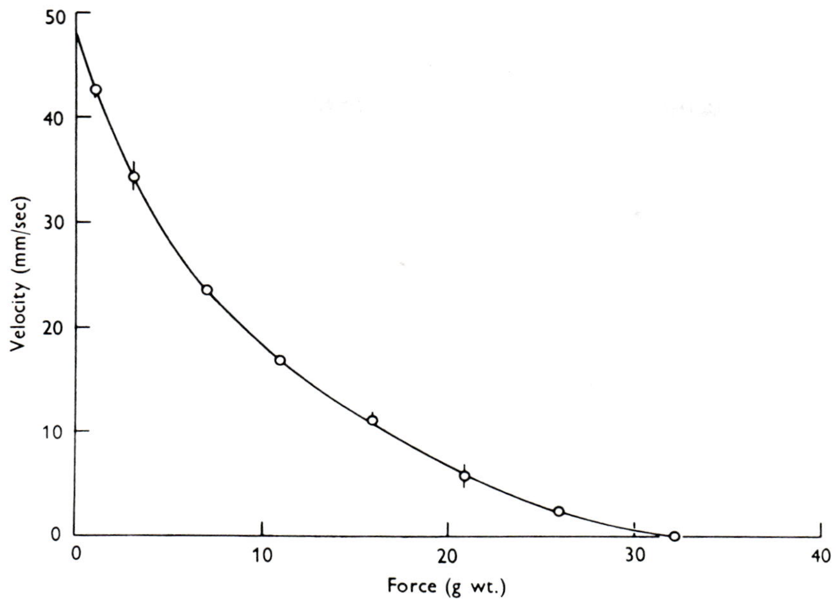

FIG. 2.14. Force–velocity relation for quick releases after full isometric force had been developed and isotonic afterloaded contractions of frog sartorius at 0°C. The results of these two types of experiments are so similar that the points usually superimpose. The vertical lines show the maximum range of variation where it exceeds the diameter of the point. In the experiment involving afterloaded shortening, there was a delay between the start of stimulation and the start of shortening. The duration of the delay was longer, the larger the isotonic load the muscle had to shorten against. Duration of the delay corresponding to each isotonic load:

Delay (ms)	P/P_0
18	0.03
24.5	0.09
34	0.22
42.5	0.35
58	0.50
84	0.66
90	0.82

Note that under afterloaded isotonic conditions, shortening started at various times after the start of stimulation, in some cases very soon, whereas in the quick-release experiments shortening did not start until much later when full isometric force had been developed. Despite the differences in the time at which shortening started, all the force–velocity results were the same. (From Jewell and Wilkie, 1958.)

stimulation as the muscle can shorten against a particular force, it does so as fast as when shortening starts later in the tetanus.

This idea, however, is not supported by the results of experiments by Cecchi *et al.* (1978), in which fibres were released at various times after stimulation had started (Fig. 2.15A). In these experiments the velocity was the controlled variable. In most cases the release occurred in two stages: a brief, rapid step, followed by constant velocity movement. This was done because, when shortening begins with a rapid step, the force is almost constant for the remainder of the shortening. In some experiments no rapid step was needed. The results were the same regardless of whether or not an initial rapid step was used. These conditions would seem to be equivalent to the afterloaded isotonic experiments of Jewell and Wilkie (1958). Figure 2.15B shows an example of the force–velocity relations Cecchi *et al.* observed at three different times after stimulation started (75, 100, and 190 ms). The tension had already developed to $0.32P_0$, $0.50P_0$, and $1.00P_0$ at the time the release started. In all cases V_{max} is the same. However, the rest of the force–velocity data depends on the time of the release. The authors conclude

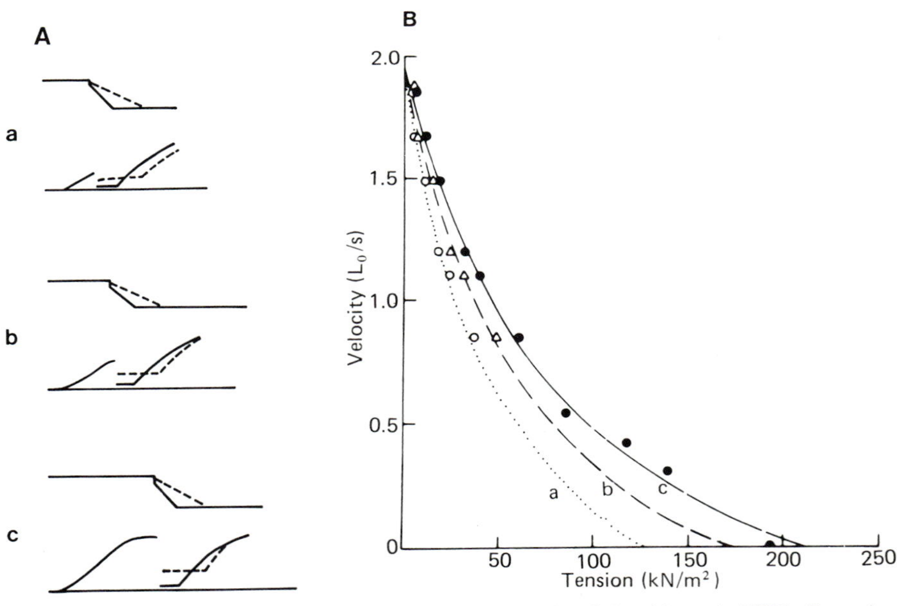

FIG. 2.15. (A) Diagram illustrating experimental methods of Cecchi *et al.* (1978). Records made during releases starting 75 ms (a), 100 ms (b), or 190 ms (c) after the first stimulus in a tetanus. In each set of traces the upper superimposed traces show the length and the lower ones show force. Shortening is indicated by downward change on length trace. Results for two different velocities of release are shown superimposed. Full lines, faster release, and broken lines, slower release. (Based on Fig. 1 of Cecchi *et al.*, 1978.) (B) *P–V* relations at various times during a tetanic contraction at 4°C. Curves a, b, and c refer to releases beginning 75, 100, and 190 ms after first stimulus, respectively. (From Cecchi *et al.*, 1978.)

that at the beginning of an isometric tetanus, the steady force for a given velocity (less than V_{max}) increases with time. In simple mechanistic terms, the number of crossbridges operating increases with time and increases the force produced at each velocity of shortening. However, at late times, but before isometric tension is fully developed, the velocity reaches the values it has during releases from the plateau of a tetanus. Thus, in contrast to the results at earlier times in the tetanus, the optimum force during shortening can be produced before the maximum isometric force is produced. The underlying mechanism for this effect is not known.

The results of Jewell and Wilkie (1958) and Cecchi *et al.* (1978) agree in showing that shortening at V_{max} can occur very early in a tetanus. However for velocities less than V_{max} and forces less than $0.4P_0$, there is an obvious discrepancy between their results (Figs. 2.14 and 2.15B). Although there are some differences in technique (whole sartorius from *Rana temporaria* was used at 0°C by Jewell and Wilkie, whereas Cecchi *et al.* used single fibres from semitendinosus of *Rana esculenta* at 4°C), it seems unlikely that these factors could be responsible for the difference in the results.

d. Factors That Change V_{max} and the Force–Velocity Relation. There are a number of experimental conditions that change the force–velocity curve. The conditions described here influence the maximum velocity of shortening (in some cases, P_0 also) and/or the shape of the curve (a/P_0).

Temperature. Values in Table 2.III show that the maximum velocity of shortening is affected by temperature more than P_0 (maximum isometric force) or a/P_0 (the curvature of the force–velocity relation).

An interesting observation made by Lannergren *et al.* (1982) is that between 10 and 15°C larger changes in V_{max} occur than between 15 and 20°C. They suggest that there may be a "critical" temperature in this range at which these properties abruptly change. These results emphasize the importance of knowing the temperature range that was used when Q_{10} values were calculated.

Hypertonic and hypotonic Ringer's solution. The mechanical properties of muscles or fibres in hyper- or hypotonic Ringer's solution have been studied by Howarth (1958), who observed the effects on V_{max}, and Edman and Hwang (1977), who observed complete force–velocity curves. Their results for hypertonic solutions are in reasonable agreement in showing that both the isometric force and V_{max} are reduced as tonicity is raised (Fig. 2.16). The Ringer's solutions were made hypertonic mainly or completely by increasing the concentration of impermeant substances. The resulting loss of water from the fibres (1) increases the concentrations of the fibre contents and the intracellular ionic strength and (2) decreases the cell volume and the spacing between the filaments. (3) There is also

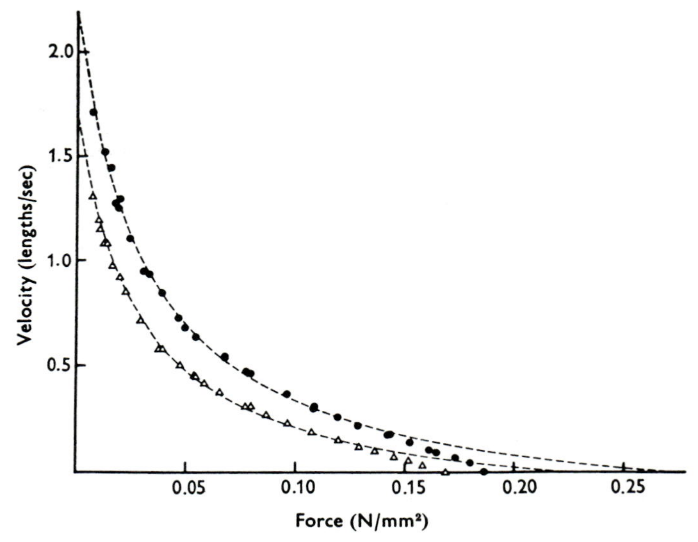

FIG. 2.16. Force–velocity relations in single muscle fibres in 1.22 times hypertonic medium (△) and in isotonic Ringer's solution (●). Hyperbolae are fitted to data truncated at $0.8P_0$. Cross-sectional area, 161×10^{-4} mm². Temperature, 0.5°C. (From Edman and Hwang, 1977.)

evidence that the release of Ca^{2+} from the sarcoplasmic reticulum is inhibited in muscles in hypertonic Ringer's solution (Homsher *et al.*, 1974; Gordon *et al.*, 1973). Two methods have been used to investigate which of these phenomena is the dominant cause of the changes in mechanical behaviour.

Edman and Hwang (1977) found that when filament spacing was reduced in normal Ringer's solution, by increasing the sarcomere length to the value believed to exist in hypertonic Ringer's, V_{max} was not changed. Thus the decreased V_{max} in hypertonic Ringer's solution is probably due to some effect other than the change in the spacing between the filaments.

Studies of P_0 by Gulati and Babu (1982) also indicate that filament spacing is not responsible for the changes in V_{max}. They found that changes in P_0 were produced by hypertonic solutions of KCl, which is permeant, and thus intracellular concentrations of these ions were increased without affecting fibre volume. Fibres were electrically inexcitable in these KCl solutions; they were stimulated in the presence of caffeine by a temperature jump technique. The changes in P_0 were the same as those observed in solutions of impermeant substances of the same tonicity relative to normal Ringer's solution. These results indicate that the increase in intracellular ionic strength or ionic concentration modifies crossbridge properties so that less force is produced. It would be interesting, in light of the decrease in both P_0 and V_{max} in experiments with impermeant substances, to know whether hypertonic solutions of permeant substances diminish V_{max} as well as P_0. Both Edman and Hwang (1977) and Gulati and Babu (1982) found that hypotonic Ringer's solutions had effects opposite to those of hypertonic solutions. It is unclear why Howarth (1958) did not detect changes in P_0 and V_{max} in hypotonic Ringer's.

Fatigue. When a muscle or fibre performs a series of isometric contractions in quick succession, the force diminishes. The extent of this type of "fatigue" (the reduced production of force) depends on the duration of stimulation, on duration of the intervals between stimulation, and on other factors. Edman and Mattiazzi (1981) investigated V_{max} during this type of fatigue and found that small decreases in P_0 ($<10\%$) occurred without any change in V_{max}; when force was reduced by more than 10%, V_{max} diminished too. They speculated that it was the increase in concentration of products of metabolism that mediate these effects, and showed that increasing extra-cellular CO_2, which would in effect cause an entry of H^+ into the fibre, also decreased P_0 and V_{max}. It remains to be shown that the hydrogen ion content (pH) of the cell is changed by the contraction patterns that they found to change the mechanical properties.

These mechanical results are interesting in that they show that P_0 can be changed, at least to some extent, without any change in V_{max} as would be expected for a factor that changes the number of active crossbridges, but not their rate of turnover. The question remains whether the initial 10% fatigue of P_0 has a different cause than subsequent change in P_0 accompanied by fatigue of V_{max}.

Degree of activation by calcium ion. A number of substances that are believed to affect Ca^{2+} release from the sarcoplasmic reticulum have been found to alter the plateau of isometric force without influencing the maximum velocity of shortening. These include nitrate (Cecchi *et al.*, 1978), dantrolene (Edman, 1979), and deuterium oxide (Cecchi *et al.*, 1981).

In Ringer's solution containing deuterium oxide, the curvature of the force–velocity curve was increased, that is, a/P_0 had a smaller value, compared to using H_2O Ringer's (Cecchi *et al.*, 1981). Since this effect was not modified by addition of nitrate or caffeine, which increase Ca^{2+} release, it was concluded that deuterium oxide can modify the kinetics of crossbridge operation in addition to changing Ca^{2+} release.

As described earlier (pp. 60–63), the experiments by Jewell and Wilkie (1958) and by Cecchi *et al.* (1978) show that the value of V_{max} is constant even during the early stages of a tetanus when the Ca^{2+} concentration is changing. All these results are in agreement in showing that in intact, living fibres, V_{max} is independent of the degree of activation by Ca^{2+}. The situation is much less clear in skinned fibres (Podolin and Ford, 1983).

2.2.2.3 *Relation of Force and Velocity during Stretching*

It has been known for many years that more force is produced by stimulated muscle while it is being stretched than can be developed during contraction under isometric conditions (see for example, Gasser and Hill, 1924). This aspect of the mechanical behaviour of muscle has continued to attract interest and investigation (A. V. Hill, 1938; Katz, 1939; Aubert, 1956; Sugi, 1972; Curtin and Davies, 1973; Flitney and Hirst, 1978; Edman *et al.*, 1978).

Katz (1939) was the first to investigate systematically the relation between force and velocity during stretching as well as shortening. Shortening velocity was measured after quick releases to various fixed loads from the plateau of the tetanus. His stretch experiments were isotonic as well; forces greater than P_0 were suddenly imposed on the tetanized muscle. Muscle lengthening followed a complex time course as seen in Fig. 2.17A. Katz described three phases in these records (indicated in Fig. 2.17A):

a. Instantaneous lengthening of elasticity believed to be outside the contractile elements in the series elastic component.
b. Relatively rapid lengthening of fibres. He called this "give" and did not consider it to be the converse of shortening. This phase of lengthening became more prominent at higher isotonic forces and above $1.8P_0$ it continued to the end of lengthening. Consequently, Katz does not consider further results for $P > 1.8P_0$.
c. Slow lengthening, which Katz referred to as "reversible" lengthening.

When the velocity of lengthening in phase c was plotted as a function of the load, the results in Fig. 2.17B were obtained. The results lie above a smooth continuation of the force–velocity curve for shortening.

In similar experiments by Mashima *et al.* (1972) on bundles of fibres from the frog *Rana nigromaculata*, and by Lannergren (1978) on single twitch fibres from *Xenopus laevis*, the velocity diminished smoothly during the stretching rather than in two distinct phases, b and c, as in Katz' experiments. It is therefore hard to summarize their results in the form of a force–velocity curve. In slow fibres from *Xenopus*, Lannergren did obtain a clear distinction between phases b and c. The velocities in phase b lay approximately on a continuation of the force–velocity curve for shortening. The velocities in phase c were much less than in b and, as in Katz' experiments, were well above the curve.

In other studies, isovelocity stretches have been used; that is, velocity of lengthening was the controlled variable while muscle force was observed. Examples of records from this type of experiment are shown in Fig. 2.18. At intermediate velocities of lengthening (B) the force increases rapidly and then comes to a relatively constant value. With slower lengthening (A) the

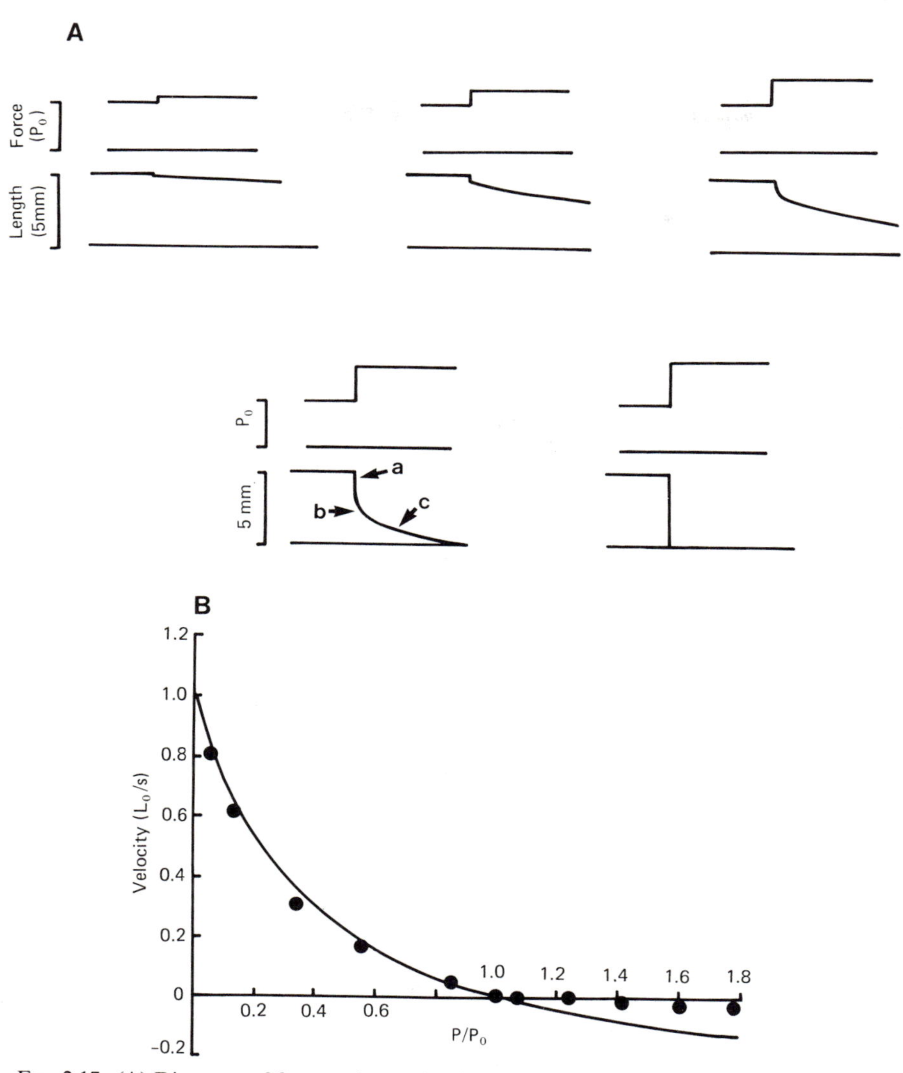

FIG. 2.17. (A) Diagrams of force and records of muscle length during isotonic stretches. The muscle was stimulated tetanically and produced force P_0 (development of force not shown). An isotonic load greater than P_0 (range shown 1.2–1.8P_0) was then imposed as shown diagramatically in the upper traces. The resulting change in muscle length (lower traces) were recorded. Downward movement indicates stretching. Three phases of stretching are seen (a, b, and c) in some records. (Based on Katz, 1939.) (B) Force–velocity relation during isotonic shortening or stretching of frog muscle. The solid line is from the Hill equation using $a/P_0 = 0.37$ and $b = 0.36L_0$/s. Points for stretch show velocity during phase c. Positive values for velocity are for shortening, and negative values for stretching. (Based on Katz, 1939.)

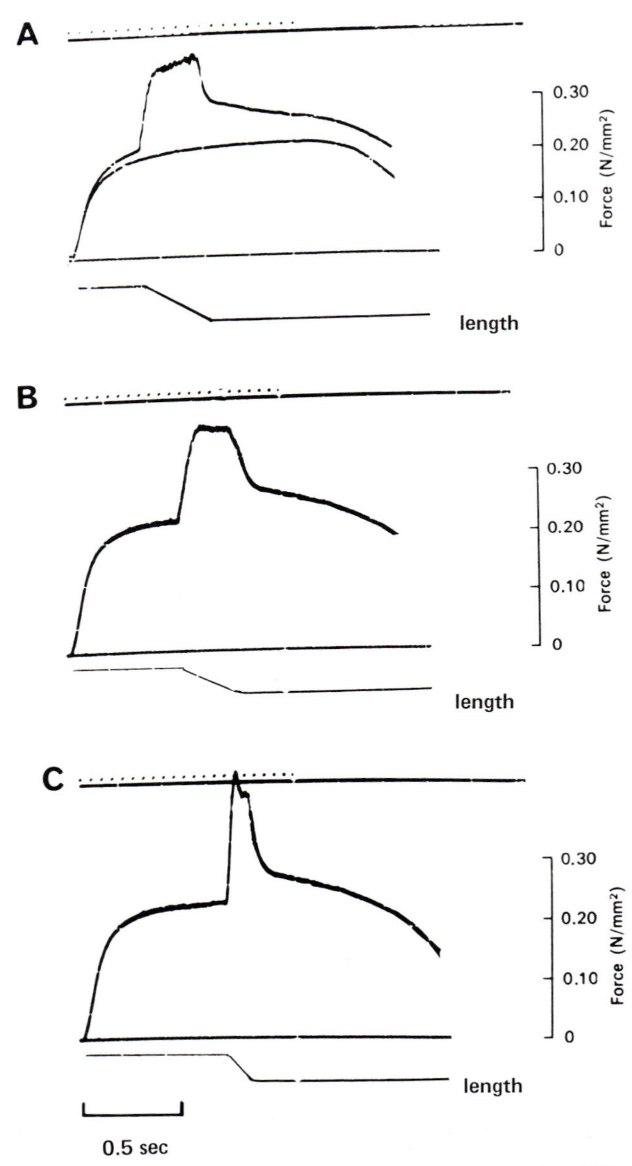

FIG. 2.18. Force and displacement records from single fibres stretched at different velocities during a tetanus. Stimulus markers are indicated at the top of each record. (A) Stretch from 2.41 μm sarcomere length to 2.54 μm, and isometric at 2.54 μm. (B) and (C) Stretches from 2.30 to 2.40 μm sarcomere length. Temperature, 2.75°C. Cross-sectional area: (A) 6.95×10^{-3} mm^2; (B), (C) 8.59×10^{-3} mm^2. (From Edman *et al.*, 1978.)

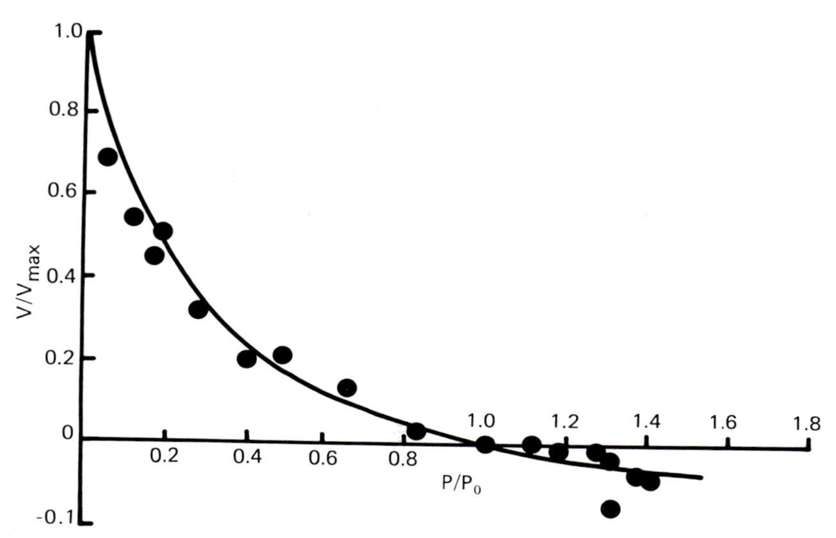

FIG. 2.19. Force–velocity relation for frog sartorius muscle at 0°C. Isovelocity stretches and shortenings were controlled by a Levin–Wyman ergometer. The force that is produced at L_0 during stretch is shown. The line was calculated from the Hill equation with $a/P_0 = 0.246$. (From Curtin and Davies, 1973.)

force does not reach a steady value but continues to increase during the lengthening. At higher velocities (C) the force rises to a peak, and then falls as lengthening continues. Obviously the shape of a force–velocity curve constructed from these results would depend on how the records were measured. Figures 2.19 and 2.20 are examples of force–velocity curves for different ways of measuring force as explained in the legends. At low velocities of stretching, the experimental points are above a continuation of the force–velocity curve for shortening; in this respect the results are like those of Katz. However, at high velocities of stretching the points cross the curve so that less force is being produced than would be expected from the curve. The discrepancy at high velocities would probably be the case even if the maximum force during stretching were plotted.

The results of a number of experiments are compared in Fig. 2.21, which includes two sets of observations (from Edman *et al.*, 1978; and Flitney and Hirst, 1978) that did not include observations during shortening. For the purpose of comparison, all values in this figure are expressed relative to P_0 and V_{max} to remove, for example, differences due to muscle or fibre size. The line shows the relation predicted by the Hill equation using values for the constant $a/P_0 = 0.25$, which is typical for frog sartorius at 0°C ($a/P_0 = 0.25$). In contrast to the results of the experiment by Katz, some of

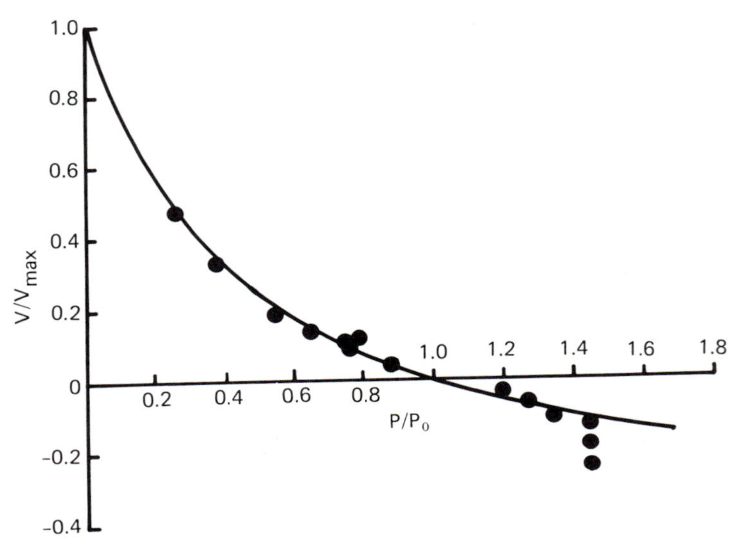

FIG. 2.20. Force–velocity relation for frog sartorius muscle at 0°C. Isovelocity stretches and shortenings were controlled by a Levin–Wyman ergometer. The value of the force that is plotted was obtained by extrapolation of records to the start of the stretch. L_0 was 28 mm. The muscle was stretched from 26 to 28 mm. (From Aubert, 1956.)

the other observations lie on or below the line predicted by the Hill equation. From these results, it would not be concluded that more force is produced than predicted by the equation. (This is possibly not a completely fair comparison, however, because the predicted line is an average one rather than based on the properties during shortening of the particular muscle or fibre used to obtain the data for stretching.) However, independent of problems about the absolute value, all the results for isovelocity stretches diverge from the results of Katz for isotonic stretches as the values of P/P_0 increase. This phenomenon needs to be investigated by performing both types of experiment (isotonic and isovelocity) on the same preparation. If the difference is really due to the type of experiment, this may be because in the isotonic experiments, the force increases very rapidly at the start of the stretch, causing what Katz referred to as "give." After recovery from this period of rapid lengthening, muscle may well behave differently than if "give" had not occurred.

Muscles can clearly exert forces much larger than P_0 while they are being stretched. It seems, however, that the force exerted may depend on more factors than the instantaneous value of the velocity of lengthening. What these other factors are and how their effects are exerted are questions requiring further investigation.

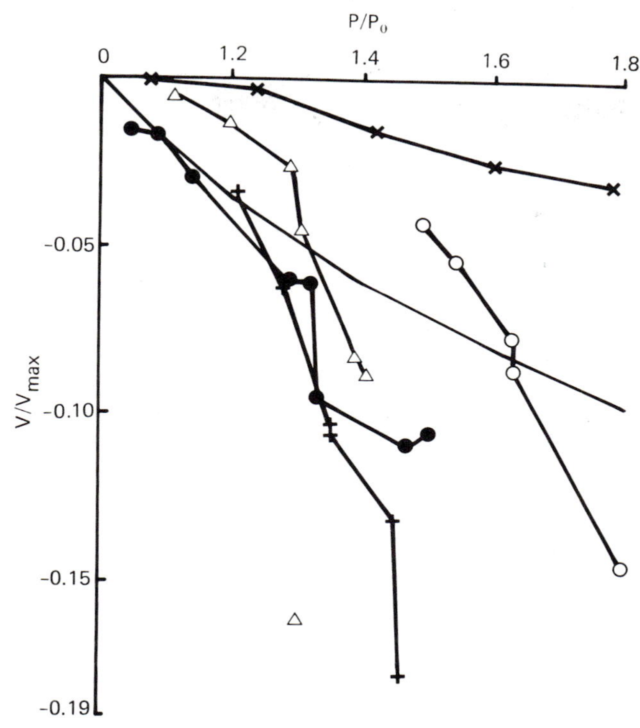

FIG. 2.21. Relation between force and velocity of stretching. In all experiments except those of Katz, the stretches were isovelocity. The line without points was calculated from the Hill equation with $a/P_0 = 0.25$. (●) Flitney and Hirst (1978) frog sartorius muscle at 0°C. Authors do not give V_{max}; it has been assumed to be $1.33L_0$/s. (○) Edman *et al.* (1978) single fibre from frog semitendinosus muscle at 1.5°C. $V_{max} = 2.13L_0$/s. (△) Curtin and Davies (1973) from Fig. 2.19. $V_{max} = 2.28L_0$/s. (+) Aubert (1956) from Fig. 2.20. $V_{max} = 1.52L_0$/s. (×) Katz (1939) from Fig. 2.17B. $V_{max} = 1.18L_0$/s.

2.2.3 AFTEREFFECTS OF LENGTH CHANGE

2.2.3.1 *Effect of Shortening on Isometric Force Production*

Both the amount of shortening and the velocity of shortening in contracting muscle can alter subsequent isometric force development. This suggests that the crossbridges have some "memory" of their past behaviour. Figure 2.22A compares a record of force development in an isometric tetanus of frog sartorius muscle with that in another contraction in which the muscle was first tetanized, then allowed to shorten at about $\frac{1}{2}V_{max}$. The isometric force exerted after shortening is 10% less than that produced

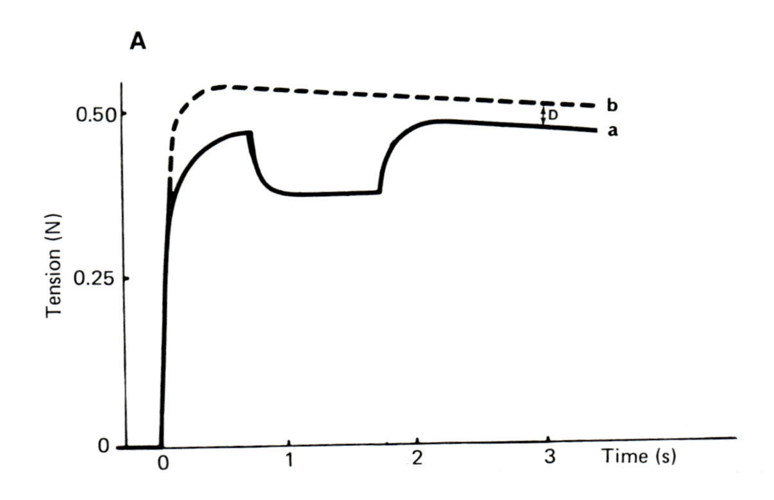
A
Tension (N)
0.50
0.25
0
0
1
2
3
Time (s)
D
b
a

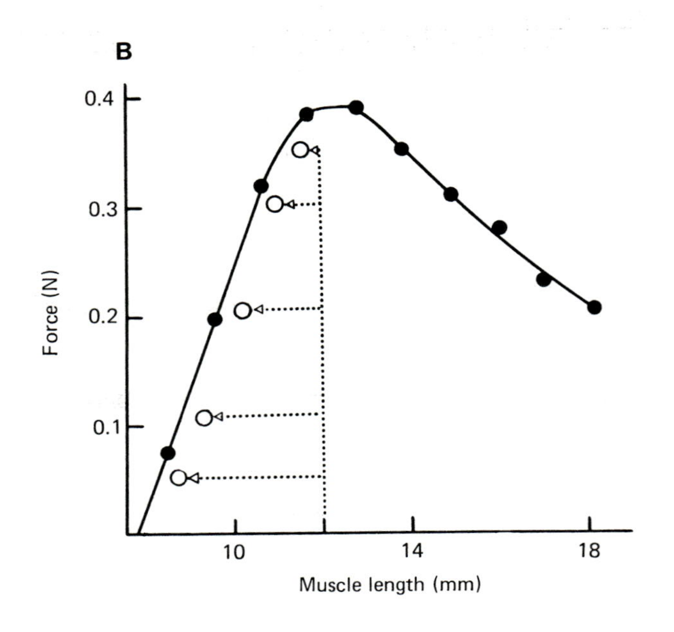
B
Force (N)
0.4
0.3
0.2
0.1
10
14
18
Muscle length (mm)

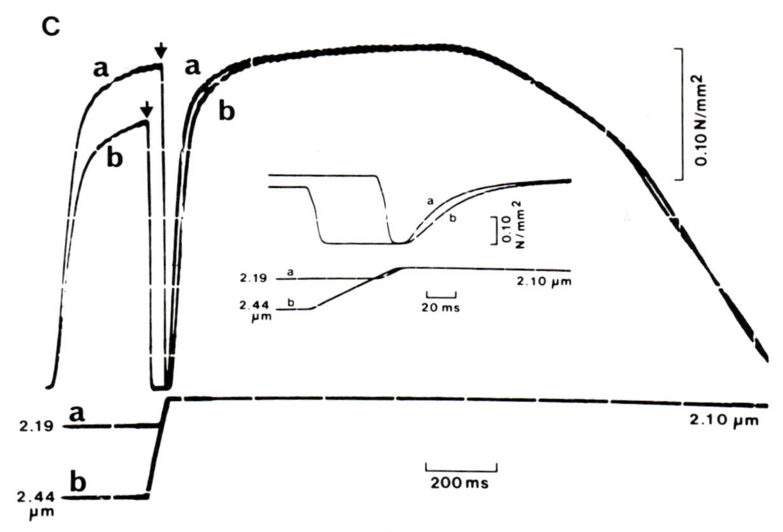

FIG. 2.22. (A) The effect of shortening on isometric force development. (a) A record of force during a tetanus in which the muscle was released after 0.8 s of stimulation and allowed to shorten at a constant velocity, then force redeveloped as stimulation continued. The period of shortening is indicated by the heavy horizontal bar. (b) A record of force produced during an isometric tetanus at the final length reached during shortening in the other tetanus. (Based on Fig. 2 of Marechal and Plaghki, 1979.) (B) A frog semitendinosus muscle at 0°C was isometrically tetanized at different initial lengths (●). It was then brought to an initial length of 12 mm and afterloaded with different loads (indicated by the dotted horizontal line), tetanized, and allowed to shorten to the shortest possible length (○). At each afterload the distance the muscle shortened was less than that expected on the basis of the force–length curve. The apparent depressant effect of shortening was not due to muscle fatigue since the point of maximum shortening at 0.05 N afterload was reached after only 1.2 s of stimulation. (Based on Fig. 4 of Deleze, 1961.) (C) The effect of rapid shortening on the rate of force redevelopment. Force records from a fibre whose contraction began at 2.19 (a) or 2.44 μm (b). At the point designated by the arrow the fibre rapidly shortened to 2.10 μm (displacement shown in the lower records) and then redeveloped force, with continued stimulation. The inset is a record with a faster time base and shows that although the isometric force redevelopment is the same in both contractions, the rate of force redevelopment is reduced in the muscle that had shortened the greater amount. (Based on Fig. 3 of Edman, 1980.)

in the completely isometric tetanus. Similar behaviour has been observed in whole muscles from frog, toad, and dogfish (Abbott and Aubert, 1952). The same effect is seen in a somewhat different way in Fig. 2.22B, which shows for isotonic contractions the relation between force (the afterload) and final muscle length, that is, the length at which shortening ceases.

Recently, further studies of whole muscles from frog have extended these results (Marechal and Plaghki, 1979). These experiments were performed as shown in Fig. 2.22A; the deficit in the force, D, is the difference between the isometric force at the short muscle length and the force produced after shortening. The following results were found.

1. The deficit in force was independent of whether shortening started at the beginning of tetanic stimulation, or 0.8 or 1.6 s later.
2. The deficit in force was directly proportional to the distance shortened (ΔL); the equation relating the two variables is $D = k \cdot \Delta L$, where ΔL is in fractional muscle length, and k is a constant. Over the sarcomere length range 1.96–2.56 μm, the value of k is independent of muscle length.
3. The deficit in force was inversely related to shortening velocity (V); i.e., the greater the shortening velocity the smaller the effect, and after shortening at V_{max} there is almost no force deficit. The equation relating the two variables is $D = Ae^{-V/a}$ where A and a are constants. The value of A indicates that the maximal amount of force depression possible is equal to about 15% of the control isometric force at the short length.
4. If shortening occurs partly at V_{max} and partly at a slower velocity during a continuous period of shortening, it is only the distance shortened at the slower velocity that has any effect on subsequent force production.
5. The deficit in force remains for a relatively long time, at least for several seconds, after the end of shortening, but can be abolished if stimulation is stopped and the muscle allowed to relax to almost zero force before stimulation is resumed.

The post-shortening depression of isometric force development is also seen in single fibres that have been allowed to shorten at velocities less than V_{max} (Edman, 1966, Fig. 4; Gordon *et al.*, 1966b, Fig. 9; Julian and Morgan, 1979b; Fig. 1B). As in whole muscles, there is little or no tension deficit after shortening at V_{max}.

Marechal and Plaghki (1979) have developed an explanation of this behaviour which is based on a stress-dependent inhibition of the G-actin monomers, which enter the zone of filament overlap during shortening. Alternatively, Julian and Morgan (1979b) give evidence that suggests that the tension deficit after shortening from the right-hand side of the length–force curve is associated with sarcomere inhomogeneities and could be explained by these effects. However, the fact that the deficit is also observed when shortening ends on the left-hand side of the plateau of the length–force curve (Marechal and Plaghki, 1979) suggests that the effect cannot be wholly explained in this way. Further experiments to test these ideas are needed. Edman (1980) found that although shortening (at or near V_{max}) greatly affected subsequent production of force in twitches and unfused tetani, it had only a very small effect, $<5\%$, during fused tetani. There was, however, a reduction in the *rate* of force development.

Figure 2.22C shows a comparison of the rate of force development after shortening different distances at a velocity near V_{max}. Since the force redeveloped is the same, there is no evidence that the depressant effect of shortening lasts longer than about 100–200 ms. This effect on the rate of force development is unlikely to be due to sarcomere inequalities because the *maximal* shortening velocity is independent of sarcomere overlap in this sarcomere length range. Additional characterization of this effect and examination of possible involvement of activation processes is needed.

2.2.3.2 *Effect of Lengthening on Isometric Force Production*

When a muscle (from frog, toad, dogfish, or cat) is stretched at a velocity between 0.05 and $2.0V_{max}$ during a maintained tetanus, the force may rise to values in excess of $2P_0$ (see Force and Velocity, Section 2.2.2). After stretching has ceased, there are two distinct phases in the recorded force (see Fig. 2.18).

First, there is a fairly rapid decline of force over a period of a second or so in frog muscle, or fibre, at 0°C. Second, the force is maintained at a more steady level, which may be well in excess of that produced under isometric or fixed-end conditions at the same muscle or fibre length as that reached in the stretch.

The first phase has been studied by L. Hill (1977), Julian and Morgan (1979b), and Edman *et al.* (1982). All these authors agree that it is characterized by greater stability of sarcomere spacing than in fixed-end contractions. For example, Julian and Morgan (1979b) point out that the phenomenon of "progressive collapse of the ends" (that is, the very large amounts of shortening that occurs at the ends) does not occur as it does in fixed-end contractions. So it seems that the crossbridges are affected by the stretch, at least for a time. Julian and Morgan (1979b) suggest that, at the end of stretch, all the crossbridges have been "stretched" and produce more force than "normal" bridges, but that when a stretched bridge detaches and reattaches it returns to its normal state. We are concerned here primarily with the second phase, in which the excess force, P_e, can be maintained for long periods of time (>5 s) with continuing stimulation (Abbott and Aubert, 1952; Curtin and Woledge, 1979b; Edman *et al.*, 1982) and increases as the distance lengthened increases (Abbott and Aubert, 1952; L. Hill, 1977; Edman *et al.*, 1979). There is general agreement that excess force is observed only when the muscle is stretched to sarcomere lengths greater than 2.25 μm (L. Hill, 1977; Edman *et al.*, 1978; Julian and Morgan, 1979b) as shown in Fig. 2.23. When a muscle is stretched to a sarcomere length *less* than 2.25 μm, the force subsequent to lengthening falls to the value obtained in a fixed-end isometric tetanus (Fig. 2.23).

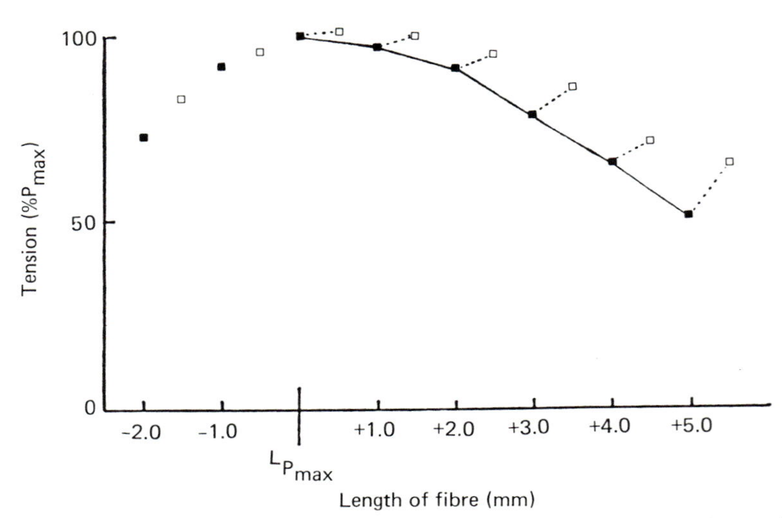

FIG. 2.23. The effect of fibre length on isometric force development in isometric tetani (■) and in tetani in which the muscle was first tetanized at a given length and, during continued stimulation, stretched by 0.5 mm and allowed to reach a stable isometric force after the stretch (□). The dotted lines connect the initial and final muscle lengths. Note that at muscle lengths less than that at which the force is optimum, the isometric and stretched muscle results are superimposed. At lengths greater than this (i.e., at sarcomere lengths > 2.25 μm) the results from stretched muscles fall above the isometric data points. (Data from L. Hill, 1977.)

Julian and Morgan (1979b) have presented evidence that stretching of some sarcomeres and shortening of others occur after the end of a stretch of a fibre. This also occurs in a fixed-end tetanus. On the basis of their results they conclude that the extra force after stretch, compared to that in fixed-end tetanus, is due to a difference in sarcomere lengths during movement under these two conditions. They measured the movement of markers on the surface of the centre of the fibre during lengthening of resting and tetanized single fibres. On resting fibres, the change in distance between the markers was proportional to the applied length change. However, they consistently found that when the tetanized fibre was stretched, the change in distance between the markers was greater (by about 15% on average) than that seen during stretch of resting fibres. Thus after stretch of active fibre, the sarcomere spacing at the fibre ends will be less (and they will, therefore, have more overlap) than when the same fibre length is reached by stretching of the resting fibre. Records of marker separation after stretch show that the centre of the fibre continues to lengthen; presumably the ends are shortening since the overall length of the fibre is constant. In a fixed-end tetanus, of a fibre stretched before stimulation, a similar

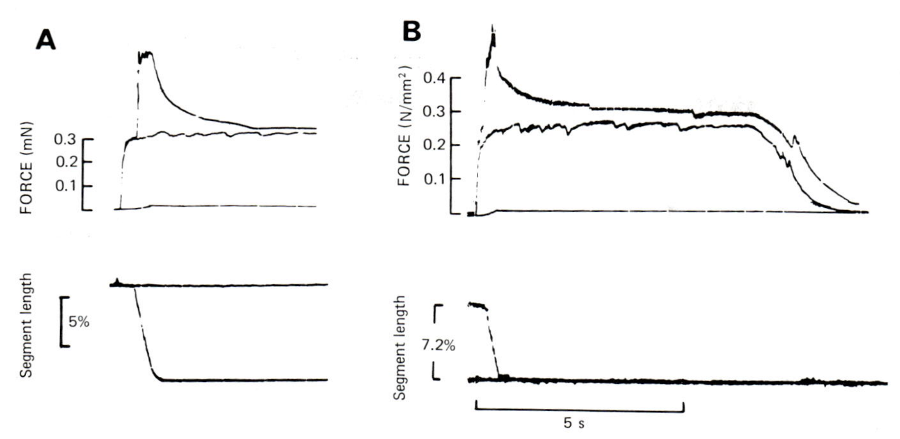

FIG. 2.24. Force production during a tetanus with length clamp. In (A) and (B) the upper records are of force and the lower records show the clamped length of the segments. In each case three sets of records were superimposed: (1) the unstimulated fibre was stretched and very little change in force occurred; (2) the fibre was stretched (downward deflection of the segment length) after the beginning of the tetanus. After the stretch the segment length was clamped at a constant value (isometric); (3) in (A) the fibre was tetanized under isometric (length-clamped) conditions at initial length for the stretch; in (B) the isometric contraction was at the final length reached during the stretch. Note that except during the stretching, segment length was constant. Single fibre from anterior tibialis at 25°C. (Based on Fig. 10 of Edman *et al.*, 1982.)

increase in marker separation occurs, indicating stretching of the middle by shortening of sarcomeres at the ends. It is suggested that because the end sarcomeres are shorter after stretch of an active fibre than in a fixed-end tetanus of a fibre stretched before stimulation, they exert more force.

In contrast to the findings of Julian and Morgan (1979b) are those of L. Hill (1977), who observed the striation pattern, by interference microscopy, during contractions which included a stretch. She measured the widths of the A bands and of the I bands during stretch of both resting and active fibres. No length changes in the A band were observed. The changes in the I band were identical in resting and active fibres. Sarcomere lengths remained uniform along the fibre during a tetanus including a stretch.

Similar evidence has also been obtained by Edman *et al.* (1982). They refer to the extra steady force produced after stretching as "residual force enhancement." They present evidence that this occurs even in the absence of the movements within the fibres, which are an essential feature of the phenomenon described by Julian and Morgan (1979b). Edman *et al.* detected length changes by observing either laser diffraction patterns or the separation between several consecutive markers along the fibre. They did not detect shortening at any of the locations that they observed. (However, shortening could have occurred at the very ends of the fibre, which were not examined.) More importantly, they found from measurements of marker separation that the length of some segments did not change at all after the end of the imposed movement. Thus, segments were producing extra force that could not be due to Julian and Morgan's explanation because the segment length was constant. Furthermore, when a length clamp was used to prevent change in segment length, there was still more force after stretch than in an isometric tetanus at the same fibre length (Fig. 2.24B).

In summary, the results of Julian and Morgan (1979b) show that shortening and lengthening can occur simultaneously in different parts of a fibre and this can affect the force produced. On the other hand, it seems from the work of L. Hill (1977) and Edman *et al.* (1982) that extra force can also be produced when no length changes are occurring within the fibre.

However, in all cases only part of the fibre has been observed while the behaviour of the rest was deduced, taking into account the length of the entire preparation. It seems unlikely that questions about events after stretching will be settled until an experiment is done in which length changes of all the individual segments of a fibre are observed and balanced with the overall length changes.

The nature and mechanism of the changes responsible for the extra tension are not clear. Edman *et al.* (1982) present evidence that it is not due to a passive, parallel elastic structure. Also, the fact that force is not greater than that produced under isometric conditions at optimal overlap would seem to rule out the idea that a simple modification of the bridge properties occurs resulting in more force being produced per bridge. Julian and Morgan (1979b) showed that stiffness does not increase as a result of stretch, and this result is taken to indicate that the number of attached bridges is not increased by stretch.

2.2.4 RAPID TRANSIENT CHANGES IN FORCE AND LENGTH

The phrase *tension transient* is used to refer to the changes in tension observed during and after a length step, that is, a sudden and fast change in the length of a muscle fibre. The principle of length step experiments is similar to that of earlier studies on whole muscle (e.g., Gasser and Hill, 1924; Jewell and Wilkie, 1958), but they have a much better time resolution. The length step lasts only a few tenths of a millisecond, and the tension changes during the step and in the next few milliseconds are observed.

A general description of the tension transients is given by Ford *et al.* (1977). It is based on studies of single fibres from anterior tibialis muscle of the frog that were stimulated tetanically and produced force, T_i, before the step change in length. The fibre was subjected to a release or stretch that was complete in 0.2 ms and varied in distance (see below), but was always small enough that the sarcomere length was in the range 2.0–2.2 μm. In other words, filament overlap was kept constant; the effects of varying overlap were investigated in another study (Ford *et al.*, 1981). A length clamp (described in Section 2.1.2) was used to control the length of a segment of the fibre and thus to eliminate effects due to the ends of the fibre and to the mechanical properties of the tendons. A length step results in the following four phases of change in tension. They occur in sequence and have the indicated durations when the temperature is about 0°C (Fig. 2.25).

Phase 1. The change in tension occurring *during* the change in length. The extreme tension that is reached is referred to as T_1.

Phase 2. A recovery of tension back toward its initial value T_i; this phase lasts for 2–5 ms and the tension reached is referred to as T_2.

Phase 3. The slowing or reversal of the recovery of tension that occurred in phase 2; the duration of phase 3 is between 10 and 50 ms.

Phase 4. The further recovery toward the original tension; this occurs much more slowly than the recovery in phase 2.

The first two of these phases have been investigated in more detail than the others. Some specific features of phases 1 and 2 are described below, along with more general conclusions arising from this work about cross-bridge operation.

An alternative experiment is to suddenly change the force and observe the resulting changes in fibre length. This was done by Podolsky (1960) and Civan and Podolsky (1966), and transients corresponding to those described above were observed (see Fig. 2.26). However, because of the technical complexity of controlling force, this type of experiment has been largely superseded by those in which fibre length is the controlled variable. See Podolsky and Tawada (1980) for an account of results obtained in force clamp experiments.

2.2.4.1 *Phase 1*

It has been concluded that phase 1 is due to an elasticity that is *linear*; that is, the extreme force reached during the step is a linear function of the extent of the step change in fibre length (see Fig. 2.27, dashed curve). In other words, the slope of this curve, which is referred to as the instantaneous stiffness, is a constant, independent of the distance released.

The experimental values of the extreme force T_1, particularly for large releases, are higher than expected for a linear elasticity, but this can be explained by the onset, during the release itself, of a recovery of tension that

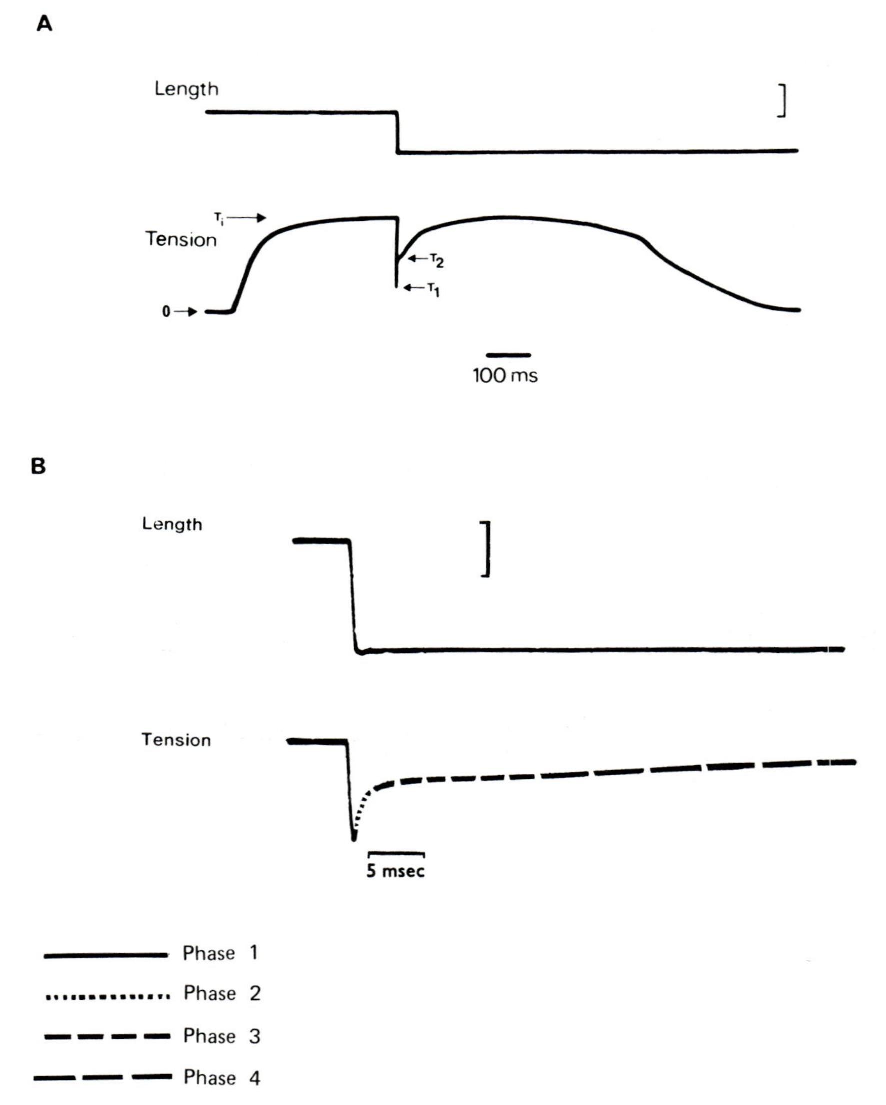

FIG. 2.25. Diagrams of tension during a tetanus (lower traces) in which a "step" release occurred (length is indicated by upper trace). The bars on the length traces correspond to a length change of 3.3 nm per half-sarcomere. (A) Records made on slow time scale to show entire time course of tension in the tetanus. T_i, Tension just before release; T_1, extreme tension reached during the release; T_2, tension approached during early recovery phase. (From Huxley, 1980.) (B) Records on a faster time scale to show the phases of tension change resulting from a step release. (Based on Huxley and Simmons, 1970a.)

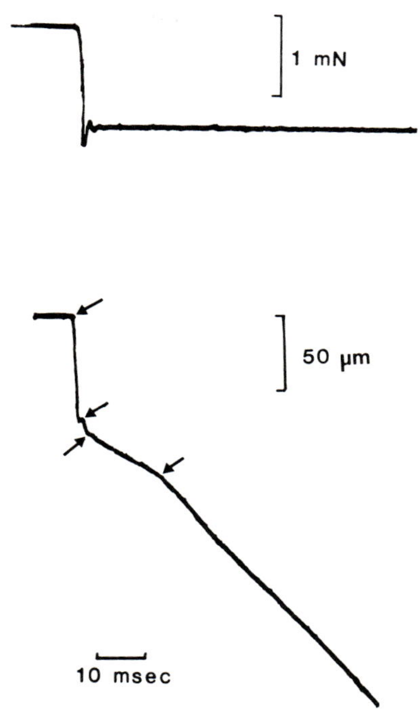

FIG. 2.26. Response of frog muscle fibres to changes in load. The upper trace shows load that was reduced at about 500 ms after stimulation began. Recording started after full isometric force had developed and shortly before the change of load. Bundle of seven fibres. Diameters not recorded; fibre length excluding tendon, 15 mm; tendon length, 3 mm; sarcomere length, 2.4 μm; temperature range, 2.7–3.2°C; stimulus, 15 pulses/s. P_0, about 5 mN. Lower trace shows the transient changes in length. Arrows mark the start of successive phases of the length transient. (From Civan and Podolsky, 1966.)

continues in phase 2 and by the limited time resolution of the tension transducer. The early onset of recovery prevents the force reaching the expected minimum value. When account is taken of the properties of the tension transducer and also of the recovery process (assuming it has the characteristics described below for phase 2), the time course of tension change during phase 1 fits that expected of an elasticity obeying Hooke's Law.

Another characteristic of phase 1 is that it is due to a *pure* elasticity having no viscous properties. This means that the stiffness (change in force per change in length) is independent of the velocity and the direction of movement. The evidence for this is that the T_1 values for stretches and releases lie on a single straight line (Fig. 2.27). In contrast, a viscosity would produce a force of the opposite sign for stretch and release, so a combination of

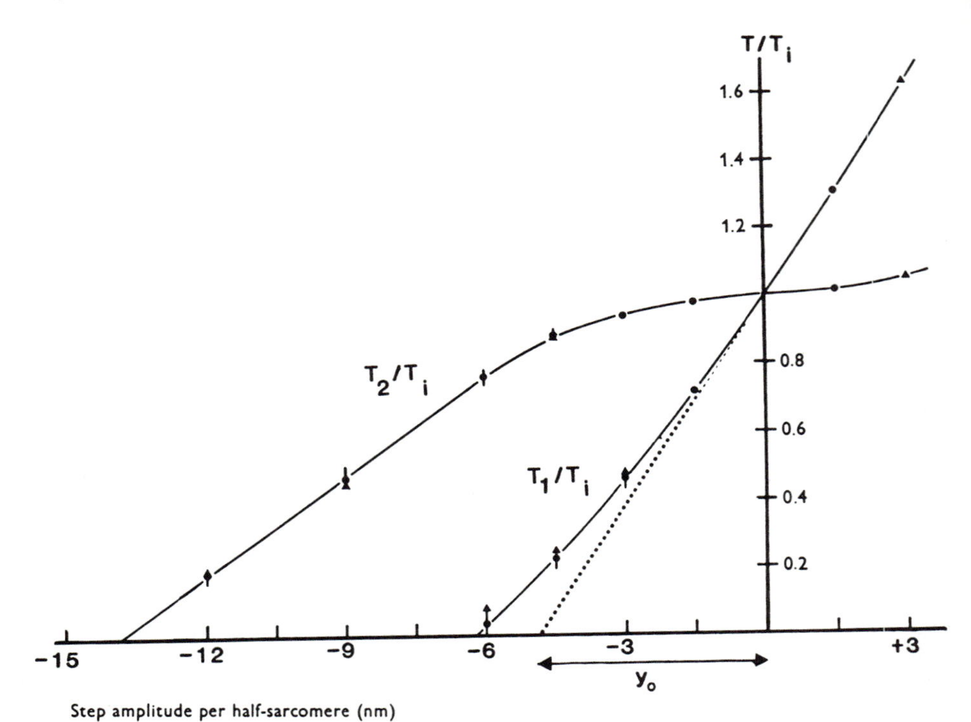

FIG. 2.27. Curve of T_1 (extreme force) and T_2 (tension approached during early recovery phase) in length-control steps of various amplitudes, both expressed as fractions of T_i, the isometric force immediately before the step. Negative values of step size indicate releases; positive values stretches. See Fig. 2.25. The dashed line shows stress–strain curve of the elastic structure thought to be responsible for the drop of tension in phase 1. Its intercept on the horizontal axis is referred to as y_0. (Based on Ford *et al.*, 1977.)

elastic and viscous properties would give a step in the T_1 curve between stretches and releases (Fig. 2.28).

The behaviour of a *resting* fibre shows that it *does* have viscous properties, because the amount of force produced during a movement depends on the velocity (Fig. 2.36B). The idea that an active fibre might have similar viscous properties that contribute to the force in phase 1 was examined by testing the hypothesis that the observed change in force has two components, one due to a viscosity equal to that present in the fibre at rest, and the other due to a linear elasticity (Fig. 2.28). If the hypothesis is correct, then after subtracting the viscous component the relation between change in force and change in length should be more linear than the original data. In fact, the corrected relation is less linear than the original (Fig. 2.29). Thus, it seems

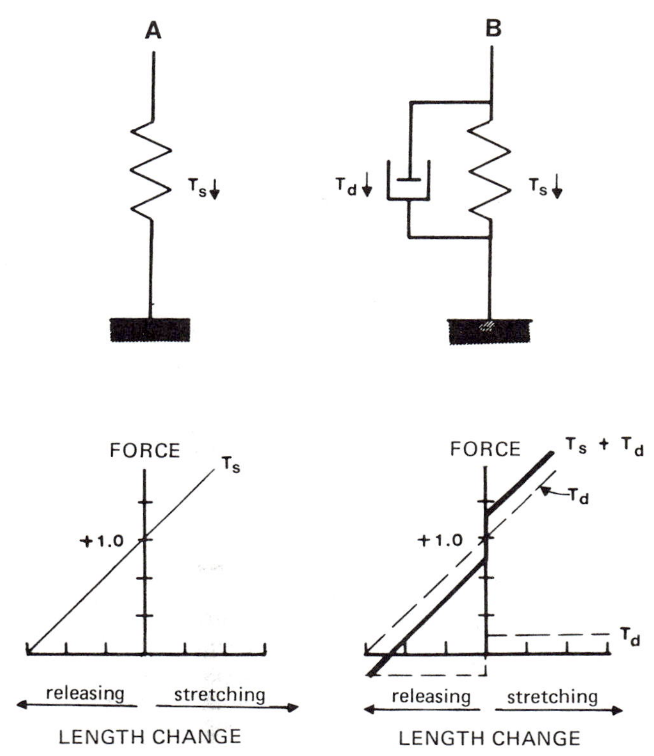

FIG. 2.28. An illustration of the behaviour of an undamped spring (A) and a damped spring (B) when stretched or released, at finite velocity, from an initial length at which force of +1.0 unit is exerted. The arrows indicate the direction of the forces regarded as positive in the graphs. The force in the damped spring is the sum of the force in the spring T_s and in the dashpot T_d. T_d depends on the velocity of length change, but not on the extent of the length change.

that the viscosity present in resting fibres does not exist in stimulated fibres.

As mentioned above, an elasticity and a viscosity can be distinguished by the effect of the velocity of movement on force. The force due to a viscosity is dependent on velocity, whereas force due to an elasticity is not. Comparison of the T_1 curve for steps lasting about 1 ms (Huxley and Simmons, 1971a) with that for faster steps (0.2 ms; Ford *et al.*, 1977) shows that they are not identical. However, the difference can be explained by the early onset of a tension recovery process that truncates the fall of force during phase 1 and has a larger effect in the slower releases. Additional evidence from experiments with even faster releases (complete in less than 0.2 ms) would be useful on this point.

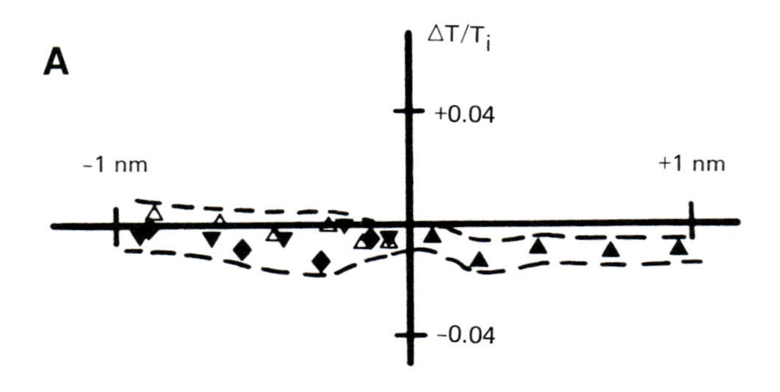

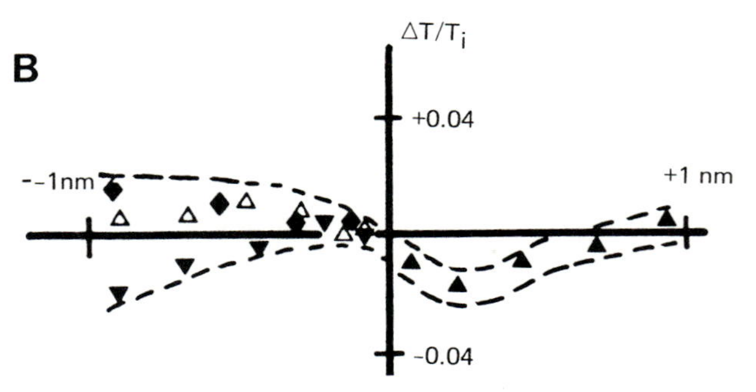

FIG. 2.29. Ordinate: deviation of force from linear dependence on length change during step itself. Forces are expressed as a fraction of T_i, the force just before the change in length. (A) Original data; (B) after correction for hypothesized viscosity. Stretch of 1.5 nm (▲); release of 1.5 nm (▼); release of 3 nm (△); release of 6 nm (◆). All steps were complete in 0.2 ms and therefore the velocities were proportional to the size of the step. The dotted "envelope" has been drawn to emphasize the greater nonlinearity in (B). (Based on Ford *et al.*, 1977.)

The structure responsible for the elastic behaviour in phase 1 can exert "negative tension." This means that it can resist further shortening (by pushing on the force transducer) when the extent of release is greater than that required to reduce the tension to zero. This effect is apparent on records of force during large releases ($\gg 4$ nm per half-sarcomere) in which the tension drops below the baseline; the fibre starts to push on the transducer rather than pull (Fig. 27 of Ford *et al.*, 1977). In terms of structure, this means that the component responsible for this force (believed to be attached crossbridges, see below) does not become floppy (like a released rubber band); it can resist compression.

Involvement of Crossbridges. The elasticity responsible for phase 1 is mostly, perhaps entirely, in the attached crossbridges. Evidence for this comes from a study of the effects of filament overlap on rapid transients (Ford *et al.*, 1981). The basis of the argument is that crossbridges are independent force generators distributed on the filaments as described in Section 2.2.1. Consequently, force due to crossbridges depends on overlap as shown in Fig. 2.8. If crossbridges are completely responsible for phase 1, stiffness (change in force per change in length) should depend on overlap in the same way as isometric force, and stiffness should be directly proportional to the isometric force T_i. The distance of release that brings force to zero (y_0 in Fig. 2.27) should be a constant, independent of overlap. On the other hand, the structures in series with the bridges, including the Z line and parts of the thick and thin filaments outside the overlap zone, may also have significant compliance. Their contribution to the elastic properties of a fibre depends on filament overlap in a different way than that due to crossbridges, and y_0 would not be independent of overlap (see Fig. 2.30C). Ford *et al.* (1981) find that y_0 hardly changes at all with sarcomere length in the range 2.2–3.2 μm. This result, which is shown in Fig. 2.31, indicates that most of the compliance is in the crossbridges. They conclude that at the "very least" 80% of the compliance is in the crossbridges in a contracting muscle at a sarcomere length of 2.2 μm.

The idea that phase 1 is due to crossbridges has two important consequences:

1. The instantaneous stiffness (change in tension per change in length) is proportional to the number of attached crossbridges. Therefore, the effects of mechanical, biochemical, or other interventions on the number of attached bridges can be assessed by how they change phase 1, during a release sufficiently rapid that the recovery of tension during the release is negligible.
2. It gives a new view of events that are occurring as tension rises at the start of an isometric tetanus. According to the traditional two-component model, the contractile element is shortening during this time and stretching the series elastic element; force is less than its maximum value because of the shortening. However, Huxley and Simmons (1971a) found that when step releases were given during the early part of a tetanus, the value of T_1 (and T_2 also) was proportional to the tension just before the step. Having accepted that the instantaneous stiffness gives a measure of the number of attached bridges, this result shows that the isometric tension increases because more bridges attach, and argues against the two-component model, according to which the number of attached bridges does not increase at this time.

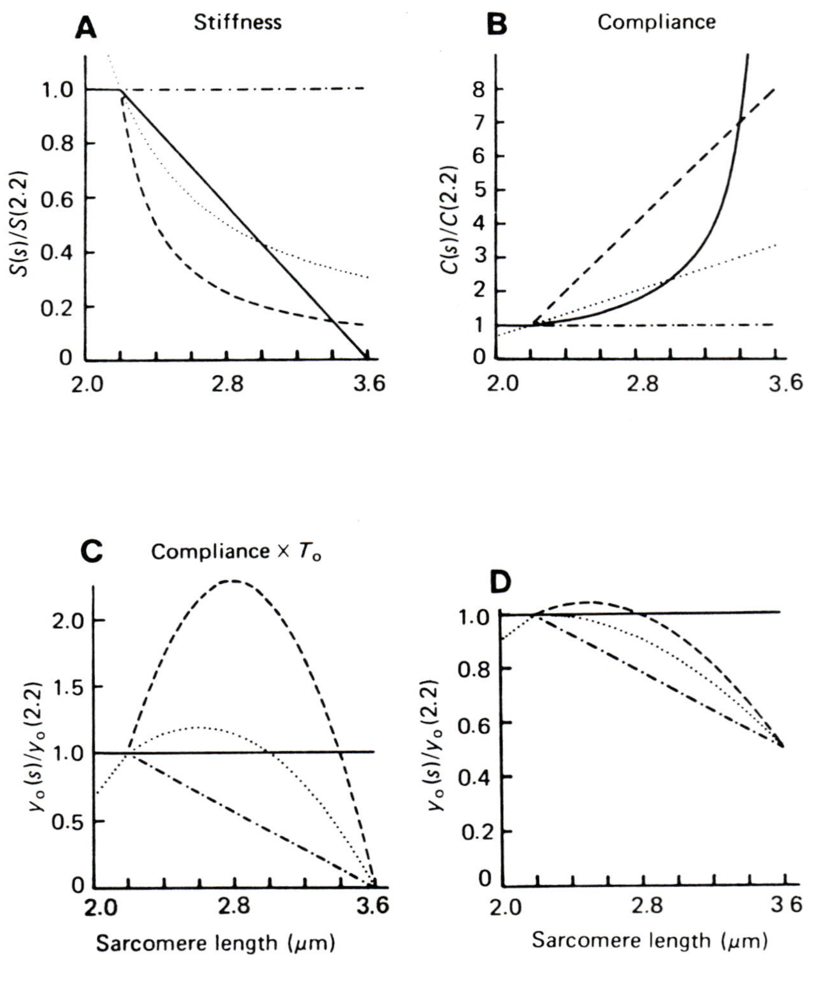

FIG. 2.30. Graphs showing various measures of stiffness plotted against sarcomere length in an idealised sliding filament system. In (A), (B), and (C), all except one of the four named structures are assumed completely rigid, compliance existing only in the crossbridges (continuous lines), the thick filaments (dashed line), the thin filaments (dotted line), or the Z lines (dots and dashes). In (D), the continuous line is the same as in (C), and the other lines give the values when the cross-bridges are compliant and also the indicated structure is compliant enough to halve the total stiffness when the sarcomere length is 2.2 μm. Stiffness, S; compliance, C (reciprocal of stiffness); product of compliance and isometric tension (assumed proportional to overlap) equals amount of sudden shortening (y_0) required to bring tension from its isometric value to zero. All expressed as fractions or multiples of the corresponding value at full overlap (2.2 μm). Filament lengths assumed: thick filaments, 1.6 μm including 0.2 μm bare zone; thin filaments, 1.0 μm on either side of Z line. (Based on Fig. 1 of Ford *et al.*, 1981.)

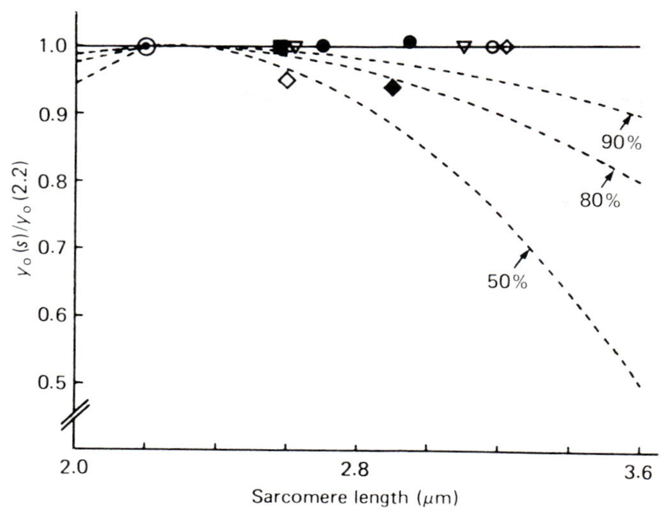

FIG. 2.31. Sarcomere–length dependence of y_0, the amount of instantaneous shortening per half-sarcomere needed to bring tension to zero from its isometric value. Curves show the ratio of y_0 at each sarcomere length to its value at sarcomere length 2.2 μm. Curves were calculated assuming that the crossbridges contribute the indicated percentages of the compliance at sarcomere length 2.2 μm. (Based on Fig. 13 of Ford *et al.*, 1981.)

2.2.4.2 *Phase 2*

a. Nonlinear Features. During this phase the tension recovers toward its initial value, T_i. The extent of this recovery can be expressed as the tension reached during phase 2 as a fraction of T_i. Figure 2.27 shows that the extent of this recovery, T_2/T_i, has a nonlinear relation to the change in length. Recovery from small releases, amounting to up to about 3 nm per half-sarcomere, was almost complete, but after larger releases, recovery was less complete the larger the release. Extrapolation of the curve indicates that no recovery would occur after a release of about 14 nm per half-sarcomere.

The time course of tension recovery in phase 2 was found to depend on the distance and direction of the length change in a complex way. In qualitative terms, the recovery was more rapid from large releases than from small ones, and it was even slower from small stretches and slowest from large stretches. These effects can be seen in Fig. 2.32A, which shows examples of the time course with which tension increases toward the T_2 tension following releases of various distances (note that the force is here expressed relative to the absolute amount of recovery, $T_2 - T_1$, which depends on the size of

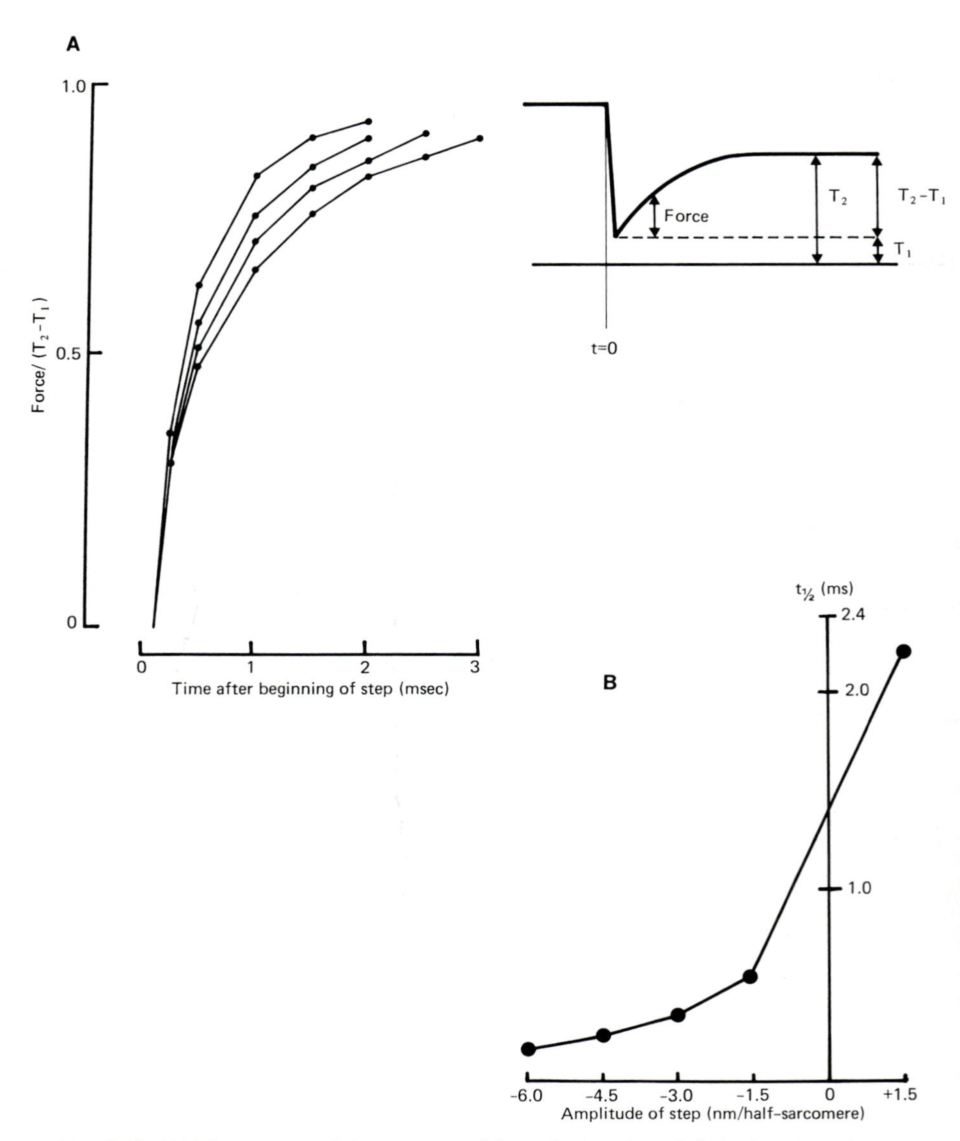

FIG. 2.32. (A) Time course of the recovery of force during phase 2 following step release of various amplitudes (reading from left to right, the amplitudes are 6, 4.5, 3, and 1.5 nm per half-sarcomere). Force is expressed as a fraction of $T_2 - T_1$. The step change in length started at time 0 and was complete in 0.2 ms. The inset is a diagram of a record to show how the measurements were made. (Based on Fig. 20 of Ford *et al.*, 1977.) (B) The half-time of recovery of force during phase 2 plotted against the amplitude of the step change in length (positive values indicate stretch and negative values release). (Based on Fig. 24 of Ford *et al.*, 1977.)

the length change). If the system behaved in a simple, linear way, the curves would have superimposed.

A mathematical description of the time course of tension change in phase 2 was devised which included four exponential terms. Only one parameter, the half-time for recovery of tension in this phase ($t_{1/2}$), depended on the distance of the length change, as shown in Fig. 2.32B. This parameter gives a simple measure of the speed of recovery.

b. Involvement of Crossbridges. The experiments of Ford *et al.* (1981) on the effects of overlap provide evidence that crossbridges are responsible for tension recovery in phase 2. For sarcomere lengths ranging from 2.2 up to 3.2 μm, the extent of the recovery, expressed as T_2/T_i, was essentially the same. In other words, for a particular size of length change, the tension always recovered to the same fraction of the initial tension.

c. Crossbridges Remain Attached during Phases 1 and 2. As already described, the evidence suggests that crossbridges, rather than structures in series with them, are responsible for both phase 1 and phase 2 of tension change after a length step. An important and related point is that the number of attached bridges is the *same* in these two phases. This idea was tested by giving two length changes in succession (Ford *et al.*, 1974). The first was a release, which was followed by the usual recovery toward the initial tension. The second length change was a stretch given when the recovery of force in phase 2 was almost complete. The instantaneous stiffness (change in force per change in length) observed in the second step, i.e., during the recovery, was only very slightly less than that in the first step. If it is accepted that instantaneous stiffness gives a measure of the number of attached bridges, the result shows that this number does not change from phase 1 to phase 2. A simple interpretation would be that the same bridges remain attached and the changes in force are due to transitions of existing attached bridges among states in which they exert different amounts of force. Huxley and Simmons (1971a) have developed a theory which uses this idea to explain the nonlinearities of phase 2 (see Chapter 5).

d. Effect of Temperature. Ford *et al.* (1977) investigated the effect of temperature in the range 0–8°C on phase 1 and phase 2 of the tension transient. They found that the absolute values of T_i and T_1 increased with temperature, but T_1 increased more than T_i. Thus the value of T_1/T_i for a particular step size was greater at a higher temperature. When the results are summarized as in Fig. 32 of their paper, in which T_1/T_i is shown as a function of step amplitude, the curves are less steep at higher temperature; the instantaneous stiffness (slope) decreases with increasing temperature with a Q_{10} of 0.8. Corresponding to this change in instantaneous stiffness, increasing temperature results in an increase in the value of y_0, the amount of instantaneous shortening per half-sarcomere that would bring tension to zero.

Phase 2 is also affected by temperature. The speed of the recovery of tension during this time increases with temperature with a Q_{10} of 1.85. The value of $t_{1/2}$ decreases with increasing temperature. Ford *et al.* (1977) considered and rejected the idea that the faster recovery of tension in phase 2 and greater truncation of the initial fall in tension produced an apparent change in T_1, rather than there being a genuine change in phase 1 with temperature.

2.2.4.3 *Later Phases*

Phase 3 is the slowing or reversal of the recovery of tension that occurs in phase 2 (Fig. 2.25). It has a complex time course which has not been fully characterized. Phase 3 is most obvious and has its longest duration after a step stretch and is less obvious and more brief after a small release. After large releases, it may appear only as a slowing of recovery, rather than a reversal (Ford *et al.*, 1977).

Phase 4 is the gradual recovery of tension as it approaches T_i. Although this point has not been fully studied, published records suggest that the final tension in phase 4 would reach T_i if stimulation were continued long enough (Ford *et al.*, 1977). In other words, there would be no deficit of tension. The time course of the early part of phase 4 after a large release is faster than the corresponding part of the rise of tension at the start of a tetanus. This is similar to findings of Jewell and Wilkie (1958) for recovery of tension after slower releases of whole muscle (Fig. 2.5B). The later part of phase 4 (tensions greater than $0.8T_i$), however, does have a similar time course to the initial development of tension. The dependence of the time course of phase 4 on the extent of length change is complex and requires further study.

With only a few modifications, the description of crossbridge operation by A. F. Huxley (1957) can explain phases 3 and 4 of tension recovery following a step change of length. During phase 3, both detachment and attachment of bridges occur; but detachment predominates and is responsible for the pause in the recovery of tension. As detachment becomes less important than attachment, the transition to phase 4 occurs (A. F. Huxley, 1974). One would expect therefore that the number of attached bridges would fall during phase 3 leading to a decrease in the instantaneous stiffness. No experiments on this point have yet been published.

2.2.4.4 *Comparison with Experiments on Whole Muscle*

In experiments on whole muscle in which changes in length have been less sudden and slower than the step changes in fibre length that have been discussed in this section, phases 1 and 2 cannot be completely resolved. However, Blange and Stienen (1979), using releases complete in about 1 ms, have obtained records from whole sartorius muscles in which all the four phases of tension transients are clearly seen. There is reasonable agreement about the distance of release that is required to bring the isometric force to zero and prevent its recovery; values reported for whole muscle are 1–1.5% of the muscle length (Jewell and Wilkie, 1958; Bressler and Clinch, 1974), and 1.75% (Blange and Stienen, 1979). These values are close to the value of 14 nm per half-sarcomere (equivalent to 1.3% of fibre length) found in the rapid tension transient experiments from the intercept of the T_2 curve (see Fig. 2.27). The whole muscle experiments would be expected to give slightly higher values because of tendon compliance.

2.2.5 RELAXATION AFTER TETANIC STIMULATION

Relaxation is defined for this discussion as the mechanical events (force and length changes) after the last stimulus of a tetanus. As described below, a number of factors influence the rate and time course of relaxation. As a starting point, Fig. 2.33A shows the tension during and after a brief tetanus of a nonfatigued muscle at low temperature. There are three phases in the record of relaxation. Initially, (1) force is maintained at its plateau value for a brief time. After a short tetanus tension may continue to rise during this time. Then, (2) force declines at a fairly constant rate. At the start of the last phase, (3) the rate of decline increases markedly, producing a "shoulder" or "hump" in the record. The force then follows an approximately exponential time course as it returns to the baseline.

This "shoulder" in relaxation after a tetanus was recognized many years ago by Hartree and Hill (1921). It is now known that it is related to movement within the fibres. Two independent methods have been used to observe these movements, which involve shortening of some regions of the muscle fibre and lengthening of others. Cleworth and Edman (1969, 1972) and Edman and Flitney (1982) detected length changes by observing the diffraction of laser light by the striations. Huxley and Simmons (1970b) and Julian and Morgan (1979a) observed displacement of markers on the fibre. The start of movement coincided with the "shoulder" in the force record and the more sudden the movement, the more obvious the "shoulder" (Huxley and Simmons, 1970b). The idea that the "shoulder" was actually *caused* by the transition from a relatively isometric condition to one involving

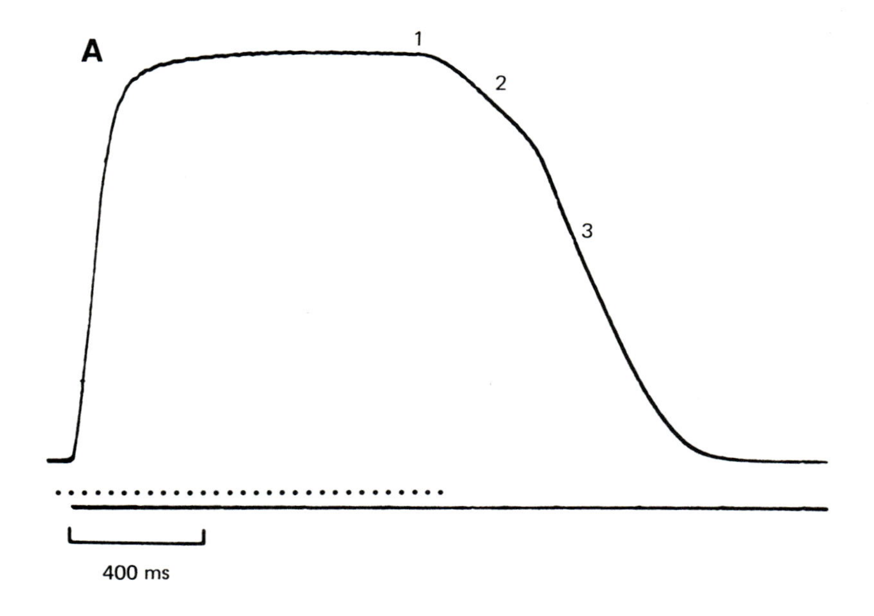

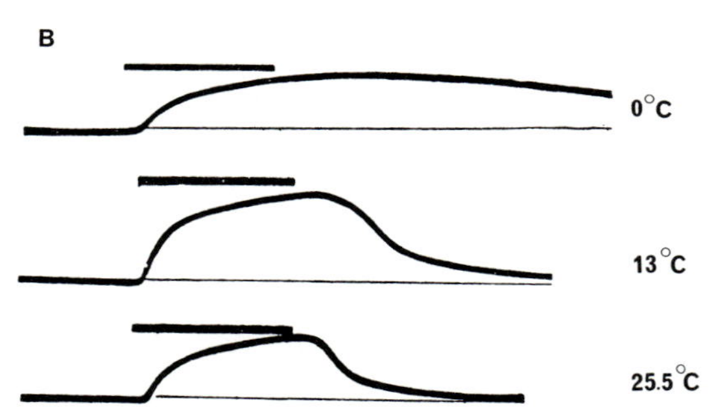

FIG. 2.33. (A) Time course of force during and after a 1-s isometric tetanus of a single fibre from anterior tibialis of *R. temporaria*, 2.2°C. The dotted line shows the period of stimulation. The numbers indicate the phases of relaxation described in the text. (From Fig. 1a of Edman and Mattiazzi, 1981.) (B) Records of force during an isometric tetanus and subsequent relaxation. Frog sartorius muscle at 0, 13, and 25.5°C. The bars show the duration of stimulation (0.2 s). (From Fig. 12A of Hill and Hartree, 1920.)

simultaneous shortening and lengthening was further supported by the later experiments of Huxley and Simmons (1973). They used a length clamp to prevent changes in the length of the middle 50% of the fibre when changes occur outside this controlled region. Under these conditions the second phase, the slow linear decline of force, continued beyond the time at which the shoulder appeared when the length clamp was not operating. Thus it seems that different regions of the fibre relax at different rates, and that the third, exponential phase of relaxation starts when the discrepancies between regions become significant. Although the time constant of the exponential phase is presumably related in some way to the kinetics of turnover of the crossbridge, the relation is certainly not a simple one. Since at any instant some bridges are shortening and others are being forcibly stretched, the exponential phase of relaxation must reflect a combination of the characteristics of bridges in quite different states. Obviously, the time constant of this phase of relaxation cannot be taken as a measure of the rate of detachment of isometrically operating bridges (Huxley and Simmons, 1970b), as had previously been suggested (A. F. Huxley, 1957).

It seems plausible that the rate of relaxation is determined by the kinetics of crossbridge turnover and/or the rate of removal of calcium ions from troponin as a result of the removal of free calcium from the sarcoplasm by the SR. Often the results of experiments on relaxation are discussed in terms of whether one of these processes can be identified as the rate-limiting one that controls the whole, or part, of relaxation. This question, however, has become less meaningful in the light of evidence that these two phenomena can affect each other. For example, experiments with aequorin have shown that changes in the free calcium concentration are influenced by muscle length (Blinks *et al.*, 1978), by shortening and lengthening (Allen, 1978), and by the internal shortening and lengthening at the "shoulder" of relaxation (Cannell, 1982). Studies on isolated proteins show that the regulatory functions of the troponin system are modified by attachment of bridges (Weber and Murray, 1973), and experiments on skinned fibres have also been interpreted in this way (Section 2.3).

Factors That Influence Relaxation after Tetanic Stimulation

A number of factors have been found to affect relaxation after isometric (fixed-end, without length clamp) contractions. (There do not seem to have been any systematic studies of relaxation after tetani with shortening.) It is worth noting that a particular variable may have a striking influence on relaxation under one set of conditions, but be less important under other conditions, and may influence the different phases of relaxation in distinct ways, although this point has not been investigated thoroughly.

1. *Species and type of muscle fibres.* Rate of relaxation varies widely from one species of animal to another and from one type of muscle to another. There is a rough correlation between the rate of relaxation and the rate of heat production. However, the existence of this correlation probably does not indicate anything about the nature of the rate-limiting process in relaxation (see Chapter 4, Section 4.2.6).

2. *Temperature.* Hill and Hartree (1920) examined relaxation of frog sartorius muscle after stimulation (single shocks and tetani of duration up to 1.6 s) at 0, 13, and 25.5°C. The example of their records for 0.2-s tetani in Fig. 2.33B shows that relaxation is very sensitive to temperature; the time for half-relaxation (the interval from the end of stimulation until force had declined to half its peak value) decreased about 2.2 times with a rise temperature of 10°C. Edman and Flitney (1982) have found that all the phases of relaxation have temperature coefficients in the range 2.2–2.5.

3. *Duration of tetanus.* Relaxation becomes slower as the duration of tetanus increases (Fig. 2.34), but eventually this effect saturates and relaxation becomes independent of duration. This behaviour was found by Abbott

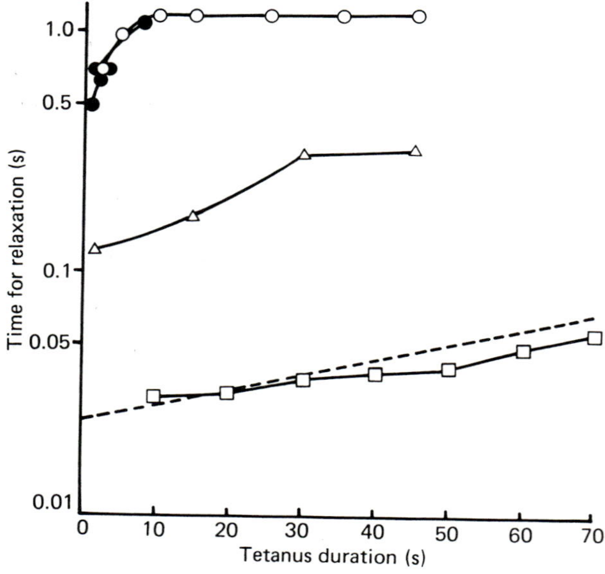

FIG. 2.34. The dependence of time required for relaxation on the duration of isometric tetanus. (●) Aubert (1956), frog sartorius (0°C). Time for tension to drop to 50%. (○) Abbott (1951), toad sartorius (0°C). Time for tension to drop to 95%. (△) Edwards *et al.* (1975a), mouse soleus (25°C), cyanide treated. Half-time of exponential phase of relaxation. (□) Edwards *et al.* (1975b) human quadriceps (about 37°C). Voluntary contractions of sufficient intensity to occlude circulation. Half-time of exponential phase of relaxation.

(1951) in sartorius muscle of toad (*Bufo bufo*), by Aubert (1956) in frog sartorius, and by Blinks *et al.* (1978) in fibres from anterior tibialis of *R. temporaria* and *X. laevis*. Surprisingly, Hill and Hartree (1920) did not find any effect of tetanus duration (up to 1.6 s) on the time to half-relaxation in frog sartorius muscle. This may have been because of the high frequency of stimulation (90 Hz), or the rather restricted range of tetanus duration they studied. In tortoise rectus femoris muscle there is no change in relaxation with tetanus duration. An effect of tetanus duration was found by Edwards *et al.* (1975a) in mouse soleus muscle which was metabolically inhibited, and by Edwards *et al.* (1975b) in human quadriceps muscles. These effects in mammalian muscle appear rather different than that in amphibian muscle, in that much larger amounts of activity are required to change relaxation rate and short tetani have relatively little effect compared with long tetani. The change in relaxation only occurs in anaerobic conditions (Edwards *et al.*, 1975a; D. A. Jones, personal communication). This may mean that the effects have a different origin in mammalian muscle than in amphibian muscle.

In amphibian muscle, Blinks *et al.* (1978) report that the fall in free calcium ion level after a tetanus becomes slower as the duration of the tetanus increases. This probably causes some or all of the slowing of relaxation with increased tetanus duration. One idea that may explain why calcium ion uptake is more rapid after a short tetanus is that some of the calcium is removed from the sarcoplasm by binding to parvalbumin and that these binding sites would be saturated after a long tetanus (see Chapter 4).

4. *Previous contractions.* It has been known for many years that the mechanical properties of muscle change during the progress of a series of contractions when the interval between them is short enough. For example, Bronk (1930) found, in a study of the relation between heat production and mechanics in frog muscle, that a pattern of stimuli that initially produced individual twitches could, if applied continuously, cause summation of the responses and eventually a "fused tetanus." Bozler (1930) made similar observations with the smooth, retractor pharynx muscle of the snail and he pointed out that the effect was due to a prolongation of relaxation.

For small amounts of stimulation this effect may have the same origin as the slowing of relaxation occurring during a single tetanus. Curtin (1976) has shown using frog muscle at 0°C that relaxation after a short "test" tetanus is prolonged when it is preceded by a single 5-s "conditioning" tetanus. The extent of the effect was independent of the muscle length during conditioning. The same slowing of relaxation was produced by conditioning at $1.7L_0$ as at $1.0L_0$, even though a quarter as much ATP would be hydrolysed, which suggests that the effect is not caused by any change in metabolite level. Recovery from the conditioning required at least 40 min at 0°C.

A number of studies have shown that during a *series* of tetanic contractions there is a progressive slowing of relaxation accompanied by other mechanical changes. These include decrease in the force produced during stimulation (Edman and Mattiazzi, 1981; Dawson *et al.*, 1980) and decrease in the maximum velocity of shortening (Edman and Mattiazzi, 1981). It is possible that all these mechanical effects have a common cause, possibly related to the levels of metabolites in the muscle.

5. *Metabolic state.* After inhibition of ATP resynthesis in muscle (with cyanide and iodoacetate) the ATP level falls (even if the muscle is not stimulated). This is accompanied by a slowing of relaxation (Edwards *et al.*, 1975a).

Production of force and calcium pumping both depend on the energy from ATP hydrolysis. The free energy for any reaction changes according to the concentration of reactants and products. As was first pointed out by Kushmerick and Davies (1969), the free energy available from ATP splitting will therefore change during a tetanus (Fig. 4.36). Changes in relaxation after metabolic inhibition, or during a contraction or a series of tetani, can be correlated with the changes in the supply of free energy (Edwards *et al.*, 1975a; Dawson *et al.*, 1980). It has been suggested that this correlation might arise because the behaviour of the crossbridge, or alternatively the behaviour of the SR, is affected by reduction in the free energy supply, but there is no clear evidence which (if either) of these explanations is correct.

6. *Intracellular pH.* The presence of CO_2 + bicarbonate buffer in the extracellular solution at low pH causes a slowing of relaxation (Bozler, 1930; Edman and Mattiazzi, 1981). It also causes reductions in force and in V_{max} similar to those that accompany a slowing of relaxation in a series of tetani (Edman and Mattiazzi, 1981). These effects are probably due to entry of CO_2 into the cells, where it forms hydrogen ion. The resulting change in intracellular pH is, in frog muscle, very long lasting (Bolton and Vaughan-Jones, 1977). There are a number of mechanisms that could produce a change in relaxation in this case; the reaction of hydrogen ions with specific amino acids in contractile proteins may modify their function, the change in intracellular pH may change the rates of various reactions, or there may be a change in the free energy from ATP hydrolysis since hydrogen ions participate in this reaction.

7. *Muscle length.* Increasing the length of a muscle fibre slows relaxation both from tetani (Hartree and Hill, 1921; Aubert, 1956) and from twitches (Jewell and Wilkie, 1960). Aubert (1956) found, in experiments using muscle length less than the optimum for tension development (L_0), that the main effect is on phase 1 of relaxation, the time for which tension is maintained before starting to fall. Edman and Flitney (1982) performed experiments at muscle lengths mostly beyond L_0 and show that in this length range both phases 2 and 3 are affected (they did not investigate phase 1).

2.2.6 UNSTIMULATED MUSCLE

2.2.6.1 *Passive Tension*

If the ends of an isolated muscle fibre are not held, its sarcomere length is about 2.0 μm. When it is stretched beyond 2.0 μm and then released, it shortens back to its original length. If the sarcomere length is brought to less than 2.0 μm, for example by allowing the stimulated muscle to shorten, the fibre reextends to this length during relaxation. If this reextension is prevented, for example by holding the fibre in a gel (Gonzalez-Serratos, 1971) the myofibrils become wavy, indicating that they lengthen, although the sarcolemma has not lengthened. There must be some forces exerted in unstimulated muscle fibres to cause these movements. The observations of Gonzalez-Serratos show that at least part of the forces that cause reextension is within the myofibrils themselves.

An isolated muscle fibre at a sarcomere length greater than 2.0 μm exerts a measurable force (Fig. 2.35), called the passive tension. It was suggested

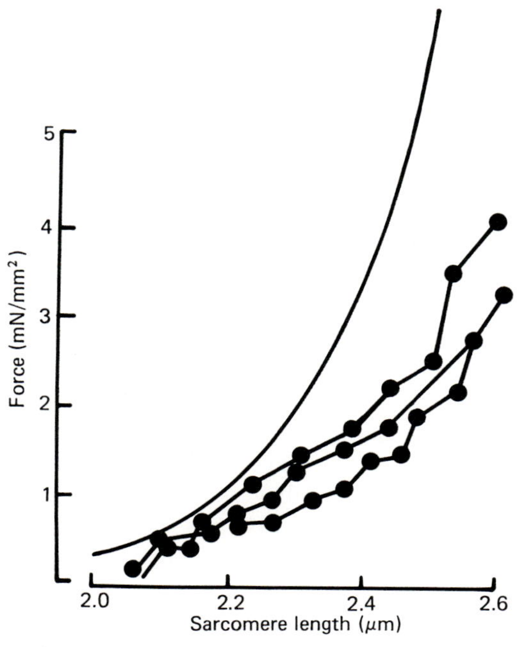

FIG. 2.35. Passive tension in single muscle fibres and whole muscle as a function of sarcomere length. (●) Single fibres from iliofibularis muscle of frog (*R. temporaria*). (From Lannergren, 1971.) Solid line, whole sartorius muscle of frog (*R. temporaria*). (From D. K. Hill, 1968.)

by Ramsey and Street (1940) that much of this tension is exerted by the sarcolemma, because an empty sarcolemmal tube (i.e., the portion left when myofibrils have been retracted from a length of a fibre by clot formation after damage) exerts a similar tension. Podolsky (1964) found that the stiffness of a skinned region of a fibre was only 20% that of an unskinned region, at sarcomere lengths between 3.2 and 3.6 μm. However, Casella (1950) and Rapoport (1972), who made more detailed studies of the mechanical properties of the sarcolemma, concluded that it can account for only a small part of the resting tension at sarcomere lengths up to 3.7 μm. Thus, neither the isolated sarcolemma nor the fibre without its sarcolemma has been shown to be responsible for the tension in the resting intact fibre. Perhaps this is because the passive tension results from some interaction between the surface and the contents of the fibre.

In whole muscles the resting tension, expressed per unit cross section of muscle, is greater than in single fibres, particularly at longer sarcomere lengths (Fig. 2.35). It seems likely that this is because, in whole muscle, part of the tension is due to connective tissue strands running parallel to the muscle fibres. It should also be remembered that the single fibres on which resting tension has been measured may not be representative of the whole muscle.

D. K. Hill (1968) suggested that part of the passive tension might be due to attached crossbridges. This idea arose from the observation that passive tension (in whole sartorius muscle) is increased in hypertonic Ringer's solution. If the hypertonic solution was acting on connective tissue and/or sarcolemma, the increment in tension would increase steeply with muscle length. However, the results show that the increment is almost independent of muscle length. The evidence for the involvement of crossbridges in this phenomenon is not strong and additional work is needed to test D. K. Hill's hypothesis.

2.2.6.2 *Short-Range Elasticity*

When an unstimulated muscle, or muscle fibre, is stretched a short distance the tension increases in a characteristic way (Lannergren, 1971) (Fig. 2.36A). For a distance of 2 nm per half-sarcomere the tension rises in proportion to the distance. During a stretch beyond 2 nm per half-sarcomere tension does not continue to rise, but remains constant for the remainder of the stretch. A release (Fig. 2.36A) produces a tension record like that for a stretch, but inverted. After the length change is complete tension returns to its original level over the next 1 or 2 min.

The component responsible for these tension changes has been called the short-range elastic component (SREC). Its stiffness, that is, the ratio of

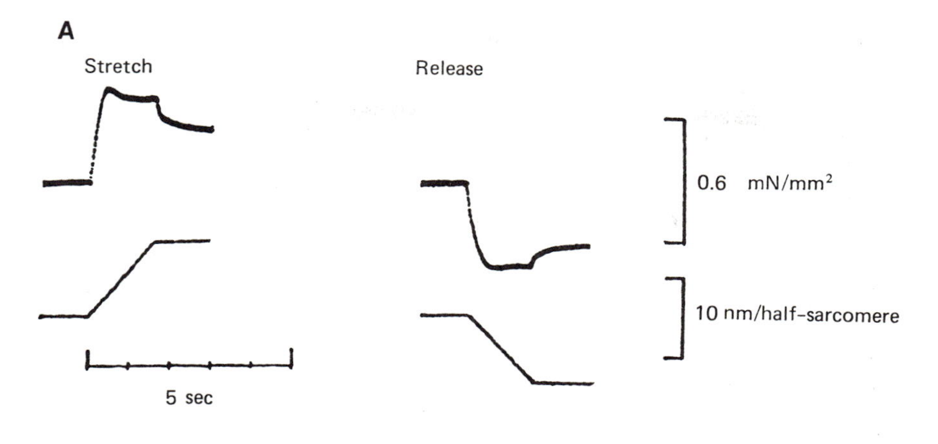

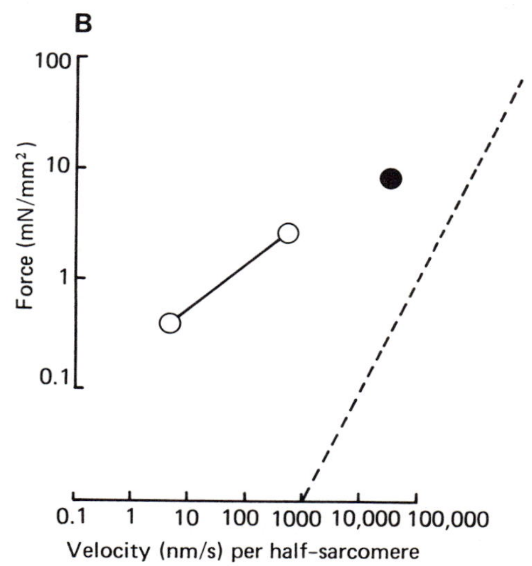

FIG. 2.36. (A) Records of tension (above) and length change (below) during stretch and release of an unstimulated single fibre from iliofibularis muscle (*R. temporaria*). The initial sarcomere length of the fibre was 2.2 μm. (From Lannergren, 1971.) (B) Force produced by resting muscle when stretched, plotted against the velocity of stretch. Log scales on both axes. Viscous behaviour, that is, force proportional to velocity, would be represented on this plot by a line of unit slope (dotted line). Measurements (●) from Ford *et al.* (1977) and (○) from Lannergren (1971). In both studies frog single fibres were used.

tension change to length change in the initial, linear part of the record, is 0.2 kN/mm^3 per half-sarcomere, which is about 0.25% of the stiffness during a tetanus. The maximum force exerted during these length changes is about 0.13% of the force produced in an isometric tetanus.

The constant force exerted during most of the stretch, such as that in Fig. 2.36A, depends on the velocity of lengthening as shown in Fig. 2.36B. It is increased fivefold by a 100-fold increase in velocity to 500 nm/s per half-sarcomere. This is not the behaviour expected of a viscosity, for which the force would increase in proportion to the velocity. At the much higher velocity of 28,000 nm/s per half-sarcomere the force exerted is increased to about 10 mN/mm^2 (Ford *et al.*, 1977) and at this high velocity the behaviour is viscous, that is, force is proportional to velocity.

The stiffness of the SREC increases as sarcomere length is increased from 2.0 to 2.6 μm (Lannergren, 1971). It cannot, of course, be observed below 2.0 μm as the fibres are then slack. This increase with sarcomere length is less steep than that of the steady resting tension. From his observations on whole muscles in hypertonic solution, D. K. Hill (1968) suggested that the SREC is due to crossbridges. If this is so, it is surprising that the stiffness does not decrease as filament overlap decreases at sarcomere lengths greater than 2.2 μm. Hill supposed that this is because the decrease in filament separation that occurs compensates for the change in filament overlap. This argument, being rather contrived, is not completely convincing. However, in the absence of a better explanation, it is a reasonable working hypothesis that the SREC is due to crossbridges.

Is the SREC present in stimulated muscle? Three observations suggest that it is not:

1. Ford *et al.* (1977, p. 484) found that, if they assumed that the viscous behaviour of resting muscle also occurs in stimulated muscle, the measurements of rapid elasticity determined from releases of different amplitudes did not agree. If, however, they assumed that there was no viscosity in the active fibre, the measured elasticity was independent of the extent of release (see Fig. 2.29).
2. D. K. Hill (1968) using very slow stretches and releases was unable to find, in tetanically stimulated muscle, any component of the tension resembling the SREC. These experiments were made in hypertonic solution, and the conclusion may therefore not apply to muscles in solutions of normal tonicity.
3. Lannergren (1971) showed that the tension produced by the SREC was greatly reduced by treatment with contracture-producing agents (K^+ and caffeine) at concentrations too small to produce an increase in active tension. It seems unlikely that the SREC would reappear when these

same agents are present in high enough concentration to cause contraction.

The SREC is an intriguing phenomenon. The possibility that it represents the behaviour of the crossbridge is worth further investigation because there may turn out to be similarities with the crossbridge cycles in contracting muscle.

2.3 Fibres without Membranes

2.3.1 INTRODUCTION

Skinned or demembranated fibres are those from which the surface membrane has been either removed mechanically (by dissection) or rendered permeable by chemical treatment with glycerol and/or detergents. In a solution containing MgATP these fibres will contract if calcium ion is present and will relax when calcium is removed. Some of the motives for experimenting on these preparations are (1) to learn how the mechanical behaviour is influenced by the concentration of substrate MgATP, and products ADP and P_i (these concentrations cannot readily be controlled in the intact fibre); (2) to investigate the effects of changes in the Ca^{2+} concentration; and (3) to hold Ca^{2+} ion concentration constant, so that effects due to changes in its concentration can be avoided.

How good are such skinned preparations as models for contraction of living fibres? This question can be answered by comparing the behaviour of skinned fibres (in solutions of ATP, Ca^{2+}, etc., at a concentration thought to exist in intact fibres) with the behaviour of intact fibres.

2.3.2 COMPARISON WITH INTACT FIBRES

Structural changes due to skinning. When the membrane is removed the muscle fibre swells. The cross-sectional area increases by about 50% (Godt and Maughan, 1977) because of an increase in the interfilament spacing (Matsubara and Elliott, 1972). This is presumably (1) due to a loss by diffusion of high molecular weight substances that are present within the fibre but excluded (at least in part) from the interior of the myofibrils; these substances exert an osmotic compressive force on the myofibrils; or (2) because the ionic environment is changed in a way which increases the net charge on the filaments.

The filament lattice in an intact resting fibre shows "constant volume behaviour" when the fibre length is changed, but this behaviour is lost when skinned fibres swell (Matsubara *et al.*, 1984) (see Fig. 2.37). If a high molecular

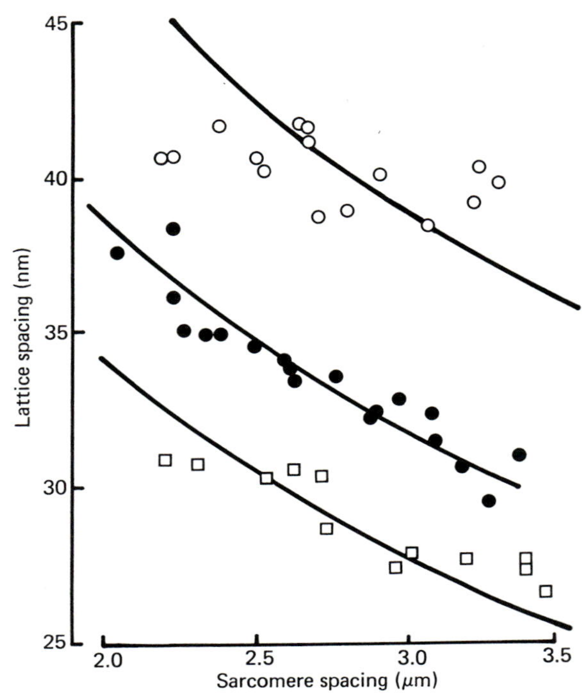

FIG. 2.37. Relationship between spacing and sarcomere length in resting frog muscle fibres. The points are experimental observation; the lines represent constant volume behaviour, that is, $(\text{lattice spacing})^2 = \text{constant/sarcomere length}$. (●) Living fibres, from Matsubara and Elliott (1972). (○) Skinned fibres; note the increase in lattice spacing and loss of constant volume behaviour. (□) Skinned fibres placed in PVP-containing solution causing, by osmotic means, a decrease in volume. Note that the constant volume behaviour is, to some extent, restored. Results for skinned fibres from Matsubara *et al.* (1984).

weight polymer (for example, dextran or polyvinylpyrolidone, PVP) is included in the solution around skinned fibres, the lattice spacing can be brought back to a value typical for the intact fibre and the constant volume behaviour of resting fibres is restored (Fig. 2.37). However, few mechanical studies have been made in the presence of PVP so it is not yet known what effect this restoration of normal filament arrangement has on many aspects of the contractile behaviour of the fibres.

The regularity of the sarcomere spacing is easily lost in skinned fibres when they contract maximally, or when they go into rigor. These changes are often irreversible. Morgan *et al.* (1982) suggest that the constant volume behaviour of intact fibres may be a factor that stabilizes the sarcomere regularity and so its loss in skinned fibres may contribute to the ease with which the sarcomeres become disordered. It would be a good practice to monitor the sarcomere spacing during experiments with skinned fibres so that the experiment can be ended when severe irregularities appear. Because it is difficult to obtain reproducible results in a fully activated skinned fibre, many experiments are done on preparations that are not fully active. Thus, it is likely that the results will be, to some degree, different from those in intact fibres either because of some loss of the orderly arrangement of the contractile material or because the fibres are not fully active.

Force. The force produced by muscle fibres is usually normalized to the cross-sectional area of the fibres. As a result of swelling, the cross-sectional area after skinning is about 50% greater than before. It is not surprising therefore that force, when normalized by the cross section after skinning, is less than that in intact fibres. After allowing for this factor (by multiplying by 1.5), the force per unit cross section in two studies of skinned fibres from frog was 222 kN/m^2 at 1°C (Ferenczi *et al.*, 1984; mechanical skinned fibres) and 218 kN/m^2 at 5°C (Podolsky and Teicholz, 1970; chemically skinned fibres). These values can be compared to values for living frog fibres of 223 kN/m^{-2} at 1°C (Edman, 1979) and 265 kN/m^{-2} (Gordon *et al.*, 1966b, average of seven values). Clearly the agreement is within the accuracy of the correction for swelling.

Moss (1979) has done experiments in which he measured force developed in the same frog fibres before and after skinning. The value after skinning was 77% of that before skinning. He states that, if he had used a higher calcium concentration in the activating solution for skinned fibres, they would have produced 90% of the force of the intact fibre. Clearly the force exerted by skinned fibres can be close to that of the living fibre. Godt and Lindley (1982) have shown that, as temperature increases from 4 to 22°C, there is an 88% increase in the force exerted by skinned fibres from frog. A similar effect of temperature is also found in intact muscle (Table 2.III).

Mammalian skinned fibres usually produce less force than frog skinned fibres. For example, Julian *et al.* (1981) give values of 63 kN/m^2 (at 10°C) and 91 kN/m^2 at (15°C) for skinned fibres from rabbit soleus (these values are not corrected for swelling). There are no observations of intact soleus muscle in the same temperature range, perhaps because they are inexcitable at these relatively low temperatures. A comparison of force production by skinned and intact mammalian fibres is therefore not possible with present information.

Rate of activation. In skinned fibres the time course with which tension develops on activation (and falls on relaxation) is generally much slower than in the intact fibre. One reason for such behaviour is that, in intact fibres, calcium is liberated from (and returns to) the SR which is distributed within the fibres, but in skinned fibres calcium must diffuse from the surface to

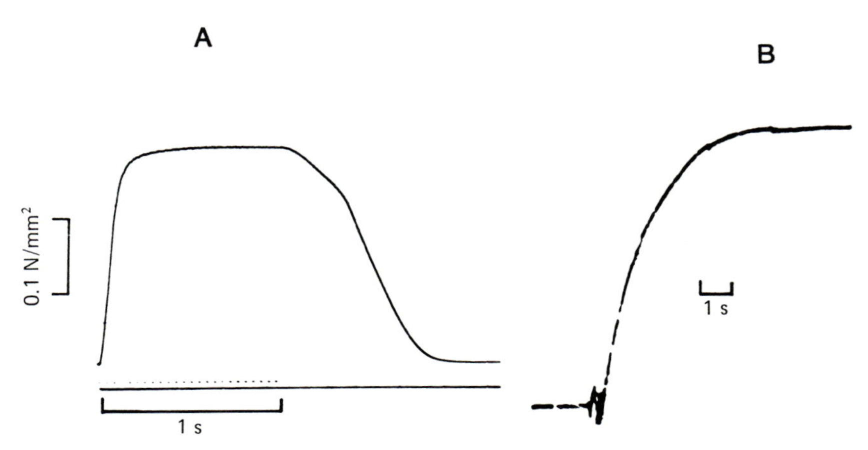

FIG. 2.38. A comparison of the rise of tension during a tetanus in a living muscle fibre (A) with that in skinned muscle fibre (B). Both observations are with fibres from frog. (A) Fibre from anterior tibialis (*R. temporaria*) at 2.2°C. (From Edman and Mattiazzi, 1981.) (B) Fibre from iliofibularis (*R. esculenta*) at 2.3°C. Record was not calibrated for force. (From Moisescu and Thieleczek, 1978.)

produce activation. Moisescu and Thieleczek (1978) describe solutions designed to minimize this delay by providing a high concentration of calcium buffer to carry calcium rapidly into the interior of the fibre. They show that, using this technique, diffusion of calcium into the interior of the fibre is no longer limiting the rate of tension development. Nevertheless the rates of rise of tension are much less than in intact fibres (Fig. 2.38). The use of this method of relatively rapid activation probably helps to preserve the regularity of sarcomere spacing.

Rate of shortening. In most experiments the maximum velocity of shortening (V_{max}) recorded for skinned fibres has been less than that in intact fibres. For example, Podolsky and Teicholz (1970) report V_{max} for mechanically skinned fibres from *R. pipiens* at 7°C to be 1.2 muscle lengths per second whereas the data of Civan and Podolsky (1966) on intact fibres suggest a value of 2.3 muscle lengths per second at this temperature. Julian and Moss (1981), using chemically skinned fibres (*R. pipiens*) at 5°C, obtain a V_{max} of 1.8 muscle lengths per second and quote (from their unpublished work) a value of 3 muscle lengths per second for intact fibres.

A notable exception to this trend is the work of Ferenczi *et al.* (1984). Their V_{max} is 2.2 muscle lengths per second (at 1°C) for mechanically skinned fibres from *R. temporaria*, similar to the values of 2.0 and 2.5 muscle lengths per second for intact fibres (Edman, 1979; Edman and Mattiazzi, 1981; Edman and Hwang, 1977). The success of Ferenczi *et al.* (1984) in obtaining

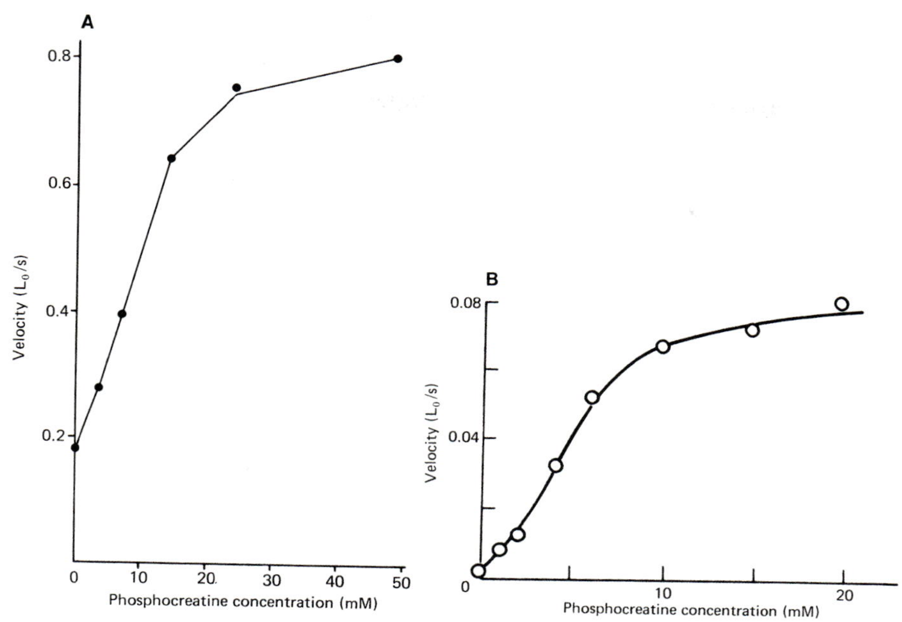

FIG. 2.39. Dependence of shortening velocity on concentration of phosphocreatine in the solution. (A) Mechanically skinned muscle fibre from anterior tibialis of *R. temporaria*, 3°C. The MgATP concentration was 0.5 m*M*. The load was 16% P_0. (From Ferenczi *et al.*, 1984.) (B) Chemically skinned fibre from rabbit psoas at 10°C. The MgATP concentration in these experiments was 0.02 m*M*. The load was zero. (From Cooke and Bialek, 1979.)

a physiological value of V_{max} may be connected with the fact that they used a high concentration of PCr and creatine kinase in the solutions. They show that shortening velocity is sensitive to the concentration of PCr (Fig. 2.39A). Cooke and Bialek (1979) report a similar result with glycerinated rabbit muscle. Since it is clear that the presence of PCr and creatine kinase influences some aspects of contractile behaviour, they ought always to be included in the solutions used for experiments with skinned fibres.

Records of isotonic shortening of skinned fibres sometimes show a progressive fall in speed as shortening progresses (Fig. 2.40A). This is not seen in intact twitch fibres (although intact slow muscles do show similar behaviour, Floyd and Smith, 1971). Ferenczi *et al.* (1984) suggest that this phenomenon is due to development of disorder within the fibre.

Force–velocity curve. The force–velocity curve of skinned muscle fibres has not been determined in as much detail as that of the intact fibres and whole muscles probably because it is not possible to make many records with one skinned fibre preparation. It is apparent that the curve has a shape

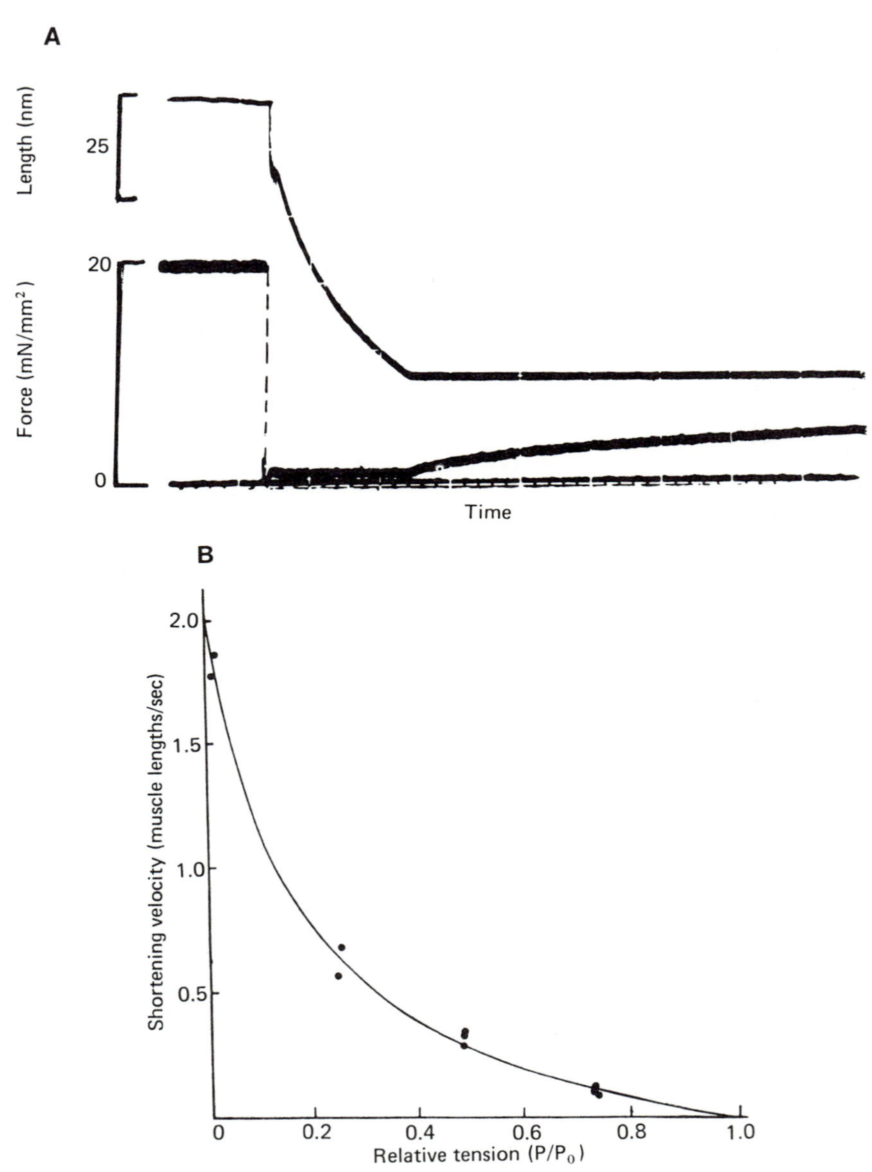

FIG. 2.40. (A) Records of length change (expressed per half-sarcomere) and force during isotonic shortening of a skinned muscle fibre from anterior tibialis of frog (*R. temporaria*) at 1–2°C. (M. Ferenczi, personal communication; see also Ferenczi *et al.*, 1984.) (B) Force–velocity curve obtained from one preparation, a mechanically skinned fibre from anterior tibialis of a frog (*R. temporaria*) at 1–2°C. The solution contained 5 m*M* MgATP and sufficient free calcium to activate the fibre maximally. (Ferenczi *et al.*, 1979.)

similar to that for intact fibres, although it is probably more curved (Fig. 2.40B). The maximum power output of skinned fibres is thus reduced (by about 10%) compared to an intact fibre. This effect might be due to a local depletion of substrate, or accumulation of products near the cross-bridges because the rate of ATP splitting is at its maximum when power output is maximum.

Tension transients. Figure 2.42A shows the tension response to a small, fairly rapid length change, in an activated skinned fibre. The record is strikingly similar to those obtained from intact fibres (compare Fig. 2.25). However, skinned fibre preparations are less stiff than contracting intact fibres (Fig. 2.45). Perhaps some of this difference is due to the damage caused to the skinned fibre segment where clips are attached to it. Because these segments are short, the damaged regions can have a relatively large effect. In principle these effects can be avoided by measuring and controlling sarcomere length in the centre region of the skinned segment. Activated skinned fibres that have been shrunken with PVP (to about the same lattice volume as an intact fibre) are much stiffer, perhaps as stiff as contracting intact fibres (Goldman and Simmons, 1978).

ATPase. Levy *et al.* (1976) have measured the rate of ATP splitting in activated skinned fibres from *R. pipiens.* Their result, 2 μmol/s per gram dry weight at a sarcomere length of 3 μm and a temperature of 4°C, is similar to the rate in whole muscle under these conditions, calculated from the results of Homsher *et al.* (1981) assuming a Q_{10} value of 3.

Conclusion. In most respects the properties of fully activated skinned muscle fibres resemble those of tetanized intact fibres in spite of the fact that they have probably lost some of the regularity of their sarcomere arrangement. However, because there are some discrepancies, inferences about mechanisms in skinned fibres cannot be accepted unequivocally as applying to intact fibres as well.

2.3.3 EFFECTS OF CHANGES IN SUBSTRATE AND PRODUCT CONCENTRATION

The MgATP concentration optimal for tension development is about 50 μ*M* and at both higher and lower MgATP concentrations less tension is produced (Fig. 2.41A). At the physiological MgATP concentration of 5 m*M*, tension is about 75% of the maximum value shown in Fig. 2.41A. Similar results have been obtained with glycerinated rabbit muscle at room temperature (Cooke and Bialek, 1979) and with mechanically skinned frog muscle at 1°C (Ferenczi *et al.*, 1982, 1984).

V_{max} varies with substrate concentration as shown in Fig. 2.41B. The relation is approximately hyperbolic, and at the physiological MgATP concentration the value of V_{max} is near its maximum. The value of MgATP

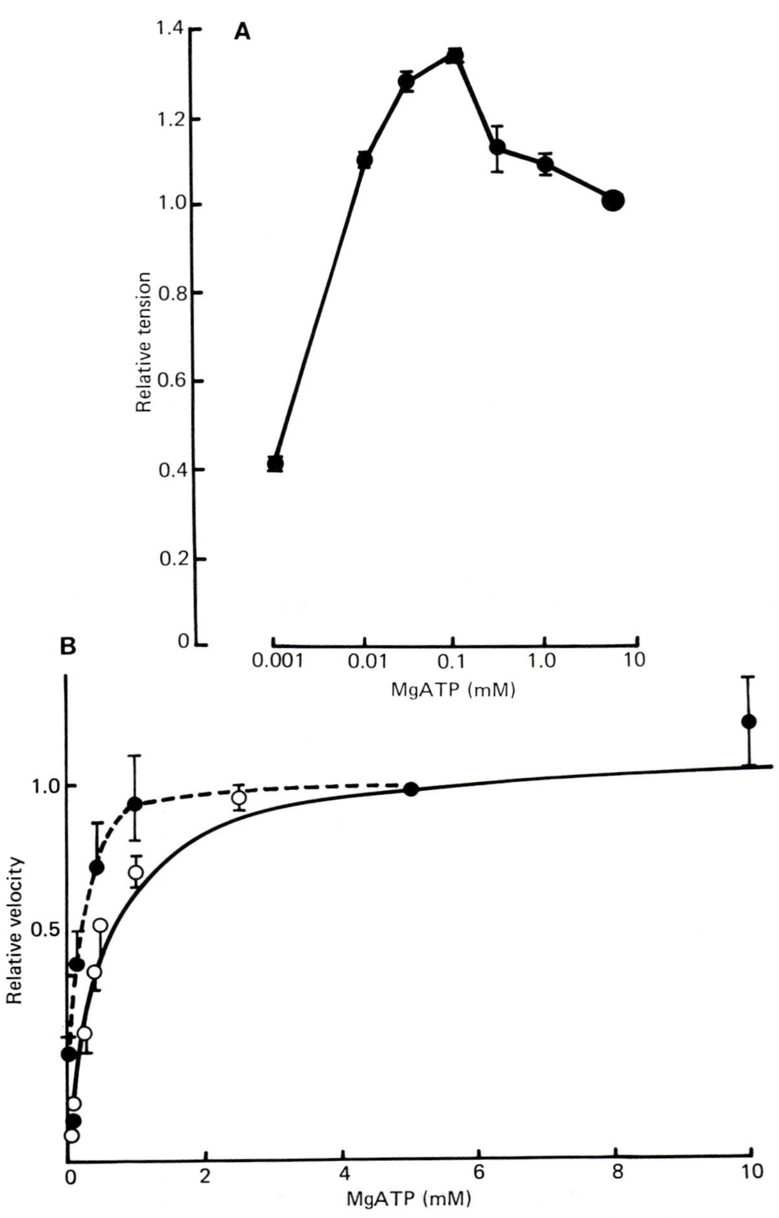

FIG. 2.41. (A) Force developed by mechanically skinned muscle fibres from frog muscle (*R. temporaria*) at different concentrations of MgATP. The force is expressed relative to the value in 5 m*M* ATP. The bars indicate ±1 SEM. (From Fig. 10.2 of Ferenczi, 1978.) (B) Dependence of maximum speed for shortening on concentration of MgATP in skinned muscle fibres. (○) Frog muscle (*R. temporaria*), 1–3°C, from Ferenczi *et al.* (1984). (●) Rabbit muscle, 10°C, from Cooke and Bialek (1979). In both cases the velocities are expressed relative to the value in 5 m*M* MgATP.

concentration at which V_{max} is half-maximal (the K_m) is 0.47 mM. Cooke and Bialek's experiments on rabbit muscle also show that V_{max} is a hyperbolic function of the MgATP concentration (Fig. 2.41B). In their experiments V_{max} had a K_m of 0.25 mM.

Ferenczi *et al.* (1984) find that the force–velocity relation is more curved at high MgATP concentration. Their values of a/P_0 are quite scattered but there is a statistically significant decrease in a/P_0 at higher MgATP concentration. Thus, for example, the value of a/P_0 at 5 mM MgATP is 0.16 ± 0.01 (mean $\pm$ SEM) and at 20 μM MgATP is 0.27 ± 0.05.

Each of the four phases of the tension response to a rapid length step is affected by MgATP concentration (Ferenczi *et al.*, 1982). This is illustrated in Fig. 2.42A, which compares a response at high MgATP concentration with one at low concentration. The behaviour of each phase is as follows:

Phase 1. The size of the drop in tension, for a particular size of step, is proportional to fibre stiffness. As the MgATP concentration is reduced from 5 mM, stiffness increases and reaches a maximum value at 50 μM. At MgATP concentrations less than 50 μM, the stiffness remains constant.

Phase 2. There is less tension redevelopment and it is slower at lower MgATP concentrations.

Phase 3 disappears at MgATP concentrations below 1 mM because it becomes so slow that it is no longer distinguishable from phase 4.

Phase 4 also becomes slower when MgATP is lower than 0.3 mM. Its dependence on MgATP concentration is hyperbolic with a K_m of 40 μM.

Figure 2.42B compares the MgATP dependence of P_0, V_{max}, stiffness (from phase 1), and the rate of phase 4 redevelopment of tension.

There are few results concerning the effects of product concentrations. However, both Cooke and Bialek (1979) and Ferenczi *et al.* (1984) note that increases in MgADP concentration, to about 1 mM, reduce V_{max}. Similar changes in P_i concentration have no effect.

Crossbridge Interpretation

A qualitative interpretation of most of these effects of MgATP concentration can be given in terms of the simple reaction scheme in Fig. 2.43. This scheme assumes the existence of one detached crossbridge state (MATP) and two attached states (AMPr and AM). It is supposed that AMPr exerts more force than AM but has the same stiffness. MgATP is required for crossbridge detachment, which occurs with a rate constant k_1 which depends on the MgATP concentration. The rate constants for the other two reactions are independent of MgATP concentration. It is also assumed that, at high MgATP concentration, k_3 is the rate-limiting step of the reaction mechanism and this rate is increased when muscle shortens.

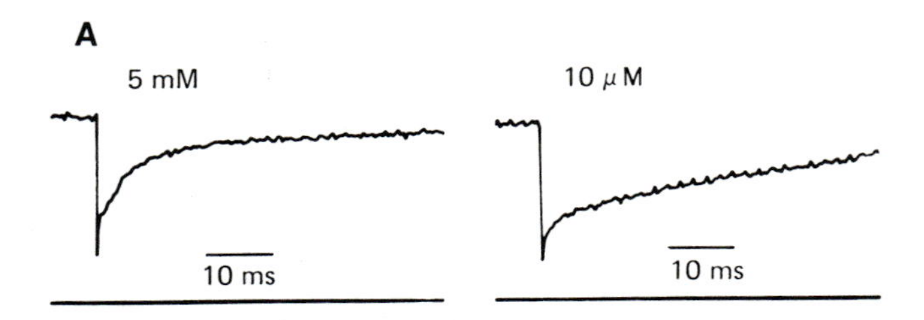

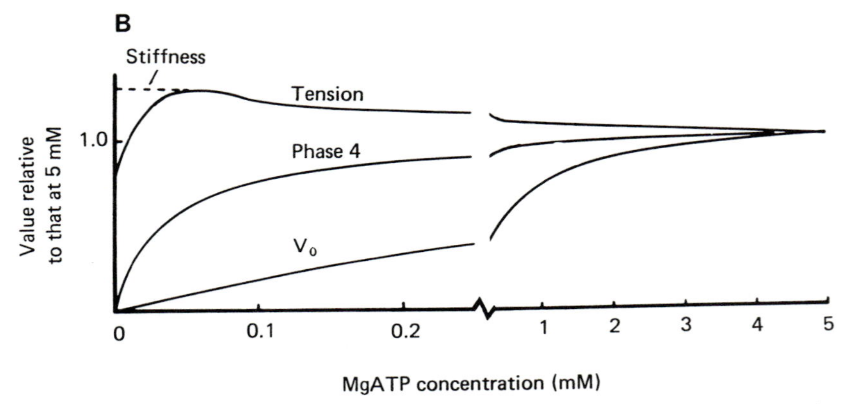

FIG. 2.42. (A) Tension responses to rapid releases at two different MgATP concentrations. Both records from the same mechanically skinned fibre from frog anterior tibialis muscle (about 1°C). (From Ferenczi *et al*., 1982, Fig. 6.) (B) Summary of the dependence on MgATP concentration of various muscle properties. Stiffness is determined from observations of phase 1 of the tension transient. Tension is that produced in a fixed-end tetanus. Phase 4 refers to the rate of rise of tension during this phase. V_0 is the maximum velocity of shortening obtained by extrapolation of the force–velocity curve. All quantities are expressed relative to that at 5 m*M* MgATP. (From Fig. 5 of Ferenczi *et al*., 1982.)

In the absence of MgATP all the crossbridges are in the low force state, because reaction 1 cannot occur. Consequently, stiffness is at its maximum, and no shortening or phase 4 tension recovery can occur after a quick release. When the MgATP concentration is raised slightly (to about 40 μM for example), some crossbridge cycling occurs, although rather slowly, and some crossbridges enter the high force state (AMPr). As a result, the force is greater than in the absence of MgATP. The value of k_1 at this MgATP level is small compared to k_2; therefore the number of detached crossbridges is small and the stiffness remains high. Because the number of detached crossbridges will also be negligible after a quick release, there is no phase 3 in the tension recovery. (Phase 3 is thought to result from an increase in the

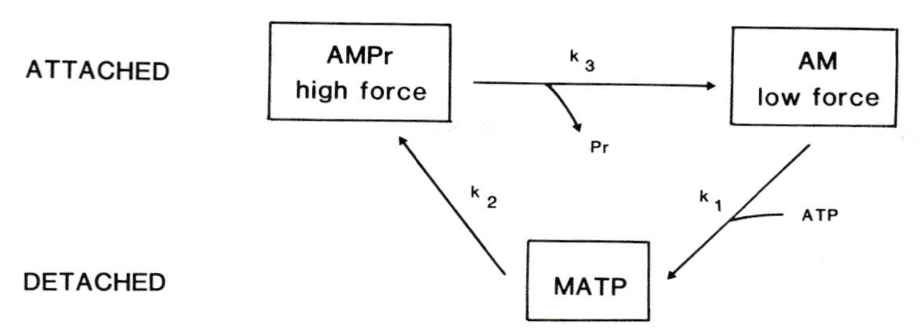

FIG. 2.43. A reaction scheme used to explain the dependence of various muscle properties on MgATP concentration. For explanation, see text.

number of detached bridges; see Section 2.2.4.) At this low MgATP concentration, the tension redevelopment after a release (phase 4) is due to crossbridges reattaching almost as soon as they have detached, and therefore the rate of tension development depends on MgATP concentration.

At moderate MgATP concentrations (e.g., around 3 m*M*) k_1 is greater than k_2. There are therefore a significant number of detached crossbridges in the steady state, and both force and stiffness are less than at lower MgATP concentrations. The number of detached crossbridges increase after a quick release, giving rise to a phase 3 in the tension recovery. Phase 4 is insensitive to MgATP concentration at this MgATP level because its rate is limited by k_2, instead of k_1. V_{max} is also insensitive to MgATP concentration because its rate is limited by k_3, instead of by k_1.

2.4 Rigor

2.4.1 EXPERIMENTAL FACTS AND THEIR INTERPRETATION

Without ATP, or some similar substance, muscle is in rigor. It is very much stiffer than relaxed muscle and, compared to living muscle, it can shorten only an extremely limited distance. It is possible that the crossbridges in rigor muscle are in one, or more, of the states they pass through as they split ATP during contraction. If this is so, it is useful to study rigor muscle, to get information about these states.

In skinned fibres, rigor can be induced simply by placing the fibres in an ATP-free solution. If a solution of low Mg^{2+} concentration is used, rigor can be produced without much loss of order, perhaps because only a very small amount of tension is developed.

In intact fibres or whole muscles, rigor can be induced by blocking the supply of ATP with dinitrofluorobenzene (DNFB) or iodoacetic acid and nitrogen, and then either waiting for the resting metabolism to use up the ATP supply, or depleting it by stimulation. As rigor develops some force is exerted; the amount varies according to the treatment used to produce rigor (Mulvany, 1975; Kawai and Brandt, 1976) but is typically about half that exerted by contracting muscle.

Interpretation of mechanical measurements is difficult unless it is known that the sarcomere arrangement remained orderly as rigor was induced. In isolated intact muscle fibres, rigor usually results in the loss of the regular sarcomere arrangement (Brocklehurst, 1973). In whole muscles, there is some increase in the dispersion of sarcomere length as rigor is induced (Mulvany, 1975), but the loss of order is minor. If such a muscle is subjected to a large stretch, most of the length increase occurs by tearing of the myofibrils.

Mulvany (1975) compared force produced during slow stretches of muscles put in rigor at different sarcomere lengths and found that rigor force declined with filament overlap. This supports the idea that rigor force is due to crossbridge interactions. Two lines of evidence indicate that most of the crossbridges are attached in rigor muscles: (1) EPR measurements show that all mobility of a chemical probe attached to myosin heads in relaxed skinned muscle is lost upon going into rigor (Thomas *et al.*, 1980; Thomas and Cooke, 1980). (2) White *et al.* (1980) measured the binding of excess free S-1 (that is, isolated myosin heads) to myofibrils in rigor. Their results indicate that all the myofibrillar myosin heads were occupying a binding site on actin.

When rigor muscle is released, the tension drop is followed by only a small tension recovery (Fig. 2.44A), which may correspond to phase 2 of tension recovery after release of a living fibre (or might come from outside the myofibrils altogether). However, there is no evidence of phases 3 and 4 of the tension response. After stretches the recovery of tension is more rapid and complete than after a release, though it is slower than in active muscle. (An example of tension responses to stretches is shown in Fig. 2.44B.) In Mulvany's experiments (1975), rigor muscles were subjected to large stretches (about 10% of their initial length) and continued to exert force. This introduces the possibility that crossbridges can (when strained) detach and reattach to a different actin site, thus allowing filament sliding to occur. However, considerable tearing of the myofibrils occurred during these large stretches and it is not clear whether or not any filament sliding actually occurred. There are no comparable experiments on skinned or isolated fibres in rigor from which it might be easier to determine whether sliding occurs.

It would be interesting to compare the stiffness of rigor muscle with that of contracting muscle, because with certain additional assumptions, the proportion of crossbridges that are attached during contraction could be

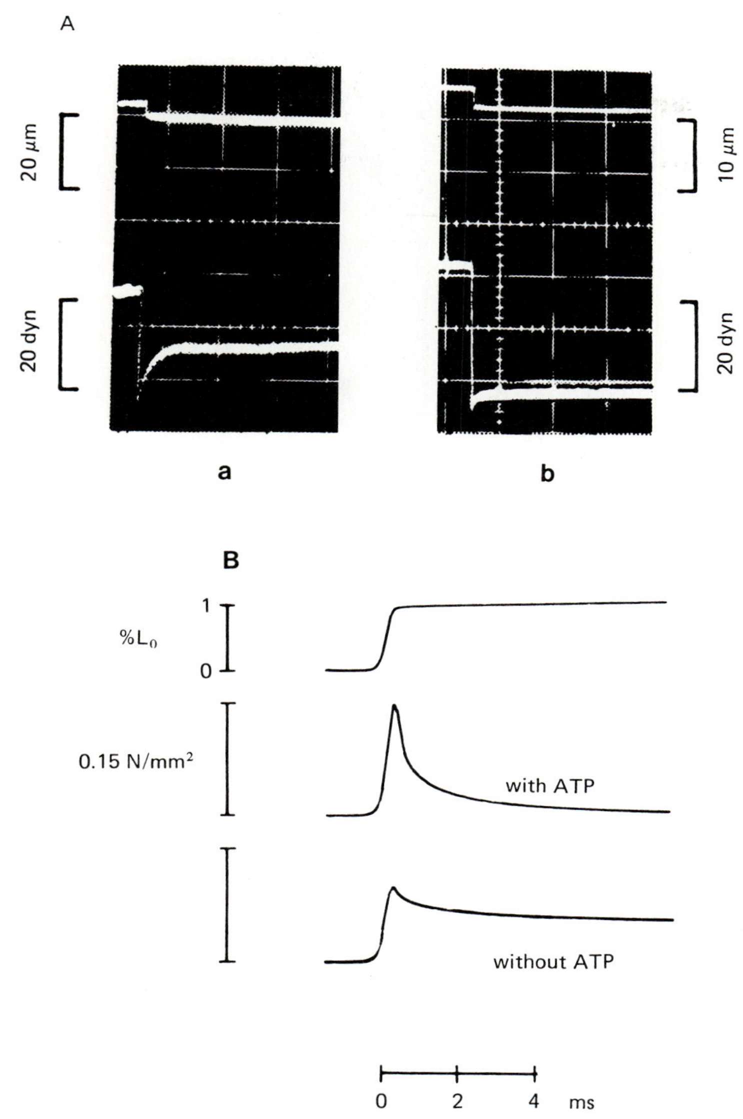

FIG. 2.44. (A) Force responses (lower trace of each pair) to length change (upper trace) recorded from a glycerinated fibre bundle obtained from a tortoise iliofibularis muscle: (a) with ATP present, at 4°C; (b) without ATP, at 20°C. Note that in the absence of ATP there is little rapid recovery of force after the length step. (From Fig. 3 of Heinl *et al.*, 1974.) (B) Tension changes (two lower lines) in response to rapid stretch (upper line), in glycerinated fibre bundle from rabbit psoas at 23°C. Note the more rapid drop in tension after the stretch in the presence of ATP. (From Fig. 4 of Guth and Kuhn, 1978.)

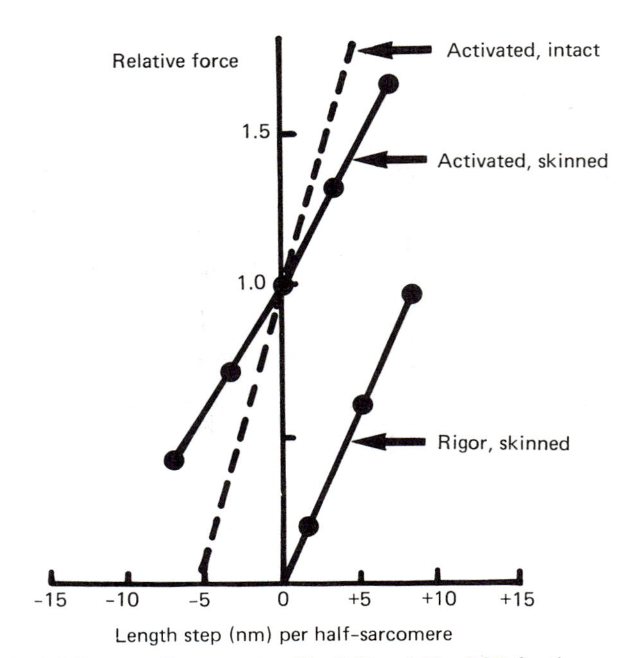

FIG. 2.45. The full lines are T_1 curves (see Fig. 2.25 and Fig. 2.27) for the same mechanically skinned fibre from frog semitendinosus muscle (*R. temporaria*) when calcium activated, and when in rigor. (Goldman and Simmons, 1978). The dotted line shows the force–extension curves obtained in intact contracting fibres by Ford *et al.* (1977).

estimated. Unfortunately there are difficulties in interpreting measurements of the stiffness of rigor muscles. The results in Fig. 2.45 illustrate the difficulty. In these experiments, the stiffness of a skinned fibre in rigor was found to be about 30% less than that of an activated intact fibre but greater than that of activated skinned fibres. It is usually stated that fibres in rigor are stiffer than contracting fibres. However, the evidence in Fig. 2.45 does not support this, unless it is assumed that the unknown factors causing the activated skinned fibre to be less stiff than the intact fibre also affect the stiffness of skinned fibres in rigor. Until there is evidence on this point, the question of whether muscle is stiffer, or less stiff, in rigor than when contracting remains unresolved.

The force exerted by whole frog muscles in rigor decays slowly with a time constant of about 10 hr at 0°C (Aubert, 1956). The stiffness remains high after the force has decayed. It is not known whether rigor force decays also in skinned fibres; it would be useful to examine this along with any structural changes that may occur.

Marston *et al.* (1976, 1979) have shown that the tension exerted by skinned rigor muscle (insect asynchronous flight muscle) is reduced when MgAMP·PNP (a nonhydrolysable analogue of ATP) is added to it. The stiffness is unchanged when comparison is made at the same tension. Similar though smaller changes are produced by adding MgADP. The changes of force are reversible, in that force increases again when the nucleotide is removed.

Interpretation. If the attached crossbridges in rigor could exist only in a single state, there would be no recovery of force after a release or stretch. As there is some recovery, at least two attached states must exist. The tension recovery would be due to a redistribution of the crossbridges between these states after a length change. Phase 2 of the tension recovery in contracting muscle is interpreted as a redistribution of this sort. However, phase 2 is much smaller and slower than in active muscle; therefore the set of attached states is not identical in rigor and in contraction. For instance, some of the states that exist in contracting muscle might not exist in rigor muscle.

Is there an equilibrium between attached and detached states of crossbridges in rigor? The answer depends on whether sliding of filaments occurs during stretches over greater distances than crossbridges are thought to remain attached. If such sliding does occur, the force exerted after stretching is evidence that crossbridges have detached from one actin and reattached to another. Presumably the detachment process would be accelerated by the large force in the crossbridge caused by stretching. If such a detachment process does occur it probably causes the slow decay of tension observed by Aubert (1956). Griffiths *et al.* (1980) have suggested that such a process, which they refer to as "crossbridge slippage," occurs not only in rigor muscle, but also in contracting muscle.

2.4.2 RELAXATION FROM RIGOR

Adding MgATP to the sarcoplasm of muscles in rigor causes relaxation if no calcium ion is present. Tension and stiffness fall to the very low values characteristic of relaxed, living muscle. The kinetics of this transition have been studied by using "caged ATP" [P^3-1-(2-nitro)phenylethyladenosine 5′-triphosphate]. This compound has no effect on muscle, but ATP can be released from it by photolysis with a rate constant of about $100\ s^{-1}$. Some results of experiments by Goldman *et al.* (1982) are shown in Fig. 2.46A and B. Relaxation is found to be surprisingly rapid; in the example shown in Fig. 2.46A it is half-complete in about 30 ms, and is even faster at higher ATP concentration. This is too fast to fit with the idea that the rate of turnover of crossbridges during isometric contraction is limited by their rate of detachment, which is one of the basic ideas of the simplest crossbridge

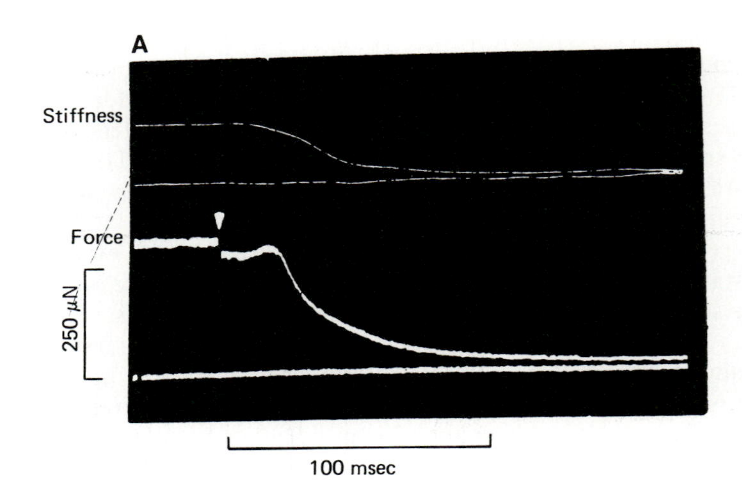

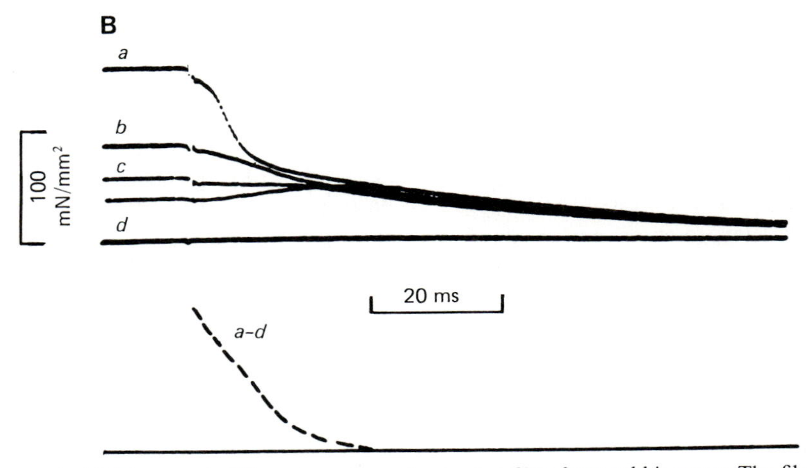

FIG. 2.46. (A) Force and stiffness in a glycerol-extracted fibre from rabbit psoas. The fibre is initially in rigor, in an ATP-free solution. At arrowhead, ATP is added (by photolysis) to a final concentration of 0.27 m*M*. (From Fig. 2 of Goldman *et al.*, 1982.) (B) Upper curves: force during relaxation from rigor caused by the addition (by photolysis) of ATP to a concentration of 480 μM. The records a–d differ in initial tension because muscle length was changed at 1 s before photolysis: (a) 0.8% stretch; (b) no change; (c) 0.4% release; (d) 0.8% release. Lower curve: difference between records a and d. (From Fig. 3 of Goldman *et al.*, 1982.)

theories. (The lifetime of a crossbridge in isometric contraction is about 0.5 s; see Chapter 5, Section 5.1, for a further discussion of this point.)

The time course of the fall of tension in the experiment in Fig. 2.46A is complex. Although stiffness declines monotonically during "relaxation," force rises for a while before it falls. The extent of this rise can be varied by changing the force at the start of the relaxation, by preceding it with a small stretch or release. Figure 2.46B shows a family of such records. It can be seen that the records are different in their early phases, but then they converge, and finally follow a common path.

An explanation of the time course of the records in Fig. 46B is suggested by the observation of Bremel and Weber (1972) that, even in the absence of calcium ion, rigor links act to activate ATPase in myofibrils. If rigor links can partially activate crossbridge turnover in skinned fibres also, the initial rise in tension (Fig. 2.46A) would be due to replacement of some of the rigor crossbridges by "active" bridges which exert more force. These newly formed crossbridges are independent of the initial length and force, and so the records in Fig. 2.46B converge as the crossbridges originally present detach and are replaced.

Another factor that may play a role in the rise of tension when ATP is added is that some rigor bridges are probably exerting negative force. If these bridges detach more rapidly than those exerting positive force (as would be expected from Huxley's 1957 theory, see Chapter 5), then tension would rise before falling. However, this phenomenon would not account for the convergence of the records in Fig. 2.46B and thus cannot provide an explanation of the entire time course of these records.

If the early decline of tension in Fig. 2.46B, upper curves, is due to the detachment of rigor crossbridges, its rate should be comparable to that of the dissociation of actin from myosin on addition of ATP *in vitro*. Figure 2.46B, lower curve, shows that the rate constant for this part of the relaxation is about 100 s^{-1}, which agrees well with the value of 120 s^{-1} for dissociation of actomyosin *in vitro* at the same ATP concentration (Chapter 3).

3
Kinetics

3.1 Introduction

Enzyme kinetics is the study of the rates of enzyme-catalyzed reactions with the aim of establishing the sequence of steps (the mechanism) by which the substrate (S) is converted to product (P). This approach has been applied to the ATPase activity of myosin and actomyosin isolated from muscle. Ideas and information have advanced rapidly during the last 15 years and have added a great deal to our knowledge of the biochemistry of these processes as they occur in solution. In this chapter the main principles and results of these studies are summarized.

How do these kinetic studies relate to other approaches toward an understanding of muscle contraction? Theories intended to account for all aspects of contraction should bring together different types of information, including kinetics. As discussed in Chapter 5, it may now be possible to assess theories in terms of their success in explaining *in vitro* kinetic experiments, as well as the mechanical and energetic aspects of contraction.

In this book we are particularly concerned with the question of how the aspect of energetics complements other ways of studying contraction. It is clear from energy balance studies (discussed in Chapter 4) that ATP splitting cannot account for all the energy produced during contraction: at least one other reaction contributes. The transitions among the various states of actin and myosin that have been characterized in the kinetic studies are possible sources of the extra energy; those with large enthalpy changes are especially attractive. This type of explanation is particularly needed to account for the energy output of muscles that are shortening rapidly (Chapter 4, Section 4.3.3).

In trying to relate the results of experiments on the kinetics of proteins in solution to those on contraction in muscle, some important differences between the systems should be recognized.

1. Muscle can produce force and convert energy from chemical reactions into work as well as heat, but in solution there is no force development nor work done, and all the free energy from ATP splitting is dissipated as heat.
2. In solution the probability of a reaction depends on concentration in a simple way. In contrast, in muscle, actin and myosin are organized into separate filaments, and the probability of the reaction between them is dependent on this structural organization and on how these structures move.
3. The ionic strength is much higher in muscle than that used in most biochemical studies, but kinetic studies of homogeneous solutions show that the ionic strength has profound effects on the reaction mechanism.

Before a full account can be given of the influence of these three factors, studies of the kinetics of structurally organized and mechanically coupled systems are needed.

3.2 The Basis of Transient Kinetics

A reaction is said to be in a steady state when the concentrations of the intermediates in the reaction are constant: only substrate and product concentrations are changing. Measurement of the velocity of the reaction (rate of product formation) and how this depends on substrate concentration is of some use in studying the reaction mechanism. However, this steady-state kinetic information cannot usually reveal much about the specific reaction mechanism. The study of the situation in which the concentrations of intermediates are changing (transient or pre-steady-state kinetics) can provide information about the minimal number of intermediates in a given reaction sequence and the rates of transition between them. We provide here a simple background to aid in understanding the development of knowledge of the interaction of muscle proteins, substrates, and ligands. For a more thorough description of transient kinetic techniques, the books by Gutfreund (1972), Roberts (1977), and Fersht (1977) can be consulted.

3.2.1 METHODS

The basic idea of transient kinetics is that a minimal sequence of reaction steps and their associated rate constants, can be inferred from the time course with which the concentration of a reactant, substrate, or intermediate changes after initiating a reaction. To use this approach techniques are required to initiate the reaction rapidly and to follow the time course of the change in concentration of substrate intermediate or products. Two specific approaches have been useful in the study of catalysis by muscle proteins: quenched flow and stopped flow analysis. Both methods have a time resolution of a few milliseconds.

Quenched flow analysis. In this technique the change in the concentration of substrates, intermediates, or products is determined by direct chemical or isotopic analysis of reaction solution. Figure 3.1A illustrates the way in which solutions can be rapidly mixed to initiate a reaction, and the reaction then halted by mixing with a quenching solution. This solution is usually an acid, base, or organic chemical that denatures the enzyme or otherwise creates a condition under which the reaction cannot proceed. The quenched solution is then analyzed. The length of time for which the reaction proceeds depends on the volume of the reaction line (ml) and the rate (ml/s) of fluid flow through it. By varying these parameters, concentrations can be measured at a variety of reaction times, and the time course of the reaction can be followed.

Stopped flow analysis. Often, a reaction produces a change in some physical property of the solution such as optical absorbance, fluorescence, light scattering, conductance, pH, heat content, etc. If this change can be monitored properly it can provide a signal giving the time course of the reaction. Figure 3.1B shows a stopped flow device in which changes in optical absorbance are used to follow the reaction. Two solutions are rapidly mixed and flow into the reaction chamber. The volume of this chamber and the lines leading to it are small (about 50 μl), and if the fluid rate is high (a typical value is 25 ml/s), the reaction will have proceeded for only about 2 ms (the "dead time") before the reaction chamber is filled with the newly initiated reaction solution.

3.2.2 SOME SIMPLE REACTION MECHANISMS

To show how transient kinetics can be used to study reaction mechanisms, we consider three types of reaction, and derive the equations describing the time course of formation of products. Then we describe how reaction mechanisms are inferred from the experimental results.

3.2.2.1 *Mechanism I. A One-Step Reaction*

Consider reaction (1) in which S reacts with E to form ES.

$$\mathrm{S} + \mathrm{E} \underset{k_{-1}}{\overset{k_{+1}}{\rightleftharpoons}} \mathrm{ES} \tag{1}$$

Here and in all the reactions which follow, the rate constants of the ith reaction proceeding from left to right are labelled as k_{+i} and those from right to left, k_{-i}. The equilibrium constant (k_{+i}/k_{-i}) is denoted by K_i. For the reaction given in Eq. (1), the rate of formation of ES ($d\mathrm{ES}/dt$, in $M\ \mathrm{s}^{-1}$)

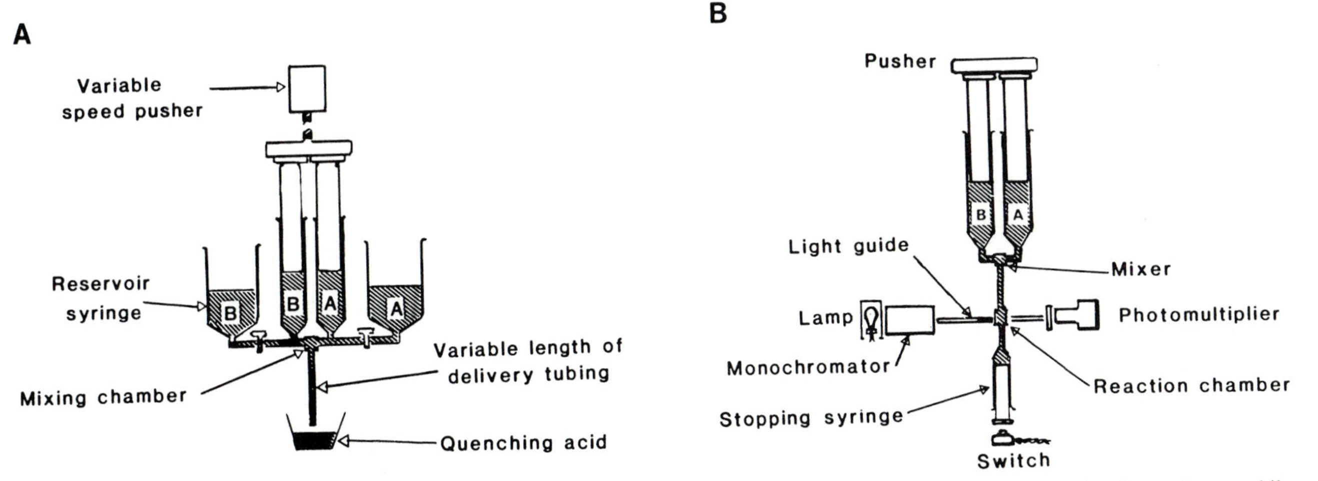

FIG. 3.1. (A) A schematic representation of a rapid quenched flow device. Two reactants, A and B, are mixed in the mixing chamber and react while flowing through the delivery tube. The reaction is stopped when the reactants enter the quenching solution. The reaction time can be varied by changing the speed of the pusher and/or the length of the delivery tube. (B) A schematic representation of a stopped flow device. The two reactants, A and B, are mixed in the mixing chamber and flow into the reaction chamber. Flow continues until sufficient reaction mixture has passed through the reaction chamber to fill the stopping syringe. The plunger of this syringe then activates a switch which causes flow to stop and triggers the sweep of the oscilloscope. By this time, sufficient reaction mixture will have passed through the chamber to wash out all the "old" solution. When monitoring of the signal produced by the reaction begins, the solution has reacted for about 2 ms, the dead time of the apparatus.

is given by Eq. (2). In this equation, and throughout this chapter, the capital letters E, S, ES, etc. represent the concentrations of the various species discussed.

$$dES/dt = k_{+1} \cdot S \cdot E - k_{-1} \cdot ES \tag{2}$$

So that, at equilibrium, when $dES/dt = 0$,

$$k_{+1}/k_{-1} = ES/S \cdot E \tag{3}$$

Note that the units of k_{+1}, the second-order rate constant, are $M^{-1}\,s^{-1}$ while those of k_{-1}, a first-order rate constant, are s^{-1}. To simplify interpretation of experimental results, the concentration of one reactant, for example S, is made large compared to E, so that S does not change significantly in the course of the experiment. Consequently, the second-order step (S + E → ES) becomes pseudo-first order. Thus, S remains at its initial value, S_0, and can be treated as a constant. Equation (2) becomes

$$dES/dt = k_{+1} \cdot S_0 \cdot E - k_{-1} \cdot ES \tag{4}$$

The total amount of enzyme E in the system, E_0, is conserved ($E = E_0 - ES$). On substitution into Eq. (4) we obtain

$$dES/dt = k_{+1}S_0 \cdot E_0 - (k_{+1}S_0 + k_{-1})ES \tag{5}$$

and the solution (see Table 3.I) for the time course of ES formation[1] is

$$\frac{ES}{E_0} = \frac{k_{+1}S_0}{k_{+1}S_0 + k_{-1}}[1 - e^{-(k_{+1}S_0 + k_{-1})t}] \tag{6}$$

3.2.2.2 *Mechanism II. Two-Step Reaction: First Step, a Rapid Equilibrium*

A reaction typical of those used in interpreting muscle protein studies is given in Eq. (7).

$$E + S \underset{k_{-1}}{\overset{k_{+1}}{\rightleftharpoons}} ES \underset{k_{-2}}{\overset{k_{+2}}{\rightleftharpoons}} EP \tag{7}$$

In this case the reaction is again made pseudo-first order. We are considering that k_{+1} and k_{-1} are much greater than k_{+2} or k_{-2}; i.e., the first step is a rapid equilibrium reaction and reaches equilibrium before the second step

[1] Equation (5) is transformed to

$$p(ES) = (k_{+1}S_0 \cdot E_0/p) - (k_{+1}S_0 + k_{-1})(ES) \tag{5a}$$

and is then solved for ES to yield

$$(ES) = \frac{k_{+1}E_0 \cdot S_0}{p[p + (k_{+1}S_0 + k_{-1})]} \tag{5b}$$

The inverse transform of Eq. (5b) (see line 3 of Table 3.I) gives Eq. (6).

TABLE 3.I

Use of Laplace transforms[a]

Original function, $f(t)$ (inverse Laplace transform)	Laplace transform
dy/dt	$p(Y)$
$dy/dt,\ y = Y_0$ at $t = 0$	$p(Y) - p(Y_0)$
$\left(\frac{k}{\pm a}\right) - \left(\frac{k}{\pm a}\right)e^{\mp at}$	$\frac{k}{p(p \pm a)}$
$k\left[\left(\frac{1}{a_1 a_2}\right) - \left(\frac{e^{-a_1 t}}{a_1(a_2 - a_1)}\right) - \left(\frac{e^{-a_2 t}}{a_2(a_1 - a_2)}\right)\right]$	$\frac{k}{p(p + a_1)(p + a_2)}$

[a] The differential equations used in this chapter have been simplified for solution using the Laplace transform method (a method described in most texts on differential equations). To facilitate the understanding of the derivations of the equations, the table of Laplace transforms of relevant functions, $f(t)$, is given above. For each class of reactions, the differential equations, the simplifying assumptions, and the equations' solutions are given in the text. The stepwise solutions to the equations using the transforms are given as footnotes. For a more complete and thorough treatment of the equations and systems described in this chapter, Capellos and Bielski (1972), Szabo (1969), Frost and Pearson (1960), or Gutfreund (1975) can be consulted.

proceeds to any significant extent. Since $E_0 = E + ES + EP$, and by definition, $K_1 = k_{+1}/k_{-1}$, it follows that

$$ES = (E_0 - EP)K_1 S_0/(K_1 S_0 + 1) \tag{8}$$

and the differential equation describing the rate of EP formation is

$$\begin{aligned} dEP/dt &= k_{+2} K_1 S_0 (E_0 - EP)/(K_1 S_0 + 1) - k_2 \cdot EP \\ &= k_{+2} K_1 S_0 \cdot E_0/(K_1 S_0 + 1) - [k_{+2} K_1 S_0/(K_1 S_0 + 1) + k_{-2}]EP \end{aligned} \tag{9}$$

The solution of Eq. (9) is[2]

$$\frac{EP}{E_0} = \frac{k_{+2}}{(k_{+2} + k_{-2}) + k_{-2}/K_1 S_0}\left\{1 - \exp\left[-\left(\frac{k_{+2} K_1 S_0}{K_1 S_0 + 1} + k_{-2}\right)t\right]\right\} \tag{10}$$

[2] The Laplace transform of Eq. (9) is

$$p(EP) + \left(\frac{k_{+2} K_1 S_0}{K_1 S_0 + 1} + k_{-2}\right)(EP) = \frac{k_{+2} K_1 S_0 E_0}{(K_1 S_0 + 1)p} \tag{9a}$$

and is solved for EP, yielding

$$(EP) = \left(\frac{k_{+2} K_1 S_0 E_0}{K_1 S_0 + 1}\right) \bigg/ p\left[p + \left(\frac{k_{+2} K_1 S_0}{K_1 S_0 + 1}\right) + k_{-2}\right] \tag{9b}$$

The inverse transform of Eq. (9b) (Table 3.I, line 3) gives Eq. (10).

3.2.2.3 *Mechanism III. Two-Step Reaction: First Step, Irreversible*

When the first step is irreversible ($k_{-1} \ll k_{+3}$ and k_{-2}) and when $S_0 \gg E$, the differential equations for the rates of change of E, ES, and EP are

$$dE/dt = k_{-1} \cdot ES - k_{+1}S_0 \cdot E \tag{11}$$

$$dES/dt = k_{+1}S_0 \cdot E + k_{-2} \cdot EP - (k_{-1} + k_{+2})ES \tag{12}$$

$$dEP/dt = k_{+2} \cdot ES - k_{-2} \cdot EP \tag{13}$$

and the solution for the time course of product formation[3] is

$$\frac{EP}{E_0} = \frac{k_{+2}}{k_{+2} + k_{-2}} \left[1 - \frac{(k_{+2} + k_{-2})e^{-k_{+1}S_0 t} - k_{+1}S_0 e^{-(k_{+2}+k_{-2})t}}{(k_{+2} + k_{-2}) - k_{+1}S_0} \right] \tag{14}$$

3.2.3 INFERRING MECHANISMS AND RATES OF REACTION

The above equations can be used to infer reaction mechanisms from experimental results. Suppose that the enzyme E reacts with a substrate S to form the enzyme–product complex E*P, the concentration of which can be monitored, for example optically. When E reacts with S at a concentration $S_0 \gg E_0$, the formation of E*P might follow a time course like that in Fig. 3.2A, which is an exponential with a rate constant λ. The time courses

[3] Using Laplace transforms, Eq. (12) becomes

$$(ES) = \frac{k_{+1}S_0(E) + k_{-2}(EP)}{p + k_{-1} + k_{+2}} \tag{12a}$$

By conservation of mass ($E = E_0 - ES - EP$)E is replaced in Eq. (12a) to yield

$$(ES) = \frac{k_{+1}S_0 E_0}{p[p + (k_{+1}S_0 + k_{-1} + k_{+2})]} + \frac{(k_{-2} + k_{+1}S_0)(EP)}{p + (k_{+1}S_0 + k_{-1} + k_{+2})} \tag{12b}$$

Using Laplace transforms Eq. (13) becomes

$$p(EP) = k_{+2}(ES) - k_{-2}(EP) \tag{13a}$$

By substituting Eq. (12b) into Eq. (13a) we obtain

$$(EP) = \frac{k_{+1}k_{+2}S_0 \cdot E_0}{p[p^2 + (k_{+1}S_0 + k_{-1} + k_{+2} + k_{-2}) + k_{+1}k_{+2}S_0 + k_{+1}k_{-2}S_0 + k_{-1}k_{-2}]} \tag{13b}$$

Since $k_{-1} \ll k_{+1}, k_{+2}, k_{-2}$, we let $k_{-1} \to 0$ and Eq. (13b) becomes

$$(EP) = \frac{k_{+1}k_{+2}S_0 \cdot E_0}{p(p + k_{+1}S_0)(p + k_{+2} + k_{-2})} \tag{13c}$$

Using the inverse Laplace transform on line 4, Table 3.I, Eq. (13c) becomes Eq. (14).

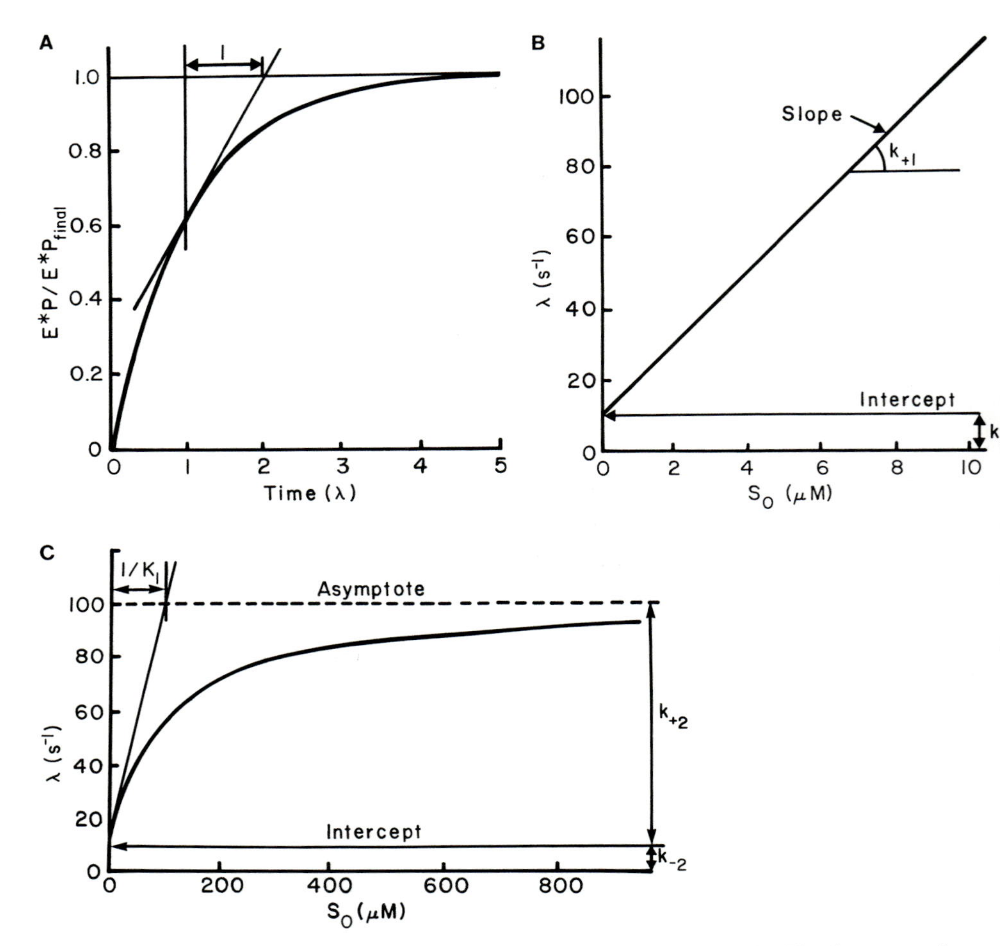

FIG. 3.2. (A) Time course of a reaction occurring by mechanism I or II. The time course is exponential. The graph has been normalized by dividing the ordinate value by the final concentration of E*P, and multiplying the time units by the rate constant λ. When normalized in this way the initial slope of the curve is unity, and a tangent drawn to the curve at any point cuts the final value one time unit beyond this point. (B) Dependence of the rate constant of a reaction (λ) on initial concentration of substrate (S_0) for a reaction occurring by mechanism I under pseudo-first-order conditions. In this example the intercept, giving k_{-1}, is 10 s^{-1}; and the slope, giving k_{+1}, is $10^7 M^{-1}$ s^{-1}. (C) Dependence of the rate constant of a reaction (λ) on initial concentration of substrate (S_0) for a reaction occurring by mechanism II under pseudo-first-order conditions. The relation is hyperbolic; the asymptote, shown by the dashed line, gives a value of $(k_{+2} + k_{-2})$. The value of K_1 can be obtained, as illustrated, from the initial slope. In this example $k_{-2} = 10\, s^{-1}$ and $k_{+2} = 90\, s^{-1}$; $K_1 = 10^4\, M^{-1}$.

of product formation by mechanism I [Eq. (6)] and by mechanism II [Eq. (10)] are of the form

$$E^*P/E_0 = \alpha(1 - e^{-\lambda t}) \tag{15}$$

where α is a constant at a given S_0. Equation (14) for mechanism III reduces to the same form if $k_{+1}S_0 \ll k_{+2} + k_{-2}$. Thus, this one experiment does not allow us to distinguish between reaction mechanisms I, II, and III. However, inspection of Eqs. (6), (10), and (14) shows that λ (the rate of E*P appearance) will be dependent on the substrate concentration, S_0, in a way that allows the three types of reaction mechanism to be distinguished. If the reaction of E with S occurred by mechanism I, then

$$\lambda = k_{+1}S_0 + k_{-1} \tag{16}$$

As S_0 is increased, λ increases (Fig. 3.2B) with a slope of k_{+1} and a y intercept of k_{-1}. Thus, for any one-step reaction mechanism, there is a linear dependence of λ on S_0. Bimolecular reactions in solution occur by collision of two molecules and are often limited in rate only by the diffusion of the molecules in the solution. In such reactions k_{+1} is of the order of 10^8–10^9 M^{-1} s^{-1} (Gutfreund, 1975; Roberts, 1977).

Equation (10) indicates that for mechanism II

$$\lambda = k_{+2}\frac{k_1S_0}{k_1S_0 + 1} + k_{-2} \tag{17}$$

In this case, the dependence of λ on S_0 is hyperbolic as shown in Fig. 3.2C. When this behaviour is observed, the reaction has at least two steps. As S_0 increases, $K_1S_0/(K_1S_0 + 1)$ approaches 1 as a limit, and λ approaches $k_{+2} + k_{-2}$. Therefore, the asymptotic value of λ is the sum of the forward and backward rate constants of the second step. When $K_1S_0 \ll 1$, $\lambda = k_{+2}K_1S_0 + k_{-2}$. Thus, at very low concentrations of S_0, the slope of a graph of λ against S_0 is $k_{+2}K_1$ and the y intercept is k_{-2}. When k_{-2} is known or is nearly zero, it is convenient to rearrange Eq. (17) as

$$\frac{1}{\lambda - k_{-2}} = \left(\frac{1}{K_1k_{+2}}\right)\left(\frac{1}{S_0}\right) + \frac{1}{k_{+2}} \tag{18}$$

$k_{+2}K_1$ can be determined from the slope of a plot of $1/(\lambda - k_{-2})$ against $1/S_0$.

It is important to remember that for all values of S_0, the time course of formation of E*P is exponential for a reaction occurring by mechanism II.

In the case of mechanism III, although results (Fig. 3.3A) are superficially similar to those for mechanism II (Fig. 3.2C), the relation between S_0 and λ is not hyperbolic. At low values of S_0, λ is a linear function of S_0, and at high values it is independent of S_0. At low values of S_0 the rate

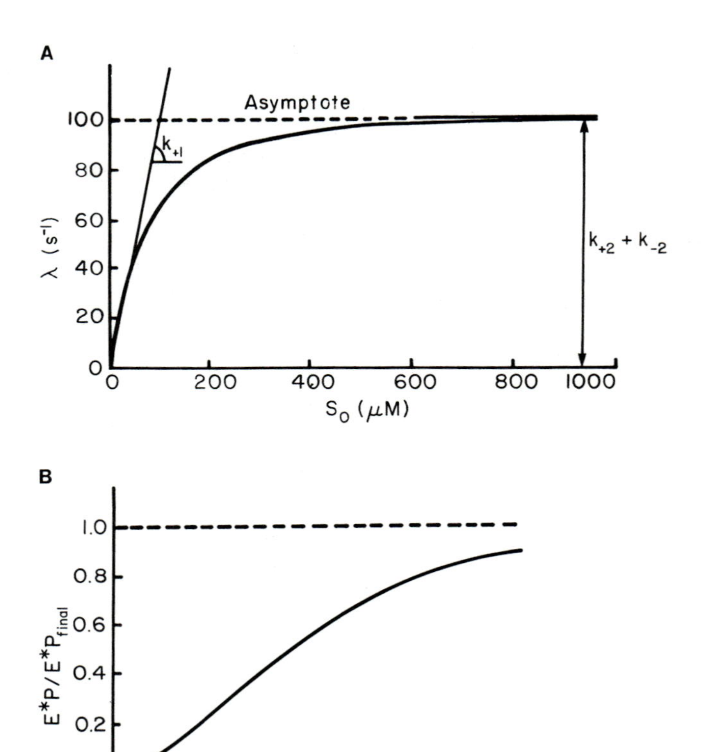

FIG. 3.3 (A) Dependence of the rate constant of a reaction (λ) on initial concentration of substrate (S_0) for a reaction occurring by mechanism III under pseudo-first-order conditions. The relation is not hyperbolic; compare Fig. 3.2C. The asymptote, shown by the dashed line, gives the value of $(k_{+2} + k_{-2})$ in Eq. 21. The initial slope gives a value of k_{+1}. (B) Time course of a reaction occurring by mechanism III, under conditions that produce a lag phase (see text). The time course is not exponential (compare Fig. 3.2A). The graph has been normalized by dividing the ordinate value by the final concentration of E*P and dividing the time units by the constant θ (see text).

of the reaction (λ) is limited by the first step, and when $k_{+1}S_0 \ll (k_{+2} + k_{-2})$, Eq. (14) becomes

$$\frac{E^*P}{E_0} = \frac{k_{+2}}{k_{+2} + k_{-2}}(1 - e^{-k_{+1}S_0 t}) \qquad (19)$$

Hence the relation between λ and S_0 at small values of S_0 is

$$\lambda = k_{+1}S_0 \qquad (20)$$

At high values of S_0, the first step of the reaction is very fast so that the term $(k_{+2} + k_{-2})e^{-k_{+1}S_0 t}$ is near zero and Eq. (14) becomes

$$\frac{E^*P}{E_0} = \frac{k_{+2}}{k_{+2} + k_{-2}} [1 - e^{-(k_{+2}+k_{-2})t}] \tag{21}$$

In this case λ is independent of S_0.

Mechanisms II and III can also be distinguished on the basis of the time course of formation of E*P. Like mechanism II, mechanism III gives an exponential time course when $k_{+1}S_0$ is much less than or much greater than $(k_{+2} + k_{-2})$. However, at intermediate values of S_0, when $k_{+1}S_0 \cong (k_{+2} + k_{-2})$, there is a lag phase. The solution for mechanism III for this case is

$$\frac{E^*P}{E_0} = \frac{k_{+2}}{k_{+2} + k_{-2}} [1 - e^{-(k_{+2}+k_{-2})t} - (k_{+2} + k_{-2})te^{-(k_{+2}+k_{-2})t}] \tag{22}$$

and, letting $\theta = (k_{+2} + k_{-2})$, Eq. (22) becomes

$$\frac{E^*P}{E_0} = \frac{k_{+2}}{\theta}\left(1 - \frac{1 + \theta t}{e^{\theta t}}\right) \tag{23}$$

Since $e^{\theta t}$ is approximately equal to $(1 + \theta t)$ when θt is <0.3, then, for a period early in the reaction time course, E^*P/E_0 is zero. Consequently the time course of E^*P/E_0 is S-shaped, as shown in Fig. 3.3B. In summary, mechanism III can be distinguished from mechanism II by the nonhyperbolic dependence of λ on S_0 and by the presence of a lag phase in the time course of E^*P/E_0 formation at intermediate values of the substrate concentration.

These examples show that, even for two-step mechanisms, the analytical solutions can be complicated. As the number of reaction intermediates increases, so does the number of exponential terms. The fitting of experimental data to analytical solutions then becomes difficult and arbitrary. Although numerical integration of the rate equations, that is, computer simulation of the reaction mechanism, can sometimes help, it is usually necessary to obtain other signals from further intermediates before the behaviour of such systems can be understood.

3.2.4 ADDITIONAL KINETIC TECHNIQUES

3.2.4.1 *Displacement Reactions*

This technique is useful for studying enzyme–ligand interactions that are not accessible to direct observation. In these experiments an enzyme–ligand complex EX (whose dissociation rate or equilibrium constant is sought) is mixed with a second ligand Y, which displaces X, forming the EY complex [Eqs. (24) and (25)]. To use this technique the concentration of EY, unlike that of EX, must be readily measured by some physical technique.

$$\mathrm{EX} \underset{k_{-1}}{\overset{k_{+1}}{\rightleftharpoons}} \mathrm{E} + \mathrm{X} \tag{24}$$

$$\mathrm{E} + \mathrm{Y} \underset{k_{-2}}{\overset{k_{+2}}{\rightleftharpoons}} \mathrm{EY} \tag{25}$$

Two extreme conditions are considered.

1. The rate constant for dissociation of EX, k_{+1}, can be easily measured when $k_{+2} > k_{+1}$, both k_{-1} and k_{-2} are much smaller than k_{+1}, and $\mathrm{Y}_0 \gg \mathrm{E}_0$. The equation describing the time course of EY formation is then[4]

$$\frac{\mathrm{EY}}{\mathrm{E}_0} = 1 - \frac{e^{-k_{+1}t}}{1 - k_{+1}/(k_{+2}\mathrm{Y}_0)} + \frac{e^{-(k_{+2}\mathrm{Y}_0)t}}{(k_{+2}\mathrm{Y}_0/k_{+1}) - 1} \tag{26}$$

As Y_0 increases, $k_{+2}\mathrm{Y}_0$ becomes much greater than k_{+1} and thus $k_{+1}/k_{+2}\mathrm{Y}_0$ becomes very small and $k_{+2}\mathrm{Y}_0/k_{+1}$ very large. Consequently, Eq. (26) becomes

$$\mathrm{EY}/\mathrm{E}_0 \cong (1 - e^{-k_{+1}t}) \tag{27}$$

In the absence of X the formation of EY from E and Y occurs [see Eq. (16)] with a rate constant of $(k_{+2}\mathrm{Y}_0 + k_{-2})$. The formation of EY from EX and

[4] Since $k_{-1}, k_{-2} \ll k_{+1}, k_{+2}$, we let $k_{-1}, k_{-2} \to 0$ and the differential equations describing $d\mathrm{E}/dt$ and $d\mathrm{EY}/dt$ are respectively

$$\frac{d\mathrm{E}}{dt} = k_{+1}\mathrm{EX} - k_{+2}\mathrm{Y}_0\mathrm{E} \tag{24a}$$

$$\frac{d\mathrm{EY}}{dt} = k_{+2}\mathrm{YE} \tag{25a}$$

By conservation of mass $\mathrm{EX} = \mathrm{E}_0 - \mathrm{E} - \mathrm{EY}$, therefore substituting for EX and using Laplace transforms, Eq. (24a) becomes

$$(\mathrm{E}) = \frac{k_{+1}\mathrm{E}_0 - k_{+1}(\mathrm{EY})}{p + (k_{+1} + k_{+2}\mathrm{Y}_0)} \tag{24b}$$

The Laplace form of Eq. (25a) is

$$p(\mathrm{EY}) = k_{+2}\mathrm{Y}_0(\mathrm{E}) \tag{25b}$$

Substituting Eq. (24b) into Eq. (25b) and solving for EY gives the Laplacian

$$(\mathrm{EY}) = \frac{k_{+1}k_{+2}\mathrm{Y}_0\mathrm{E}_0}{p(p + k_{+1})(p + k_{+2}\mathrm{Y}_0)} \tag{25c}$$

Table 3.I, line 4, shows that the inverse Laplace transform of Eq. (25c) is Eq. (26).

excess Y is slower, the rate being determined by k_{+1}. Thus the displacement of X by Y can be used to determine the rate of X dissociation from the EX complex.

2. The equilibrium constant for the reaction of E and X can be determined when the first step [Eq. (24)] is a rapid equilibrium reaction, that is, when both k_{+1} and $k_{-1}X_0 \gg k_{+2}Y_0$. As $X_0 \gg E_0$

$$K_1 \approx \frac{E \cdot X_0}{EX} \tag{28}$$

and since $EX = E_0 - E - EY$

$$E = \frac{E_0 - EY}{(X_0/K_1) + 1} \tag{29}$$

The differential equation to be solved is thus

$$\frac{dEY}{dt} = k_{+2}Y_0 \cdot E - k_{-2} \cdot EY \tag{30}$$

Substituting Eq. (29) into Eq. (30) produces

$$\frac{dEY}{dt} = k_{+2}Y_0\left(\frac{E_0 - EY}{(X_0/K_1) + 1}\right) - k_{-2} \cdot EY \tag{31}$$

of which the solution is[5]

$$\frac{EY}{E_0} = \frac{k_{+2}Y_0}{(X_0/K_1) + 1}\left(1 - e^{-\{k_{-2} + (k_{+2}Y_0)/[(X_0K_1) + 1]\}t}\right) \tag{32}$$

Thus, the rate constant λ for the formation of EY is a linear function of Y_0:

$$\lambda = k_{-2} + k_{+2}\frac{Y_0}{(X_0/K_1) + 1} \tag{33}$$

Since k_{-2} and k_{+2} can be measured by reacting Y and E in the absence of X, a graph of λ against Y_0 can be used to measure K_1.

[5] The Laplace transform of Eq. (31) is

$$p(EY) = \frac{k_{+2}Y_0 \cdot E_0}{p[(X_0/K_1) + 1]} - \left[\frac{(k_{+2}Y_0)}{(X_0/K_1) + 1} + k_{-2}\right](EY) \tag{31a}$$

Solving Eq. (31a) for EY, Eq. (31b) is obtained.

$$(EY) = \left[\frac{k_{+2}Y_0 \cdot E_0}{(X_0/K_1) + 1}\right] \Big/ p\left[p + \frac{k_{+2}Y_0}{(X_0/K_1) + 1} + k_{-2}\right] \tag{31b}$$

The inverse Laplace transform of Eq. (31b) is of the form in line 3 of Table 3.I and yields Eq. (32).

3.2.4.2 *Single-Turnover Experiments*

For the reaction shown in Eq. (34), an approximate value of K_2 $(= k_2/k_{-2})$ can sometimes be obtained by a simple experiment.

$$\mathrm{E + S} \underset{k_{-1}}{\overset{k_{+1}}{\rightleftharpoons}} \mathrm{ES} \underset{k_{-2}}{\overset{k_{+2}}{\rightleftharpoons}} \mathrm{EP} \xrightarrow{k_{+3}} \mathrm{E + P} \tag{34}$$

An amount of S is mixed with a large excess of E so that E is essentially constant ($\mathrm{E_0}$). If k_3 is small compared to the other rate constants, an equilibrium of the first two reactions can be established before a significant amount of P is formed. If E is sufficiently larger than S, the concentration of free S will be small, compared to ES + EP. Thus

$$\mathrm{ES} \approx \mathrm{S_0} - \mathrm{EP}$$

If ES and EP are in equilibrium, then

$$\frac{d\mathrm{EP}}{dt} = k_{+2} \cdot \mathrm{ES} - k_{-2} \cdot \mathrm{EP} = 0$$

$$K_2 = \frac{k_{+2}}{k_{-2}} \approx \frac{\mathrm{EP}}{\mathrm{S_0} - \mathrm{EP}} \tag{35}$$

Thus, if EP can be measured, K_2 can be estimated.

3.2.4.3 *Some Considerations about Steady-State Kinetics*

When a reaction is in a steady state, only the concentrations of reactants and products are changing: the concentrations of all intermediates are constant. In a two-step reaction,

$$\mathrm{E + S} \underset{k_{-1}}{\overset{k_{+1}}{\rightleftharpoons}} \mathrm{ES} \xrightarrow{k_{+2}} \mathrm{E + P} \tag{36}$$

In the steady state, $d\mathrm{ES}/dt = 0$, thus

$$0 = \frac{d\mathrm{ES}}{dt} = k_{+1}\mathrm{E} \cdot \mathrm{S} - (k_{-1} + k_{+2})\mathrm{ES} \tag{37}$$

Under pseudo-first-order conditions $\mathrm{S} = \mathrm{S_0}$. Since $\mathrm{E} = \mathrm{E_0} - \mathrm{ES}$, we obtain by substitution:

$$0 = k_{+1}(\mathrm{E_0} - \mathrm{ES})\mathrm{S_0} - (k_{-1} + k_{+2})\mathrm{ES} \tag{38}$$

$$\mathrm{ES} = \frac{k_{+1}\mathrm{E_0} \cdot \mathrm{S_0}}{k_{+1}\mathrm{S_0} + k_{-1} + k_{+2}} \tag{39}$$

Since the rate of product formation (V) is given by $k_{+2} \cdot \mathrm{ES}$

$$V = \frac{k_{+1}(k_{+2}\,\mathrm{E}_0)\mathrm{S}_0}{k_{+1}\mathrm{S}_0 + (k_{-1} + k_{+2})} \tag{40}$$

The maximum velocity of the reaction is V_{max}. V_{max}/E_0, where E_0 is the total enzyme concentration, is often called k_{cat}. For the reaction given in Eq. (36), $V_{max} = k_{+2}\mathrm{E}_0$ and $k_{cat} = k_{+2}$. Thus

$$V = \frac{k_{+1}V_{max}\mathrm{S}_0}{k_{+1}\mathrm{S}_0 + k_{-1} + k_{+2}} \tag{41}$$

$$= \frac{V_{max}\mathrm{S}_0}{\mathrm{S}_0 + [(k_{-1} + k_{+2})/k_{+1}]} \tag{42}$$

If K_m is defined as the concentration of S at which V is half of V_{max}, it can be shown that

$$V = V_{max}\mathrm{S}_0/(K_m + \mathrm{S}_0) \tag{43}$$

As shown in Fig. 3.4A the relation between V and S_0 is hyperbolic; as S_0 increases, V approaches V_{max} asymptotically. To obtain values of V_{max} and K_m the linear form of Eq. (43) can be used.

$$\frac{1}{V} = \left(\frac{K_m}{V_{max}}\right)\left(\frac{1}{\mathrm{S}_0}\right) + \frac{1}{V_{max}} \tag{44}$$

Figure 3.4B shows an example: the y intercept is $1/V_{max}$ and the x intercept is $-1/K_m$.

The value of K_m and the rate constants that determine it depend on the reaction mechanism. For the reaction in Eq. (34), $K_m = (k_{+2} + k_{-1})/k_{+1}$. If $k_{+2} \ll k_{-1}$, then $K_m = 1/K_1$. However, K_m is not always equal to $1/K_1$. For example, in the following reaction mechanism

$$\mathrm{E} + \mathrm{S} \underset{k_{-1}}{\overset{k_{+1}}{\rightleftharpoons}} \mathrm{ES} \xrightarrow{k_{+2}} \mathrm{EP} \xrightarrow{k_{+3}} \mathrm{E} + \mathrm{P} \tag{45}$$

the steady-state rate (V) is again given by Eq. (43), as for the mechanism of Eq. (34), but the equations for K_m and V_{max} are

$$K_m = [k_{+3}(k_{-1} + k_{+2})]/[k_{+1}(k_{+2} + k_{+3})] \tag{46}$$

$$V_{max}/\mathrm{E}_0 = k_{+2}k_{+3}/(k_{+2} + k_{+3}) \tag{47}$$

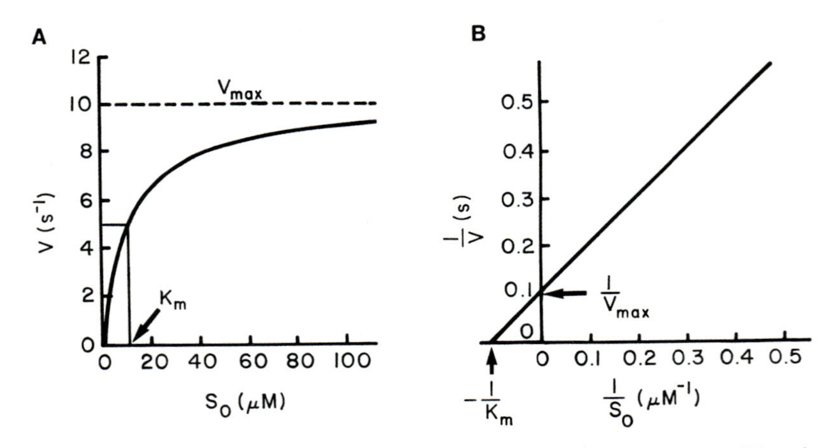

FIG. 3.4. The relation between the rate of product formation in the steady state (V) and substrate concentration S_0. In (A), V is plotted against S_0. The asymptote is V_{max}. K_m is, by definition, the value of S_0 at which V is half of V_{max} [see Eq. (43)]. In (B), $1/V$ is plotted against $1/S_0$. The values of V_{max} and K_m can be obtained as illustrated.

Thus, the individual rate constants that determine K_m and V_{max} depend on the specific reaction mechanism. Measurements of K_m and V_{max} alone cannot give information about the mechanism.

3.3 Application to Reactions of Muscle Proteins

3.3.1 ATP HYDROLYSIS BY MYOSIN

Several groups have used quenched flow analysis to study the time course of ATP hydrolysis by myosin (Weber and Hasselbach, 1954; Tonomura and Kitagawa, 1960; Lymn and Taylor, 1970, 1971). The studies of Lymn and Taylor have been the most influential; they studied the time course of ATP hydrolysis by rabbit myosin, HMM, and S-1.

For convenience in the following discussion the letter M is used to denote a "myosin head" whether it is in the native myosin, HMM, or S-1. In Lymn and Taylor's experiments, M was mixed with ATP (the terminal phosphate group of which was labelled with ^{32}P) in a quenched flow apparatus similar to that shown in Fig. 3.1A. The ATP hydrolysis was monitored by measuring the time course of $^{32}P_i$ production after acid quenching. Since the acid quench displaces P_i bound to M, it produces a solution containing the P_i that had been bound before quenching, as well as any that was free. The time course of P_i formation is shown by the data points in Fig. 3.5A. The striking feature of these results is that after mixing

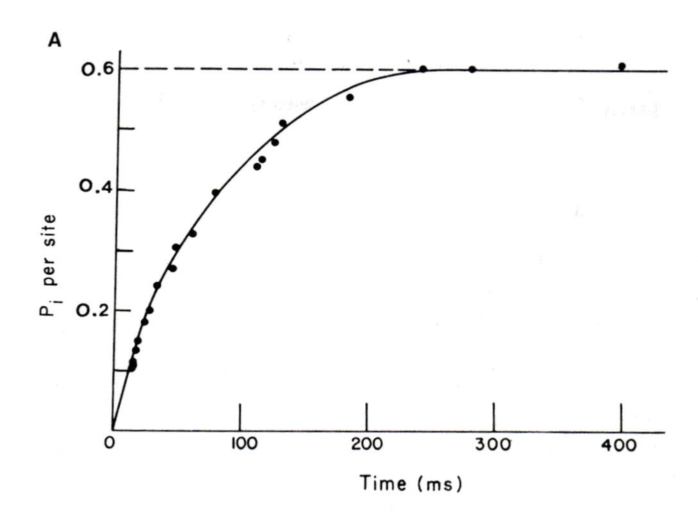

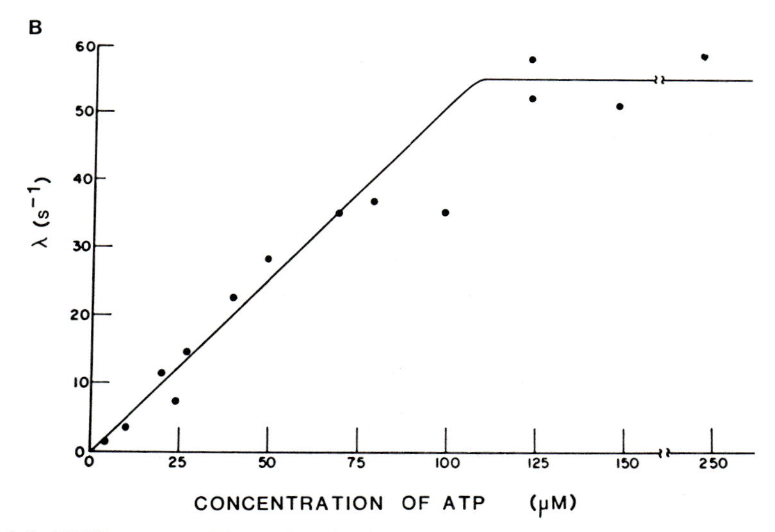

FIG. 3.5. (A) Time course of formation of P_i following mixing of ATP with myosin. Determined by Lymn and Taylor (1970) using a quenched flow apparatus. The quantity measured is the amount of P_i in the solution after quenching. The symbols (•) are the experimental results and the solid line drawn through them approximates to an exponential with a value of λ of 10 s^{-1}. The dashed line shows the steady-state rate of ATP hydrolysis. The reaction conditions for these experiments were 0.5*M* KCl, 0.1*M* Tris, 0.01*M* $MgCl_2$, pH 8.0, 20°C. The myosin concentration was 2 mg/ml and that of ATP, 32 μM. (B) The dependence on ATP concentration of the rate constant (λ) determined in experiments like that shown in Fig. 3.5A. (From Lymn and Taylor, 1970.) The results are shown by the points.

there is a very rapid production of $^{32}P_i$ (called the "phosphate burst") which is followed by a slow linear production of $^{32}P_i$, shown by the dashed line in Fig. 3.5A. The simplest mechanism consistent with this result is

$$\mathrm{M + ATP} \underset{k_{-1}}{\overset{k_{+1}}{\rightleftharpoons}} \mathrm{M \cdot ADP \cdot P_i} \xrightarrow[\text{slow}]{k_{+2}} \mathrm{M + ADP + P_i} \tag{48}$$

Upon binding to M, ATP is rapidly hydrolysed to form a ternary complex of M, ADP, and P_i ($M \cdot ADP \cdot P_i$). This is followed by a slow, rate-limiting release of products, ADP and P_i, from M. The slow linear rate of P_i production (shown by the dashed line in Fig. 3.5A) is 0.02 mol of P_i per mole of M per second, and is negligible on this time scale. The solid line drawn through the points in Fig. 3.5A is an exponential with a rate constant λ of 10 s^{-1}. Further experiments show that λ is a nonlinear function of ATP concentration (Fig. 3.5B) reaching a maximum value of 150 s^{-1}. These results are similar to Figs. 3.2C and 3.3A in that the rate of cleavage does not increase at ATP concentrations greater than 100 μM. Therefore, ATP hydrolysis is at least a two-step process:

$$\mathrm{M + ATP} \underset{k_{-1}}{\overset{k_{+1}}{\rightleftharpoons}} \mathrm{M \cdot ATP} \underset{k_{-2}}{\overset{k_{+2}}{\rightleftharpoons}} \mathrm{M \cdot ADP \cdot P_i} \tag{49}$$

The maximum rate constant of the phosphate burst (150 s^{-1}) gives a value for $(k_{+2} + k_{-2})$.

It is not possible to decide between mechanisms II and III on the basis of the data in Fig. 3.5B because the result can, within experimental error, be described by a hyperbola. Trentham *et al.* (1972) examined the time course of ATP cleavage at intermediate ATP concentrations (S_0) such that $k_{+1}S_0 = k_{+2} + k_{-2}$, and found a lag phase in the phosphate burst. This result, along with others described below, show that ATP binding is essentially irreversible and that k_{+1} is about $2 \times 10^6\ M^{-1}\ s^{-1}$.

3.3.1.1 *Chemistry of the ATP–Myosin Interaction*

What is the nature of the $M \cdot ADP \cdot P_i$ ternary complex? During hydrolysis the ATP molecule is cleaved between the oxygen and the γ-phosphate:

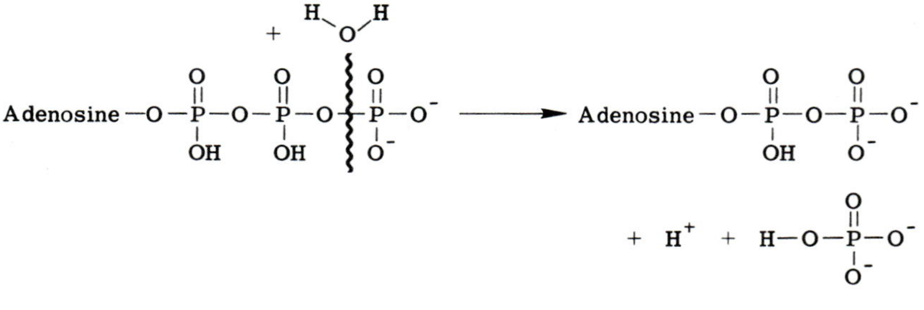

$$\text{Adenosine}-\mathrm{O}-\underset{\mathrm{OH}}{\overset{\mathrm{O}}{\overset{\|}{\mathrm{P}}}}-\mathrm{O}-\underset{\mathrm{OH}}{\overset{\mathrm{O}}{\overset{\|}{\mathrm{P}}}}-\mathrm{O} \wr \underset{\mathrm{O^-}}{\overset{\mathrm{O}}{\overset{\|}{\mathrm{P}}}}-\mathrm{O^-} + \mathrm{H_2O} \longrightarrow \text{Adenosine}-\mathrm{O}-\underset{\mathrm{OH}}{\overset{\mathrm{O}}{\overset{\|}{\mathrm{P}}}}-\mathrm{O}-\underset{\mathrm{O^-}}{\overset{\mathrm{O}}{\overset{\|}{\mathrm{P}}}}-\mathrm{O^-} + \mathrm{H^+} + \mathrm{H}-\mathrm{O}-\underset{\mathrm{O^-}}{\overset{\mathrm{O}}{\overset{\|}{\mathrm{P}}}}-\mathrm{O^-}$$

The phosphate is replaced by a hydrogen from water in the solution. In experiments in which steady-state reaction mixtures were quenched using a wide variety of denaturing agents (Sartorelli *et al.*, 1966) no phosphorylated myosin intermediates (i.e., a phosphate group covalently bound to the myosin) were found. Thus, it is likely that ADP and P_i are bound to myosin noncovalently. In this respect, the myosin is different from other ATPases (e.g., Ca^{2+}-activated sarcoplasmic reticulum ATPase and sarcolemmal Na^+–K^+ ATPase) (Glynn and Karlish, 1975; Ikemoto, 1982).

The interaction of myosin and ATP can be monitored by changes in myosin fluorescence. Myosin contains 28 tryptophan residues which, when exposed to light at 280 nm, emit light (fluoresce) at 335 nm. Normally this fluorescence is reduced (partially quenched) by interaction of the tryptophan residues with their aqueous environment. When ATP is added to myosin, the tryptophan fluorescence increases by about 12%, indicating that myosin and ATP interaction produces a change in the tertiary structure of the protein so that some tryptophan residues become further buried in the protein molecule. Consequently their fluorescence is less effectively quenched by the aqueous environment. The conformational change responsible for the fluorescence effect does not involve the 50–70% of amino acid residues making up the helical portion of the myosin molecule because Gratzer and Lowey (1969) found no significant change in the circular dichroism spectra when ATP was added. The S-1 which is formed by treatment of myosin with α-chymotrypsin is less fluorescent than myosin because it contains fewer tryptophan residues. Upon interaction of ATP with S-1 there is a greater percentage increase (36%) in protein fluorescence than with myosin (Johnson and Taylor, 1978), because of the removal of fluorescent residues which were not associated with the ATPase activity.

Studies of the fluorescence signal have been useful in analyzing the myosin ATPase mechanism. In Eq. (50), the number of asterisks (*) indicates the intensity of protein fluorescence: the more asterisks, the more intense the fluorescence.

$$\mathrm{M} + \mathrm{ATP} \underset{k_{-1}}{\overset{k_{+1}}{\rightleftharpoons}} \mathrm{M}^* \cdot \mathrm{ATP} \underset{k_{-2}}{\overset{k_{+2}}{\rightleftharpoons}} \mathrm{M}^{**} \cdot \mathrm{ADP} \cdot \mathrm{P_i} \xrightarrow{k_{+3}} \mathrm{M} + \mathrm{ADP} + \mathrm{P_i} \qquad (50)$$

The evidence for this scheme can be summarized, as follows:

1. ATP analogues such as ADP, AMP–PNP, or ATP–γ–S which bind to M, but are not cleaved at a significant rate, cause an increase of 10–16% in fluorescence. This result suggests by analogy that M*·ATP has 10–16% more fluorescence than M.

2. As seen in Fig. 3.6A, λ, the rate constant for the fluorescence change (ΔFl) is a nonlinear function of ATP concentration; this indicates that the reaction mechanism responsible for the fluorescence change has at least two steps. The fact that it is nonhyperbolic suggests that the first step is essentially irreversible.

3. At high concentrations of ATP ($>100\ \mu M$) the fluorescence change has the same time course as the phosphate burst. This suggests that the two events are results of the same reaction.

4. The change in fluorescence on adding ATP is apparently smaller at high than at low ATP concentration. At high ATP the rate of the first step is fast enough that most of the S-1 is transformed to M*·ATP during the transit from the mixer to the reaction chamber and the only signal to be observed is due to the further transformation to M**·ADP·P_i. Since the formation of M*·ATP at 200 μM ATP is largely complete within the 2 ms dead time of the stopped flow apparatus, the rate constant for its formation must be greater than $1.5 \times 10^3\ s^{-1}$ and k_{+1} must be at least $7.5 \times 10^6\ M^{-1}\ s^{-1}$ (that is, $1.5 \times 10^3\ s^{-1}$ divided by $2.0 \times 10^{-4}\ M$).

5. When M is mixed with ATP, at a concentration which exceeds 200 μM, the time course of the fluorescence change is biphasic. There is an early rapid phase associated with the formation of M*·ATP, followed by the slower formation of M**·ADP·P_i. Thus the fluorescence change occurs faster than the phosphate burst.

The mechanism given in Eq. (50) shows the minimum number of intermediates needed to explain these results. However it is probable, from comparison with other enzyme mechanisms (Gutfreund, 1975; Johnson and Taylor, 1978) that there is another enzyme–substrate (M·ATP) species formed by a diffusion-limited collision process. Thus, although there is no direct evidence for the existence of a M·ATP species, the mechanism shown in Eq. (51) can be suggested, k_{+1} is of the order of $10^8\ M^{-1}\ s^{-1}$ and both k_{+1} and $k_{-1} \gg k_{+2}$.

$$\mathrm{M + ATP} \underset{k_{-1}}{\overset{k_{+1}}{\rightleftharpoons}} \mathrm{M \cdot ATP} \underset{k_{-2}}{\overset{k_{+2}}{\rightleftharpoons}} \mathrm{M^* \cdot ATP} \underset{k_{-3}}{\overset{k_{+3}}{\rightleftharpoons}} \mathrm{M^{**} \cdot ADP \cdot P_i} \xrightarrow{k_{+4}} \mathrm{M + ADP + P_i} \quad (51)$$

Assuming this mechanism, the slope of λ versus ATP at low ATP concentration gives $K_1 k_{+2}$ as $2 \times 10^6\ M^{-1}\ s^{-1}$. Using the value of k_{+2}, $1.5 \times 10^3\ s^{-1}$ (point 4 above), K_1 must be about $1.3 \times 10^3\ M^{-1}$. Since this mechanism assumes k_{+1} to be $10^8\ M^{-1}\ s^{-1}$, k_{-1} is $10^5\ s^{-1}$. With these

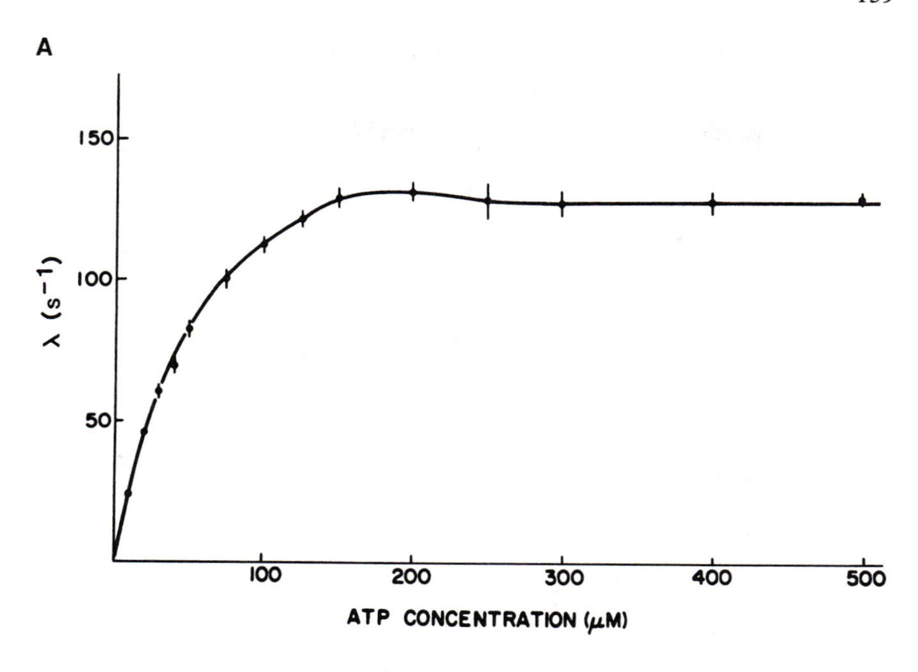

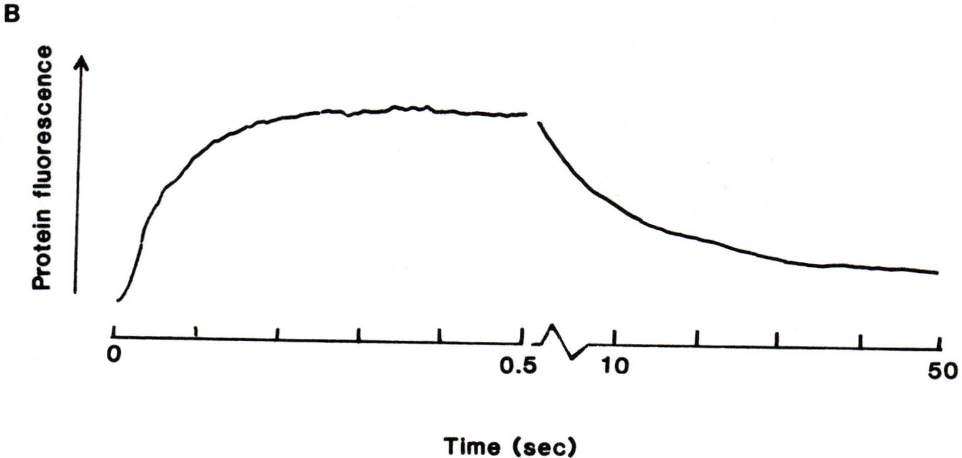

FIG. 3.6. (A) The dependence on ATP concentration of the rate constant (λ) of the fluorescence transient following the mixing of ATP with S-1. The points are the means of three or four observations and the bars show ± 1 SE. (From Johnson and Taylor, 1978.) (B) Stopped flow spectrophotometric record of protein fluorescence during a single turnover of the subfragment 1 ATPase at 21°C. One syringe contained 15 μM subfragment 1 (reaction chamber concentrations) and the other 2.5 μM ATP. Both syringes contained 5 mM $MgCl_2$, 50 mM KCl, and 20 mM Tris adjusted to pH 8.0 with HCl. (From Bagshaw and Trentham, 1973.)

values, the mechanism can account quantitatively for the results in Fig. 3.6A (Johnson and Taylor, 1978).

3.3.1.2 *The Equilibrium Constant for the Cleavage Reaction*

In the initial rapid phase of ATP splitting (Fig. 3.5A) rather less than 1 mol of ATP was split per mole of S-1, about 0.9 mol in that example. This could have been because the M solution contained some inactive protein, but, in fact, it is probably because the equilibrium constant of the cleavage reaction, K_3 in Eq. (51), is not very large, about 10. Because k_{+4} is very small compared to k_{+3} and k_{-3}, the early rapid phase of ATP splitting produces an equilibrium mixture of M*·ATP and M**·ADP·P_i [see Eq. (35)]; if $K_3 = 10$, this will consist of 9% M*·ATP and 91% M**·ADP·P_i. Bagshaw and Trentham (1973) measured the value of K_3 by mixing [^{32}P]ATP with excess S-1. The reaction mixture was quenched when a steady state of fluorescence had been reached (after 0.3 s) (see Fig. 3.6B). If the reaction occurs by the mechanism in Eq. (51), this would mean that an equilibrium mixture of M*·ATP and M**·ADP·P_i has been formed. No significant amounts of free ATP or free P_i would be expected at this time because of the low values of k_{-2} and of k_{+4}; thus the ratio of ATP to P_i in the quenched solution (1:9) gives the ratio of M*·ATP to M**·ADP·P_i [Eq. (35)] before quenching. K_3, the equilibrium ratio M**·ADP·P_i/M*·ATP, is thus about 9. Since $(k_{+3} + k_{-3})$ is known, from the maximum value of λ for the phosphate burst to be 120 s^{-1} under these conditions, we can now calculate k_{+3} (108 s^{-1}) and k_{-3} (12 s^{-1}). Values for other conditions are given in Table 3.II.

These results also allow us to specify the value of k_{+4} in Eq. (51) more accurately. The steady-state rate (V) of P_i formation by the mechanism shown in Eq. (51) is related to k_{+4} by

$$\frac{V}{M_0} = \frac{k_{+4}(M^{**} \cdot ADP \cdot P_i)}{M_0} \tag{52}$$

and since

$$M^{**} \cdot ADP \cdot P_i / M_0 = k_{+3}/(k_{+3} + k_{-3})$$

then

$$\frac{V}{M_0} = \frac{k_{+4} k_{+3}}{k_{+3} + k_{-3}} = k_{+4} \frac{K_3}{K_3 + 1} \tag{53}$$

Thus, if $V/M_0 = 0.05\ s^{-1}$ and $K_3 = 9$, then k_{+4} is 0.055 s^{-1}.

Bagshaw and Trentham (1973) also performed experiments to test the conclusion that k_{-2} is small. The experiment was started by rapidly mixing

TABLE 3.II

A summary of the kinetics of S-1 ATPase[a,b]

$$M + ATP \underset{k_{-1}}{\overset{k_{+1}}{\rightleftharpoons}} M \cdot ATP \xrightarrow{k_{+2}} M^* \cdot ATP \underset{k_{-3}}{\overset{k_{+3}}{\rightleftharpoons}} M^* \cdot ATP \cdot P_i \underset{k_{-4}}{\overset{k_{+4}}{\rightleftharpoons}} M^* \cdot ADP \underset{k_{-5}}{\overset{k_{+5}}{\rightleftharpoons}} M \cdot ADP \underset{k_{-6}}{\overset{k_{+6}}{\rightleftharpoons}} M + ADP$$

Muscle type	T (°C)	pH	I (M)	K_1k_{+2} (M^{-1} s^{-1})	$(k_{+3} + k_{-3})$ (s^{-1})	K_3	k_{+4} (s^{-1})	k_{+5} (s^{-1})	k_{-5} (s^{-1})	K_6 (M)
Rabbit–fast muscle	20	8	0.15	2×10^6	250–300	9	0.06	1.4	400	2.7×10^{-5}
	25	7	0.02	2.5×10^6	126	—	—	2.5	—	—
	20	7	0.10	2.4×10^6	125	4	—	—	—	—
	15	8	0.5	3×10^5	>300	—	—	—	—	—
	15	8	0.1	1.5×10^6	120	—	—	—	—	—
	15	7	0.1	1×10^6	55	—	—	—	—	—
	15	7	0.01	—	28	3.6	—	—	—	—
	<5	7	>0.1	2×10^5	8	1	—	0.07	—	—
Chicken—slow (ALD)	25	7	0.02	1.2×10^6	91	—	0.03	1.7	—	—
Smooth (gizzard)	25	7	0.02	3.2×10^6	40	—	0.06	0.9	—	—
Frog (skeletal)	2–5	7	0.15	1×10^5	>11	18	0.01	0.5	—	—
Cardiac (bovine)	25	8	0.16	5×10^5	100	11	0.02	0.3	—	—

[a] The results tabulated are taken from the following publications: Bagshaw and Trentham (1973, 1974); Bagshaw *et al.* (1974); Eisenberg and Kielley (1973); Ferenczi *et al.* (1978); Johnson and Taylor (1978); Lymn and Taylor (1971); Marston and Taylor (1980); Martinosi and Malik (1972); Sleep *et al.* (1978); Taylor (1977); Trentham *et al.*, (1972).

[b] Both K_3 and $(k_{+3} + k_{-3})$ increase with increasing ionic strength (I), temperature (T), and pH. K_1k_{+2} decreases with increasing ionic strength and decreasing temperature, increases with k_{+5} increased pH, and increases strongly with increasing temperature.

a molar excess of S-1 (30 μM) with [γ-^{32}P]ATP (5 μM). After about 300 ms when all the added [γ-^{32}P]ATP had been bound to the S-1, sufficient unlabelled ATP was added to raise the ATP concentration by 50- to 100-fold. If k_{-2} were large, the [γ-^{32}P]ATP would dissociate from M*·ATP and be replaced by the *unlabelled* ATP (since it is at a much greater concentration). The rate of hydrolysis of [γ-^{32}P]ATP would, therefore, fall to a very low value after the "cold chase" (the addition of unlabelled ATP). In contrast, if k_{-2} were small, the rate of [γ-^{32}P]ATP splitting would not be altered by addition of the *unlabelled* ATP. Bagshaw and Trentham did not detect any change in the rate of [γ-^{32}P]ATP hydrolysis after the "cold chase" and concluded that k_{-2} must be less than 0.02 s^{-1}.

3.3.1.3 *Post-Cleavage Steps*

The slow linear rate of ATP hydrolysis after the phosphate burst indicates the existence of a rate-limiting step subsequent to ATP cleavage. If this step followed the release of one or both products, then there would be a rapid early burst of *free* product(s) after mixing ATP and S-1. In fact there are no such early bursts, as was demonstrated by Trentham *et al.* (1972) using linked enzyme assay systems to follow the time course of the release of ADP and P_i (Fig. 3.7). For either product there is only a slow linear release at a rate equal to k_{cat}. Thus the rate-limiting step in the ATPase mechanism is either the release of one of the products or a step that precedes product release.

If reaction products are mixed with the enzyme, the rate of product binding and dissociation can be observed. If the complexes formed can be shown to be intermediates on the usual reaction pathway, these results will elucidate the steps in which products dissociate from the enzyme. Two techniques have been useful in such studies: displacement experiments (see Section 3.2.4.1) and transient kinetic analysis using the fluorescent enhancement accompanying ADP binding to S-1.

Use of the displacement techniques (Trentham *et al.*, 1972, 1976) made it possible to measure the rate of ADP release from S-1. A mixture of S-1 (11 μM) and ADP (50 μM) was reacted with ATP (either 75 or 470 μM). As seen Fig. 3.8A, ATP binds to S-1 under these conditions at a rate of only 2.3 s^{-1}, regardless of the ATP concentration. In contrast, in the absence of ADP, the rate of ATP binding is much higher (up to 350 s^{-1}). This shows that the rate of ADP release from S-1 is 2.3 s^{-1} [Eq. (27)]. If the reaction is performed so that the ADP is generated from added ATP (50 μM), the same result is obtained. Since the steady-state rate of ATP hydrolysis under these conditions is 0.05 s^{-1}, ADP release cannot be rate limiting. These

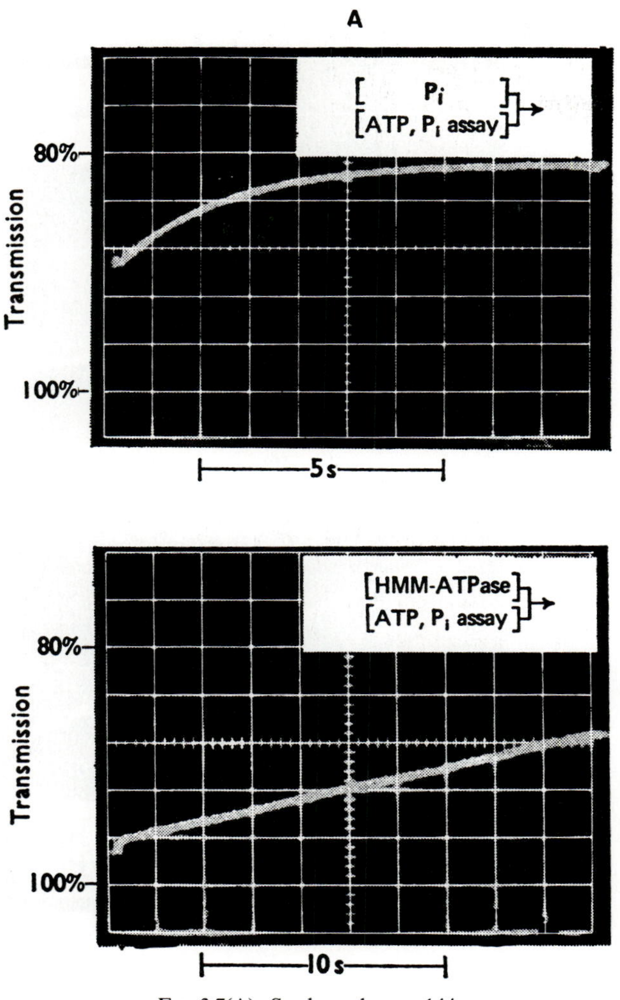

FIG. 3.7(A). See legend on p. 144.

experiments were performed at 20°C. At 6°C the rate of ADP release is much lower, about 0.025 s^{-1}, and k_{cat} is 0.014 s^{-1}. Thus, ADP dissociation *is* a rate-limiting step at lower temperatures.

When ADP binds to S-1 there is an increase in protein fluorescence. The rate of this change is a hyperbolic function of ADP concentration (Fig. 3.8B), well described by equations such as Eqs. (17) and (18). This indicates that

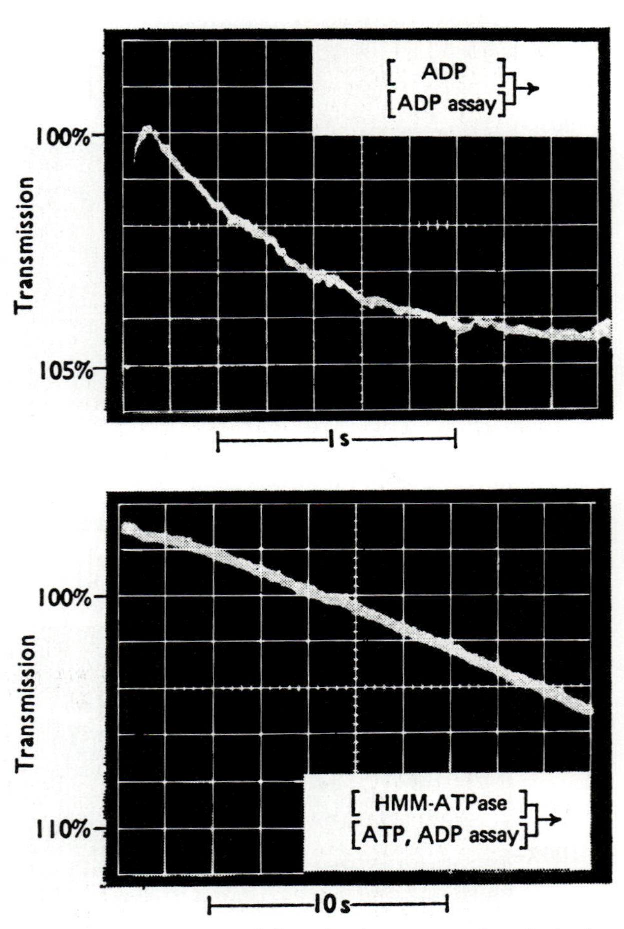

FIG. 3.7. Use of linked assay systems to follow the time course of product release from HMM during ATP hydrolysis. Upper graphs: controls; lower curves: experiments. The experiment in (A) follows P_i release. (B) follows ADP release. The experimental observations show what happens when the assay systems follow product release. In each case there is only a linear release of products; no "early bursts" are observed. The control observations demonstrate the speed of response of the assay systems when mixed with P_i or ADP, and show that these are adequate to have revealed "early bursts" had they existed. (From Trentham *et al.*, 1972.)

ADP binding to S-1 proceeds according to mechanism II (two steps, the first step a rapid equilibrium):

$$\mathrm{M + ADP} \underset{k_{-1}}{\overset{k_{+1}}{\rightleftharpoons}} \mathrm{M \cdot ADP} \underset{k_{-2}}{\overset{k_{+2}}{\rightleftharpoons}} \mathrm{M^* \cdot ADP} \tag{54}$$

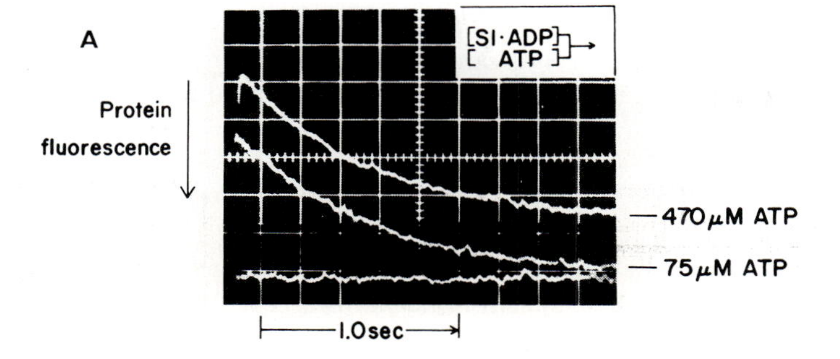

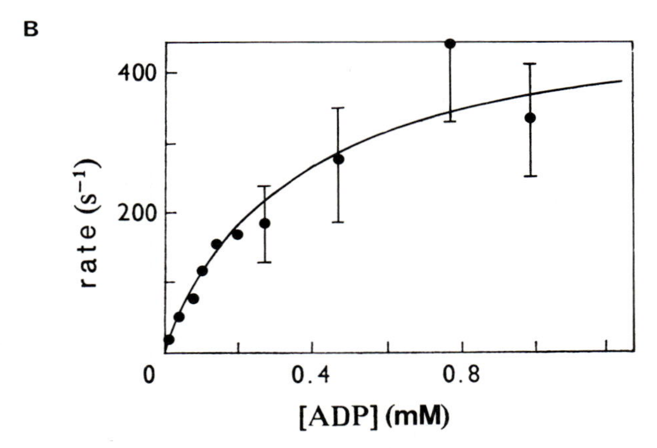

FIG. 3.8. (A) Records of the protein fluorescence obtained on mixing S-1 · ADP with ATP. The time course of the change observed is the same whether the ATP concentration is 470 μM or 75 μM. (From Bagshaw *et al.*, 1973.) (B) Plot of the rate of fluorescence change when ADP is mixed with 4.0 μM S-1. Reaction conditions 20°C, 0.1 M KCl, 5 mM $MgCl_2$, 50 mM Tris, pH 8.0. (From Bagshaw *et al.*, 1974.)

From the displacement experiments k_{-1} is 2.3 s^{-1}. The data in Fig. 3.8B indicate that $k_{+2} = 400\ s^{-1}$ and from double reciprocal plots of this data [Equation (18)] the value of k_{+1}/k_{-1} is $3.7 \times 10^3\ M^{-1}$. These results suggest that the post-cleavage mechanism for ATP hydrolysis is

$$M^{**} \cdot ADP \cdot P_i \xrightarrow{\text{slow}} M^* \cdot ADP \rightleftharpoons M \cdot ADP \rightleftharpoons M + ADP \quad (55)$$

in which either P_i release itself, or a protein isomerization prior to P_i release, is rate limiting.

This interpretation has *assumed* that the M*·ADP complex generated by addition of ADP to M is an intermediate in the normal reaction mechanism; that is, the reaction could be condensed to read

$$\mathrm{M + ATP} \xrightarrow{\text{fast}} \mathrm{M^{**} \cdot ADP \cdot P_i} \xrightarrow{k_a} \mathrm{M^{*} \cdot ADP} \xrightarrow{k_b} \mathrm{M + ADP} \qquad (56)$$

It could, however, be argued that ADP binds M to form a complex which is *not* part of the normal reaction mechanism [Eq. (57)]:

$$\begin{array}{ccccc} \mathrm{M + ATP} & \xrightarrow{\text{fast}} & \mathrm{M^{**} \cdot ADP \cdot P_i} & \xrightarrow{k_a} & \mathrm{M + ADP + P_i} \\ & & & & \downarrow k_b \\ & & & & \mathrm{M^{*} \cdot ADP} \end{array} \qquad (57)$$

If this is so, hardly any M*·ADP is present during steady-state splitting of ATP. This is because ATP binds much more strongly to M than does ADP and the concentration of ATP is much greater than that of ADP. On addition of excess ATP to M the fluorescence would rise to a high value (correspondingly to M**·ADP·P_i) which would persist until almost all the ATP had been hydrolysed. Only then would a significant amount of the less fluorescent M*·ADP form arise. The fluorescence would have the time course seen in Fig. 3.9A. In contrast, if M*·ADP is an intermediate in the reaction mixture [Eq. (55)], it will be present at appreciable concentration during the steady state, particularly at low temperature (6°C), because k_b (0.025 s^{-1}) is similar to k_{cat} (0.014 s^{-1}). On adding ATP to S-1 the fluorescence would rise due to the formation of M**·ADP·P_i, then fall to a lower value as a steady-state mixture of M**·ADP·P_i and M*·ADP is formed (Fig. 3.9B). Measurements of S-1 fluorescence during and after the approach to the steady state (Bagshaw *et al.*, 1974) give a result of this kind (Fig. 3.9C). This confirms that M*·ADP is an intermediate in the normal reaction mechanism, and that the mechanism in Eq. (57) can be rejected.

Some evidence about whether P_i release is rate limiting [Eq. (55)] or whether it is preceded by a slow protein isomerization is provided by the study of P_i binding to M·ADP. The results of Bagshaw *et al.* (1974), using displacement techniques, indicate the P_i is in rapid equilibrium with the protein and that P_i is released from M·ADP·P_i at a rate greater than 40 s^{-1}, which is much faster than the steady-state rate. However, there is no evidence that the M·ADP·P_i complex formed in the steady state can dissociate

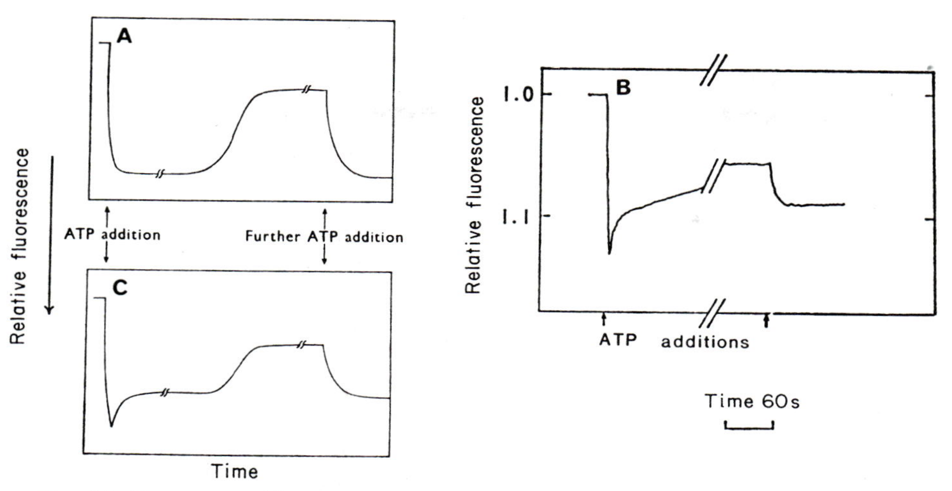

FIG. 3.9. Time course of fluorescence changes during ATP hydrolysis by S-1. (A) and (B) show hypothetical time courses corresponding to reaction schemes (57) and (55), respectively. (C) shows the time course observed. (From Bagshaw and Trentham, 1974.)

P_i at this rate. Thus the most simple mechanism for ATP hydrolysis by S-1 is that given in Eq. (58).

$$\mathrm{M + ATP} \underset{k_{-1}}{\overset{k_{+1}}{\rightleftharpoons}} \mathrm{M \cdot ATP} \xrightarrow{k_{+2}} \mathrm{M^* \cdot ATP} \underset{k_{-3}}{\overset{k_{+3}}{\rightleftharpoons}} \mathrm{M^{**} \cdot ADP \cdot P_i}$$

$$\xrightarrow{k_{+4}} \mathrm{P_i + M^* \cdot ADP} \underset{k_{-6}}{\overset{k_{+6}}{\rightleftharpoons}} \mathrm{M + ADP} \quad (58)$$

This mechanism is similar to those proposed for other enzymes (Gutfreund, 1972, 1975) in that the rate-limiting step, k_{+4}, is either an isomerization of an enzyme complex or the dissociation of the product itself.

Table 3.II summarizes values for some of the rate constants and equilibrium constants measured at 0 and 20°C for rabbit myosin and S-1. Studies of myosin from other types of muscle (avian and frog skeletal, bovine cardiac, and smooth muscle) have shown that they all hydrolyse ATP by this or a very similar mechanism. The values of the rate constants for some of these types of muscle are also given in the table.

3.3.2 ACTIN INTERACTION WITH MYOSIN AND ITS ACTIVATION OF MYOSIN ENZYMATIC ACTIVITY

Two aspects of the interaction of actin and myosin may be relevant to muscle contraction. First, in the absence of ATP, actin binds, noncovalently but very tightly to myosin, HMM, and S-1. This binding probably accounts for the stiffness of muscle in rigor. Second, actin accelerates, by more than 100-fold, the rate of ATP hydrolysis by myosin. This behaviour is thought to be the basis of contraction of living muscle. Before considering the mechanism of this catalysis, the results of studies of actin binding to myosin or S-1 are considered.

3.3.2.1 *Binding of Actin to Myosin*

In the early studies it was found that both the viscosity and turbidity (or optical density) of myosin solutions (in 0.5 *M* KCl) increased upon addition of F-actin. These changes were interpreted as indicating that the myosin heads had bound to the actin filaments, an interpretation later confirmed by similar studies of the sedimentation pattern of actomyosin and acto-S-1 solutions in the ultracentrifuge. At ionic strengths above 0.3 *M* it is possible to study the binding of intact myosin to F-actin, but at physiological ionic strengths (0.2–0.25 *M*) a lumpy precipitate is formed, making such experiments difficult. However, both S-1 and HMM are soluble at physiological ionic strengths and can be used to examine the binding of F-actin and M with various kinetic techniques.

These kinetic measurements exploit the fact that light is more effectively scattered by S-1 bound to actin than by either protein alone. Measurement of the light scattering (Fig. 3.10A) has shown that the stoichiometry of the reaction is 1:1 (F-actin monomer: S-1). The time course of the change of optical density upon mixing actin and S-1 or HMM can be used to monitor the time course of the binding reaction. The results of such studies are shown in Fig. 3.10B, in which the rate constant of S-1 binding to a fixed concentration of actin is seen to be a linear function of [S-1]. This result indicates that the reaction is a simple bimolecular collision [mechanism I, Eq. (16)].

$$\mathrm{A} + \text{S-1} \underset{k_{-1}}{\overset{k_{+1}}{\rightleftharpoons}} \mathrm{A}\cdot\text{S-1} \tag{59}$$

From which the rate of acto-S-1 formation, λ, would be

$$\lambda = k_{+1}\mathrm{A}_0 + k_{-1} \tag{60}$$

The slope of line in Fig. 3.10B shows that in this experiment (using a solution containing 40 m*M* KCl) k_{+1} is $6 \times 10^6\ M^{-1}\ \mathrm{s}^{-1}$ while the *y* intercept, k_{-1}, is only about $0.2\ \mathrm{s}^{-1}$. The ratio k_{-1}/k_{+1} is thus $3 \times 10^{-8}\ M$. This dissociation constant (K_D) is in good agreement with measurements made using equilibrium binding techniques (Marston and Weber, 1975) and fluorescence depolarization (Highsmith *et al.*, 1976). The second-order rate constant for acto-S-1 association, k_{+1}, is very sensitive to ionic strength and falls from $2 \times 10^7\ M^{-1}\ \mathrm{s}^{-1}$ at 10 m*M* KCl to only $1.3 \times 10^6\ M^{-1}\ \mathrm{s}^{-1}$ at 100 m*M* KCl.

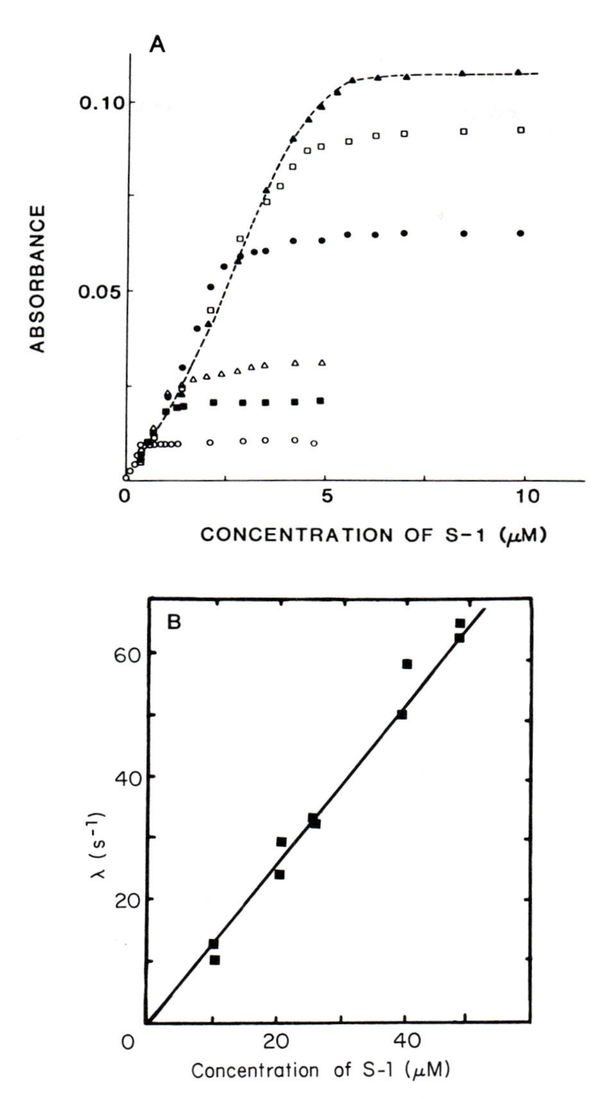

FIG. 3.10. (A) Measurement of light scattering during titrations of actin with S-1. The actin concentrations used were 0.4 (○), 0.8 (■), 1.2 (△), 2.5 (●), 4.0 (□) and 5.0 (▲) μM. (From White and Taylor (1976).) (B) The dependence on S-1 concentration of the rate constant λ for the association of S-1 and actin. The intercept on the ordinate, expected from other observations, is about 0.2 s^{-1}, which is essentially zero on the scale used here. (From White and Taylor (1976).)

3.3.2.2 *Effect of ATP on Actomyosin*

When ATP is added to an actomyosin system (in the presence of magnesium) several remarkable phenomena are observed.

In a solution containing 0.5 *M* KCl there is first a sudden decrease in optical density. This is called the "clearing phase"; it is due to dissociation of the actin and myosin. At the same time there is a modest (three- to five-fold) increase in the rate of ATP hydrolysis (as compared to myosin alone). When the ATP has been consumed, the turbidity returns to the level observed before ATP addition.

If the ionic strength of an actomyosin solution is lowered to physiological levels in the absence of ATP, the actomyosin becomes a gel. When ATP is added, the solution clears, but the extent of the change in optical density is less and the ATP hydrolysis is faster than that seen at the higher ionic strengths. Eventually, the actomyosin suspension "superprecipitates," that is, the gel contracts, squeezing out water.

3.3.2.3 *Actin Activation of Myosin ATPase*

Steady-state analysis of the ATPase of actomyosin systems has been of limited use in determining the reaction mechanism for two reasons. First, at low ionic strengths the actomyosin systems contain heterogeneous clumps of proteins in which the access of the enzyme to ATP and actin is limited. At high ionic strength, at which myosin is soluble (and forms a homogeneous solution), actin only modestly activates myosin ATPase, even at the highest usable actin concentrations. Actin + S-1 and actin + HMM are biochemical systems in which a greater degree of actin activation can be studied because they can be used at lower ionic strengths and remain homogeneous.

Minifilaments. Reisler and co-workers (Reisler, 1980; Reisler *et al.*, 1980) discovered that "minifilaments" are formed when rabbit myosin in 5 m*M* sodium pyrophosphate is dialysed against 10 m*M* citric acid buffered to a pH of 8.0 with Tris. These minifilaments are aggregates of 16–18 myosin molecules, only about 0.35 μm long but otherwise like the natural thick filaments. Minifilaments are soluble and stable in low ionic strength solutions (<40 m*M* KCl) and migrate in the ultracentrifuge as a single narrow band. Thus, minifilaments appear to be miniature species of the thick filaments that exist in living muscle. Results of parallel studies of the kinetics of minifilaments and S-1 or HMM ATPase are similar; all forms exhibit a phosphate burst of the same magnitude, and the time courses of the reactions with ATP are similar. Studies of the actin activation of myosin minifilaments and HMM show that they have the same V_{max} and K_m for actin.

S-1 and HMM. Studies of the acto-S-1 and acto-HMM steady-state hydrolysis have shown that the reaction kinetics are very dependent upon temperature and ionic strength. For example, at an ionic strength of 0.018 *M* and 20°C, V_{max} and K_m are 13 s^{-1} and 64 μM, respectively, while at 0.5°C the values are 0.55 s^{-1} and 28 μM (Wagner and Weeds, 1979; see Fig. 3.11). These results indicate that the Q_{10} of V_{max} is about 5, and that of K_m about 1.5. On the other hand, increasing the ionic strength at constant temperature has little effect on the V_{max}, but increases K_m markedly. For example, in 40 m*M* KCl (at 25°C), V_{max} for acto-HMM is 11.5 s^{-1} and the K_m is 78 μM; raising the KCl to 80 m*M* KCl does not change the V_{max}, but increases the K_m to 280 μM. The K_m for actin approximately doubles for each 20 m*M* rise in ionic strength. The highest concentration of actin that can be used is 100 μM because more concentrated solutions are too thick to be mixed effectively with other solutions. At high ionic strengths, values of

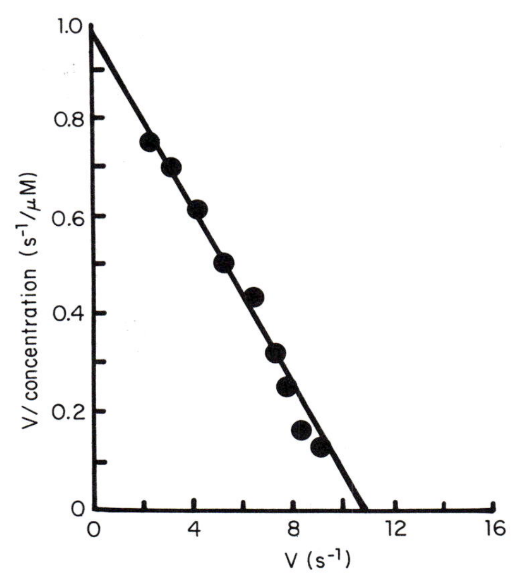

FIG. 3.11. An example of the determination of V_{max} and K_m for actin activation of S-1 ATPase. The experiments were made with a fixed concentration of S-1 (0.77 μM) and the concentration of actin was varied. V_{max} is given by the intercept on the abscissa; K_m is obtained from the reciprocal of the slope of the line. Average results obtained at different salt concentrations and temperatures from Wagner and Weeds (1979):

Salt concentration	T (°C)	V_{max} (s^{-1})	K_m (μM)
Low	20	11	19
High	20	13	64
Low	0.5	0.35	4.1
High	0.5	0.52	28

K_m and v_{max} can only be obtained by considerable extrapolation on graphs such as Fig. 3.11. For this reason most studies of acto-S-1 are made at low ionic strength. It is hard to know whether the results at low ionic strength apply to the conditions inside muscle cells, where the ionic strength is high (about 0.2 M). This is because the rate-limiting steps may be different at high ionic strength (see Table 3.III for the effect of ionic strength on the rate constants of the acto-S-1 mechanism). Consequently, extrapolation of the *in vitro* results of these kinetic studies to an intact muscle reaction scheme must be taken with a grain of salt.

3.3.2.4 *Mechanism of Actin Activation*

The fact that the acto-S-1 steady-state ATPase conforms to Eq. (43) does not reveal much about the reaction mechanism. There are at least two ways in which actin might directly alter the myosin ATPase rate: first, by interacting with the myosin after the hydrolytic step it could accelerate the rate-limiting step; second, by binding to S-1 it could form a different ATPase. Taylor and co-workers (Lymn and Taylor, 1971; Sleep and Taylor, 1976; White and Taylor, 1976) have performed experiments supporting the first of these alternatives. Eisenberg and colleagues (Eisenberg *et al.*, 1972; Stein *et al.*, 1979, 1981) have shown that, under different conditions, the second alternative is correct.

a. Evidence for Obligatory Dissociation. Lymn and Taylor (1971) studied the phosphate burst produced when acto-HMM is mixed with ATP. As shown in Fig. 3.12A, the steady-state rate is increased in the presence of actin. In contrast, the phosphate burst is similar to that in the absence of actin. This indicates that the hydrolytic step precedes the rate-limiting step. A plot of the phosphate burst rate (as determined from either P_i liberation or changes in the HMM fluorescence) as a function of ATP concentration suggests that ATP cleavage occurs by mechanism III (two-step, first step irreversible) as in Eq. (61) below.

$$\begin{array}{ccccc} \mathrm{AM + ATP} & \underset{k_{-1}}{\overset{k_{+1}}{\rightleftharpoons}} & \mathrm{AM \cdot ATP} & \underset{k_{-2}}{\overset{k_{+2}}{\rightleftharpoons}} & \mathrm{AM \cdot ADP \cdot P_i} \\ & & & & \downarrow k_{+3} \\ & & & & \mathrm{AM + ADP + P_i} \end{array} \quad (61)$$

The results in Fig. 3.12B are consistent with a second-order rate constant for ATP binding of about $2 \times 10^6\ M^{-1}\ \mathrm{s}^{-1}$ and a maximal value for $(k_{+2} + k_{-2})$ of about $150\ \mathrm{s}^{-1}$.

TABLE 3.III

A summary of the kinetics of actin-activated S-1 ATPase[a,b]

$$\begin{array}{ccccccccc} \text{ATP} + \text{AM} & \rightleftharpoons & \text{AM}\cdot\text{ATP} & \underset{k_{-7}}{\overset{k_{+7}}{\rightleftharpoons}} & \text{AM}^{**}\cdot\text{ADP}\cdot\text{P}_i & \underset{k_{-5}}{\overset{k_{+5}}{\rightleftharpoons}} & \text{AM}\cdot\text{ADP} & \underset{k_{-6}}{\overset{k_{+6}}{\rightleftharpoons}} & \text{AM} + \text{ADP} \\ & & k_{+2}\downarrow\uparrow k_{-2} & & k_{+4}\downarrow\uparrow k_{-4} & & & & \\ & & M^{*}\cdot\text{ATP} & \underset{k_{-3}}{\overset{k_{+3}}{\rightleftharpoons}} & \text{M}^{**}\cdot\text{ADP}\cdot\text{P}_i & & & & \end{array}$$

$$(k_{+7} + k_{-7}) \geq (k_{+3} + k_{-3});\ K_2 \cong K_4 \cong 1\text{–}3 \times 10^3\ M^{-1}$$

$$\text{A} + \text{S-1} \xrightarrow{k_a} \text{Acto-S-1}$$

Muscle type	T (°C)	pH	I (M)	K_1k_{+2} ($M^{-1}\,s^{-1}$)	k_{+2} (s^{-1})	$(k_{+3}+k_{-3})$ (s^{-1})	K_4k_{+5} ($M^{-1}\,s^{-1}$)	k_{+6} (s^{-1})	V (s^{-1})	K_m (μM)	k_a ($M^{-1}\,s^{-1}$)
Rabbit—fast (skeletal)	20	8	0.1	1.4×10^6	>1500	240	1.5×10^4	—	~20	~400	1.3×10^6
	20–25	8	0.04	1.5×10^6	—	—	2×10^5	—	—	—	6×10^6
	20	8	0.01	3×10^6	—	—	6×10^5	—	17	22	2×10^7
	20	7	0.04	4×10^6	—	—	2×10^5	—	13	64	—
	4	8	0.04	1×10^6	—	—	1×10^3	—	—	—	—
	0.5	7	0.018	—	—	—	—	—	0.5	10	—
	0.5	7	0.035	—	—	—	—	—	0.16	28	—
Frog (skeletal)	2–5	7	0.15	7×10^5	—	>10	—	—	—	—	5×10^4
	2–5	7	0.02	—	—	—	—	—	5	120	—
Chicken—slow (ALD)	20–25	7	0.01	6×10^6	>1100	60	1×10^5	75	4.5	35	1×10^7
	3	7	0.01	3.5×10^6	660	15	—	—	—	—	—
Smooth (gizzard)	20–25	7	0.01	2×10^6	1300	38	5×10^3	15	0.7	59	1.2×10^7
	3	7	0.01	8×10^5	400	6	—	—	—	—	—
Cardiac (bovine)	25	8	0.03	3×10^6	1500–2000	—	—	—	3	—	1×10^6

[a] References for results tabulated: Chock *et al.* (1976); Eisenberg and Moos (1968, 1970); Eisenberg *et al.* (1972); Eisenberg and Kielley (1973); Ferenczi *et al.* (1978a, b); Johnson and Taylor (1978); Lymn and Taylor (1971); Mulhern and Eisenberg (1976); Stein *et al.* (1979, 1981); Wagner and Weeds (1979); White (1977); White and Taylor (1976).

[b] K_4k_{-5}, k_a, and K_m are very sensitive to ionic strength (I). K_4k_{+5} and V_{max} are very sensitive to temperature (T).

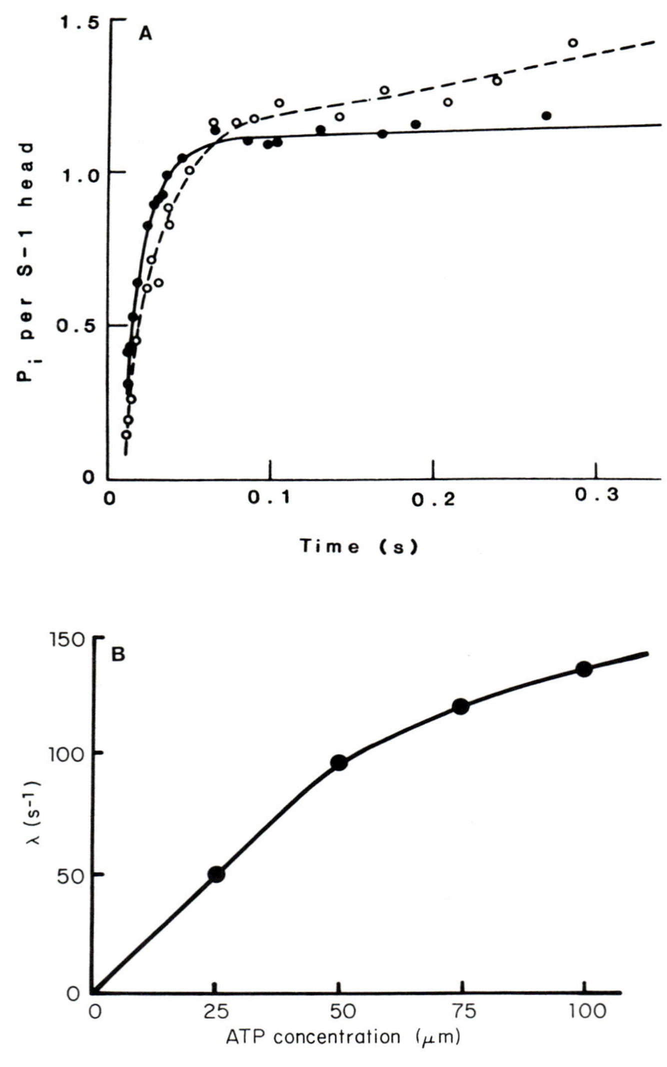

FIG. 3.12. (A) A comparison of the phosphate burst in the hydrolysis of ATP by HMM (●) and by acto-HMM (○). (From Lymn and Taylor, 1971.) (B) Rate constant for the early burst of P_i production by acto-HMM as a function of ATP concentration. Replotted from Fig. 3 of Lymn and Taylor (1971). The initial slope of the line is about 2×10^6 s^{-1} M^{-1}.

ATP also causes dissociation of actomyosin. Does this occur before or after the hydrolytic step? This question can be answered by a stopped flow experiment monitoring the optical density of an actomyosin solution upon the addition of ATP; the optical density will decline upon acto-HMM or acto-S-1 dissociation. The possible mechanisms are

$$\begin{array}{ccccc} \mathrm{AM} + \mathrm{ATP} & \underset{k_{-1}}{\overset{k_{+1}}{\rightleftharpoons}} & \mathrm{AM}\cdot\mathrm{ATP} & \underset{k_{-2}}{\overset{k_{+2}}{\rightleftharpoons}} & \mathrm{AM}\cdot\mathrm{ADP}\cdot\mathrm{P_i} \\ & & k_{+3}\downarrow\uparrow k_{-3} & & k_{+4}\downarrow\uparrow k_{-4} \\ & & \mathrm{A} + \mathrm{M}\cdot\mathrm{ATP} & \rightleftharpoons & \mathrm{A} + \mathrm{M}\cdot\mathrm{ADP}\cdot\mathrm{P_i} \end{array} \qquad (62)$$

Dissociation might occur *before* ATP cleavage via k_{+3} or *after* ATP cleavage via step k_{+4}. In the former case dissociation will occur at a faster rate than the hydrolytic step; in the latter case dissociation will occur at a rate equal to or less than that of the hydrolytic step.

In their original studies at 20°C and 0.1 *M* KCl, Lymn and Taylor (1971) found that the rate of acto-HMM dissociation increased linearly with ATP concentration so that at 200 μM ATP, dissociation occurs at a rate of 250 s^{-1}, or twice the hydrolysis rate; this suggests that the pathway through k_{+3} was being followed. At lower temperatures (3°C) and ionic strength (0.05 *M* KCl), the result is even more emphatic because the rate of ATP cleavage is slower (Sleep and Taylor, 1976). Figure 3.13 shows these results; the acto-S-1 dissociation is 10 times faster than the ATP cleavage. From these results the simplest reaction mechanism for the interaction of ATP and acto-S-1 is

$$\begin{array}{ccccc} \mathrm{AM} + \mathrm{ATP} & \underset{k_{-1}}{\overset{k_{+1}}{\rightleftharpoons}} & \mathrm{AM}\cdot\mathrm{ATP} & & \\ & & k_{+2}\downarrow\uparrow k_{-2} & & \\ & & \mathrm{A} + \mathrm{M}\cdot\mathrm{ATP} & \underset{k_{-3}}{\overset{k_{+3}}{\rightleftharpoons}} & \mathrm{M}\cdot\mathrm{ADP}\cdot\mathrm{P_i} \end{array} \qquad (63)$$

The rate of acto-S-1 dissociation is a linear function of ATP concentration to at least 700 s^{-1} with a slope of 10^6 M^{-1} s^{-1} (Fig. 3.14). Thus k_{+2} [Eq. (63)] is >700 s^{-1}.

The mechanism in Eq. (63) predicts that there should be a plateau in the relationship between the rate of dissociation of acto-S-1 and ATP concentration. Such a plateau has been observed using smooth muscle S-1, which has a very slow acto-S-1 ATPase (Marston and Taylor, 1980). Cold chase experiments have demonstrated that ATP binding to acto-S-1 is

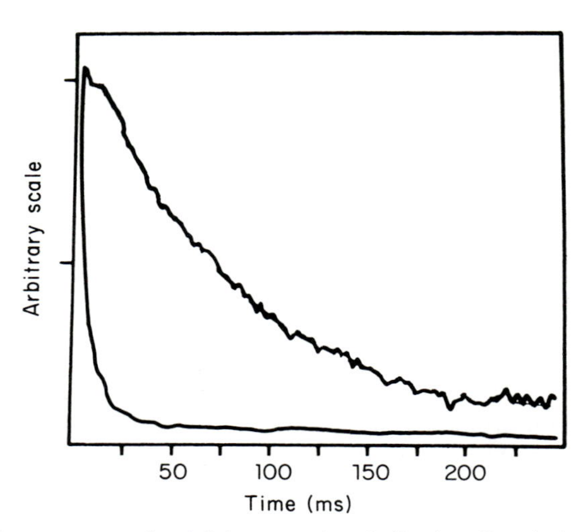

FIG. 3.13. Simultaneous records of light scattering (indicating dissociation) and change of fluorescence (indicating ATP hydrolysis) for the reaction of acto-S-1 with ATP. The upper trace is the fluorescence signal (a downward deflection indicating increase in fluorescence). The lower trace is the light scattering signal. The ordinate scales can be regarded as arbitrary. (From Sleep and Taylor, 1976.)

effectively irreversible and either k_{-2} or k_{-1} is small (Johnson and Taylor, 1978).

Assuming that the reaction of S-1 is similar to other enzyme mechanisms, Eq. (63) can be expanded to Eq. (64) (below). In this scheme the tight binding is due to a protein isomerization and the tightly bound ligand–enzyme complex is different from the collision intermediate formed from acto-S-1 and ATP

$$\begin{array}{ccccc} \mathrm{AM} + \mathrm{ATP} & \underset{k_{-1}}{\overset{k_{+1}}{\rightleftharpoons}} & \mathrm{AM} \cdot \mathrm{ATP} & \underset{k_{-2}}{\overset{k_{+2}}{\rightleftharpoons}} & \mathrm{AM}^* \cdot \mathrm{ATP} \\ & & & & k_{+3} \downarrow\uparrow k_{-3} \\ & & & & \mathrm{A} + \mathrm{M}^* \cdot \mathrm{ATP} \underset{k_{-4}}{\overset{k_{+4}}{\rightleftharpoons}} \mathrm{M}^{**} \cdot \mathrm{ADP} \cdot \mathrm{P_i} \end{array} \tag{64}$$

In this case the tight binding occurs via step 2 and the acto-S-1 dissociation via step 3. The asterisks indicate increased levels of protein fluorescence.

b. An Explanation of Actin Activation. The rate of ATP cleavage and the change in fluorescence measure only $(k_{+4} + k_{-4})$ [see Eqs. (21) and (49)].

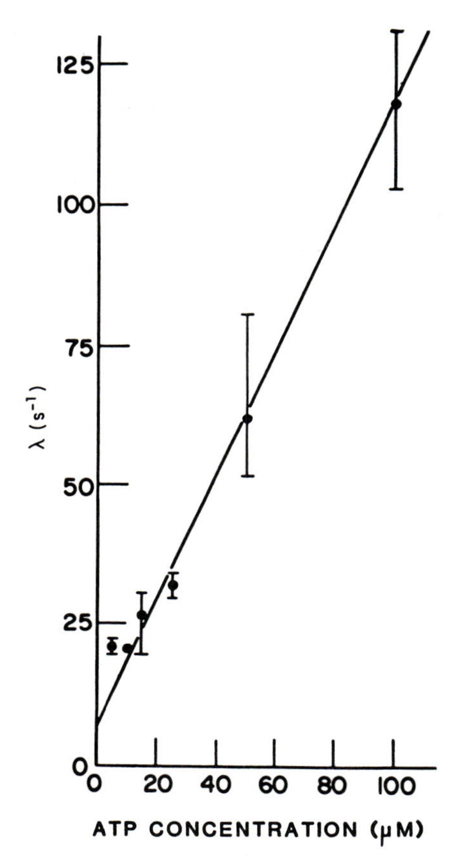

FIG. 3.14. The dependence on ATP concentration of the rate constant (λ) for the dissociation of acto-HMM. Error bars indicate the range of the observations. At the lowest concentrations of ATP the dissociation is incomplete and the ATP concentration not constant. (From Lymn and Taylor, 1971.) The results of Sleep and Taylor (1976), not shown here, indicate that the linear dependence of λ on ATP concentration continues to at least 700 s^{-1}.

As cleavage takes place on the S-1 the equilibrium constant for the hydrolytic step ought to be the same as that for S-1 alone. Consequently, at 20°C, $k_{+4} = 9k_{-4}$ and thus $k_{+4} = 100\ s^{-1}$. Since this rate (k_{+4}) is much greater than V_{max}, the rate-limiting step in the steady state must occur subsequent to ATP hydrolysis. A simple mechanism is

$$M^{**} \cdot ADP \cdot P_i + A \underset{k_{-5}}{\overset{k_{+5}}{\rightleftharpoons}} AM^{**} \cdot ADP \cdot P_i \xrightarrow{k_{+6}} AM + ADP + P_i \quad (65)$$

and either step 5 or step 6 could be rate limiting. As discussed in relation to Eq. (59) and Fig. 3.10B, the rate of actin and S-1 binding is, at low ionic strength, probably diffusion limited [k_{+1} in Eq. (59) is $2 \times 10^7\ M^{-1}\ s^{-1}$] and is thus not much affected by temperature. One might expect that binding of actin to M**·ADP·P_i would also be very rapid. White and Taylor (1976) measured the rate of this binding by following the turbidity after mixing actin with M**·ADP·P_i. The turbidity increased at a rate which was linearly dependent on the actin concentration. This rate was found to be very temperature sensitive (having a Q_{10} of about 5.4) and (as seen in Fig. 3.15A) was equal to the steady-state rate of ATP hydrolysis at the actin concentrations used ($< 10\ \mu M$). The apparent second-order rate constant for M**·ADP·P_i and actin binding was $1.3 \times 10^4\ M^{-1}\ s^{-1}$. The apparent second-order rate constant is too low by two to three orders of magnitude and has too great a temperature sensitivity to be associated with a simple binding mechanism like that shown in Eq. (59). A more likely explanation is that the step governed by k_{+5} and k_{-5} is a rapid equilibrium reaction with the bound M**·ADP·P_i state stabilized by the release of products which is governed by the rate constant k_{+6}. If this were so, the rate of M**·ADP·P_i and actin binding, λ, would be

$$\lambda = k_{+6} \frac{K_5 A_0}{K_5 A_0 + 1} \tag{66}$$

and when $K_5(A_0) \ll 1$, $\lambda = k_{+6} K_5 A_0$. If $k_{+6} = V_{max}$ or about $10\ s^{-1}$, then K_5 must be about $1 \times 10^3\ M^{-1}$. The temperature sensitivity of the reaction in Eq. (65) would thus be conferred by the temperature dependence of the rate constant k_{+6}. The mechanism given in Eq. (65) can thus quantitatively account for the data in Fig. 3.15A.

Equation (67) summarizes the acto-S-1 reaction developed to this point.

$$\begin{array}{ccccccc}
\mathrm{AM + ATP} & \underset{k_{-1}}{\overset{k_{+1}}{\rightleftharpoons}} & \mathrm{AM \cdot ATP} & \xrightarrow{k_{+2}} & \mathrm{AM^{*} \cdot ATP} & & \mathrm{AM^{**} \cdot ADP \cdot P_i} \xrightarrow{k_{+6}} \mathrm{AM + ADP + P_i} \\
 & & & & k_{+3} \downarrow\uparrow k_{-3} & & k_{+5} \uparrow\downarrow k_{-5} \\
 & & & & \mathrm{A + M^{*} \cdot ATP} & \underset{k_{-4}}{\overset{k_{+4}}{\rightleftharpoons}} & \mathrm{M^{**} \cdot ADP \cdot P_i}
\end{array} \tag{67}$$

This reaction mechanism has been particularly attractive because (as indicated in Chapter 1) it offers a simple correspondence to the physiological model of crossbridge behaviour.

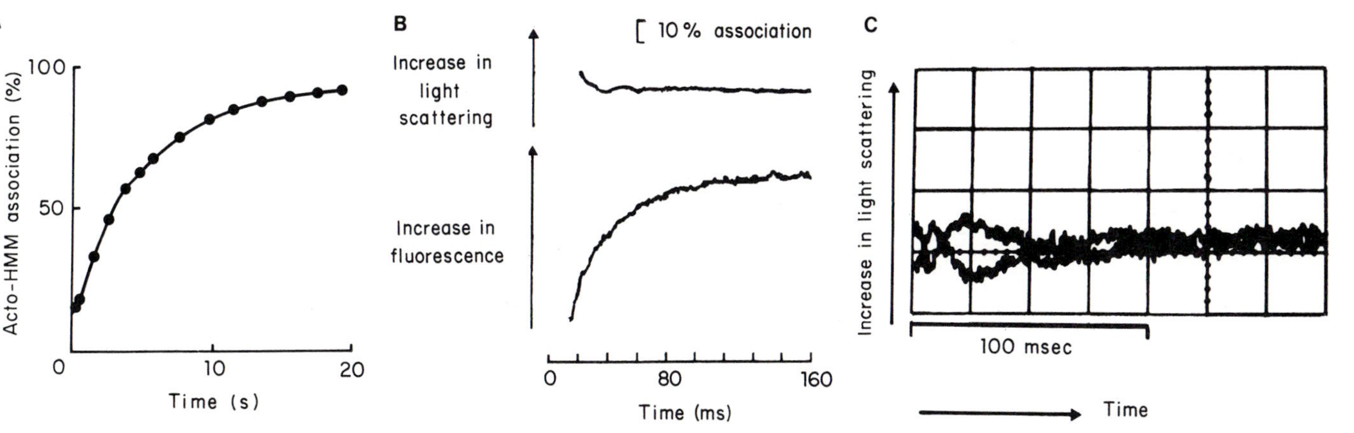

FIG. 3.15. (A) Association of actin with HMM·ADP·P_i; 20 μM ATP was added to a solution containing 10 μM actin and an equivalent amount of HMM. The degree of association was found from measurements of light scattering. The initial rapid dissociation, not shown, was followed by a slow reassociation of actin with the HMM product complex. The rate of this process can be obtained from such observations and is compared below with the steady-state ATPase rate (White and Taylor, 1976):

T (°C)	Rate of reassociation (s^{-1})	ATPase rate (s^{-1})
4	0.1 ± 0.05	0.15 ± 0.05
20	20 ± 5	25 ± 5
36	180 ± 50	200 ± 25

(B) A comparison of the changes in light scattering and changes in fluorescence on addition of S-1 to actin + ATP. The fluorescence change indicates the time course of the initial rapid burst of ATP splitting. The light scattering signal, indicating the degree of association of actin and S-1, shows that there is little if any change in association during the rapid phase of ATP splitting. The steady state of association has been reached (or nearly reached) in the dead time of the stopped flow apparatus. [From Stein *et al.*, (1979).] In contrast to the experiment shown in Fig. 3.15A, these observations were made at low salt concentration and high actin concentration. (C) Light scattering changes observed on mixing actin with a mixture of S-1 and ATP, in which most of the S-1 will be in the form M**·ADP·P_i. No consistent change in light scattering is observed, indicating that the steady-state degree of association of actin and S-1 had been reached within the dead time of the stopped flow apparatus. (From Stein *et al.*, 1979.)

c. Evidence against Obligatory Dissociation. The reaction scheme in Eq. (67) was devised to explain results obtained at actin concentrations far below those necessary to bring the steady state actin-activated ATPase near to V_{max}. The only practical way to test this scheme nearer to V_{max} is to use very low ionic strength (about 10 mM) so that V_{max} is reached at actin concentrations of about 50–100 μM. Eisenberg and co-workers (Eisenberg and Keilley, 1973; Eisenberg *et al.*, 1972; Stein *et al.*, 1979, 1981) have performed experiments under such conditions.

In Eq. (67), ATP binding to an actomyosin complex obligates the dissociation of the actin from the myosin; thereafter actin does not rebind to myosin until the bound ATP has been cleaved to form the ternary complex M**·ADP·P_i. If this is correct, when acto-S-1 is reacted with ATP, there would first be a rapid dissociation of acto-S-1 caused by the ATP binding. This step would be followed by a reassociation of acto-S-1 with a time course equal to the rate of ATP cleavage. Stein *et al.* (1981) found no evidence of such behaviour (Fig. 3.15B). Their experiments were made at low ionic strength (about 25 mM) and at an actin concentration great enough to bring the ATPase rate to a value of $0.95V_{max}$. When acto-S-1 was reacted with 2 mM ATP, a steady-state binding of actin to S-1 was reached within 2 ms (the dead time of the stopped flow machine). This result can only be explained by supposing that there is rapid equilibrium between M**·ATP and AM**·ATP, similar to that between AM*·ADP·P_i and M**·ADP·P_i. This conclusion is confirmed by studies of the time course of actin binding to M**·ADP·P_i (Fig. 3.15C). Again, the steady-state binding of actin to M**·ADP·P_i was reached within 2 ms, indicating that the reaction at this ionic strength must be a rapid equilibrium. Since the extent of light scattering by actin bound M**·ATP and M**·ADP·P_i is similar, the equilibrium constants of actin binding to the two species of myosin must also be similar: that is, k_3 and k_5 are about 10^3 M.

The rapid reversibility of reaction 3 in reaction scheme (67) has these consequences.

1. The steady-state ATPase will show a maximum rate, with respect to actin concentration, and will be inhibited at higher actin concentrations.
2. At high actin concentrations, the stoichiometry of the early rapid phase of P_i production (the size of the phosphate burst) will be reduced.
3. The rate of the phosphate burst will approach a limiting value of k_{+6} at high actin concentration.

These conclusions are derived algebraically in Table 3.IV. They can also be understood as follows, from consideration of reaction scheme (67).

1. At low actin concentrations, reaction step 5 is rate limiting and the steady-state rate therefore increases with actin concentration. At somewhat higher actin concentrations, reaction 6 is rate limiting. Increasing actin

TABLE 3.IV
Some consequences of the Lymn–Taylor kinetic scheme, Eq. (67)

If reactions 3 and 5 in Eq. (67) are in rapid equilibrium, the reaction scheme can be simplified, at high ATP concentrations, to

$$M_1^* \cdot ATP \underset{k'_{-4}}{\overset{k'_{+4}}{\rightleftharpoons}} M_1^{**} \cdot ADP \cdot P_i \xrightarrow{k'_{+6}} M_1^* \cdot ATP + P_i \quad \text{(T-1)}$$

Here M_1 stands for the equilibrium mixture of S-1 and acto-S-1, and AM is omitted because it binds ATP so rapidly that its concentration will always be negligible. k'_{+6} is separate from k'_{-4} because the step controlled by the former involves the release of P_i. The constants k'_{+4}, k'_{-4}, and k'_{+6} are related to the rate constant of Eq. (67) as follows.

$$k'_{+4}(M^* \cdot ATP + AM^* \cdot ATP) = k_{+4}(M^* \cdot ATP) \quad \text{(T-2)}$$

By definition the dissociation constant $K_3 = (A)(M^* \cdot ATP)/(AM^* \cdot ATP)$. Using this to substitute for $(AM^* \cdot ATP)$ in Eq. (T-2),

$$k'_{+4}\left[M^* \cdot ATP + \frac{(A)(M^* \cdot ATP)}{K_3}\right] = k_{+4}(M^* \cdot ATP) \quad \text{(T-3)}$$

$$k'_{+4}\left(1 + \frac{A}{K_3}\right)(M^* \cdot ATP) = k_4(M^* \cdot ATP) \quad \text{(T-4)}$$

$$k'_{+4} = k_{+4}K_3/(K_3 + A) \quad \text{(T-5)}$$

Similarly: $k'_{-4}(M^{**} \cdot ADP \cdot P_i + AM^{**} \cdot ADP \cdot P_i) = k_{-4}(M^{**} \cdot ADP \cdot P_i)$ (T-6)

$$k'_{+6}(M^{**} \cdot ADP \cdot P_i + AM^{**} \cdot ADP \cdot P_i) = k_{+6}(M^{**} \cdot ADP \cdot P_i) \quad \text{(T-7)}$$

By definition $(AM^{**} \cdot ADP \cdot P_i)K_5 = (A)(M^{**} \cdot ADP \cdot P_i)$ and substitution for $AM^{**} \cdot ADP \cdot P_i$ in Eq. (T-6) and for $M^{**} \cdot ADP \cdot P_i$ in Eq. (T-7) gives

$$k'_{-4} = k_{-4}/(K_5 + A) \quad \text{(T-8)}$$

$$k'_{+6} = k_{+6}(A)/(K_5 + A) \quad \text{(T-9)}$$

Since, as described above, $K_5 \cong K_3$, Eqs. (T-8) and (T-9) become

$$k'_{-4} = k_{-4}K_3/(K_3 + A) \quad \text{(T-10)}$$

$$k'_{+6} = k_{+6}A/(K_3 + A) \quad \text{(T-11)}$$

The steady-state rate of ATP hydrolysis, v/M_0, is therefore

$$\frac{V}{M_0} = \frac{k'_{+4}k'_{+6}}{k'_{+4} + k'_{-4} + k'_{+6}} = \frac{k_{+6}k_{+4}K_3A}{[(k_{+4} + k_{-4})K_3 + k_{+6}A][K_3 + A]} \quad \text{(T-12)}$$

In reciprocal form, for plotting $1/(v/M_0)$ against $1/A$,

$$\frac{M_0}{V} = \frac{k'_{+4} + k'_{-4} + k'_{+6}}{k'_{+4}k'_{+6}} \quad \text{(T-13)}$$

(Table continued)

TABLE 3.IV (*continued*)

Substituting for k'_{+4}, k'_{-4}, and k'_6 from Eqs. (T4-5), (T4-10), and (T4-11),

$$\frac{M_0}{V} = \frac{1}{k_{+6}} + \frac{1}{k_{+4}} + \frac{k_{-4}}{k_{+6} \cdot k_{+4}} + \left(\frac{K_3}{k_{+6}} + \frac{k_{-4}K_3}{k_{+6}k_{+4}}\right)\frac{1}{A} + \frac{A}{k_{+4}K_3} \quad \text{(T-14)}$$

Note that the last term in Eq. (T-14) indicates that, at high concentrations of actin, M_0/v will be large and thus the velocity of the reaction small. The form of Eq. (T-1) is similar to the general case given in Eq. (6) and therefore the time course of $M_1^{**} \cdot ADP \cdot P_i$ production will be given by

$$\frac{M_1^{**} \cdot ADP \cdot P_i}{M_0} = \frac{k'_{+4}}{k'_{+4} + k'_{-4} + k'_{+6}} \{1 - \exp[-(k'_{+4} + k'_{-4} + k'_{+6})t]\} \quad \text{(T-15)}$$

Since $M^{**} \cdot ADP \cdot P_i = M^{**} \cdot ADP \cdot P_i + AM^{**} \cdot ADP \cdot P_i$, and again substituting from Eqs. (T-5), (T-10), and (T-11),

$$\frac{M_1^{**} \cdot ADP \cdot P_i}{M_0} = \frac{K_3 k_{+4}}{K_3(k_{+4} + k_{-4}) + k_{+6}(A)} \left(1 - \exp - \left\{\left[\frac{K_3(k_{+4} + k_{-4}) + k_{+6}A}{K_3 + A}\right]t\right\}\right) \quad \text{(T-16)}$$

Thus the rate of the phosphate burst, λ, will be a function of actin concentration.

$$\lambda = \frac{K_3(k_{+4} + k_{-4}) + k_{+6}A}{K_3 + A} \quad \text{(T-17)}$$

Dividing the numerator and denominator of Eq. (T-17) shows that as $A \to \infty$, $\lambda \to k_{+6}$ as a limiting value. The amplitude of the phosphate burst will also be a function of the actin concentration, as seen in Eq. (T-18), and approaches 0 as $(A) \to \infty$

$$P_i^{burst} = \frac{K_3 k_4}{K_3(k_{+4} + k_4) + k_{+6}A} \quad \text{(T-18)}$$

---

concentration still increases velocity because it increases the concentration of $AM^{**} \cdot ADP \cdot P_i$. At very high actin concentration, however, reaction 4 becomes rate limiting, and increasing actin concentration lowers the velocity because it increases the concentration of $AM^* \cdot ATP$, and therefore diminishes that of $M^* \cdot ATP$ and of $AM^{**} \cdot ADP \cdot P_i$.

2. The phosphate burst is due to the fact that the initial rate of reaction 4 exceeds that of reaction 6. The restriction in the rate of reaction 4 at high actin concentration, due to accumulation of $AM^* \cdot ATP$, allows reaction 6 to keep pace with reaction 4, and prevents the build up of $M^{**} \cdot ADP \cdot P_i$. This causes the burst size to diminish at high actin concentration.

3. For the same reason, the rate of the phosphate burst decreases, and approaches k_{+6} as a limit at high actin concentration.

Stein *et al.* (1981) tested these predictions and their results are shown in Fig. 3.16. So that these results could be directly compared to the predictions

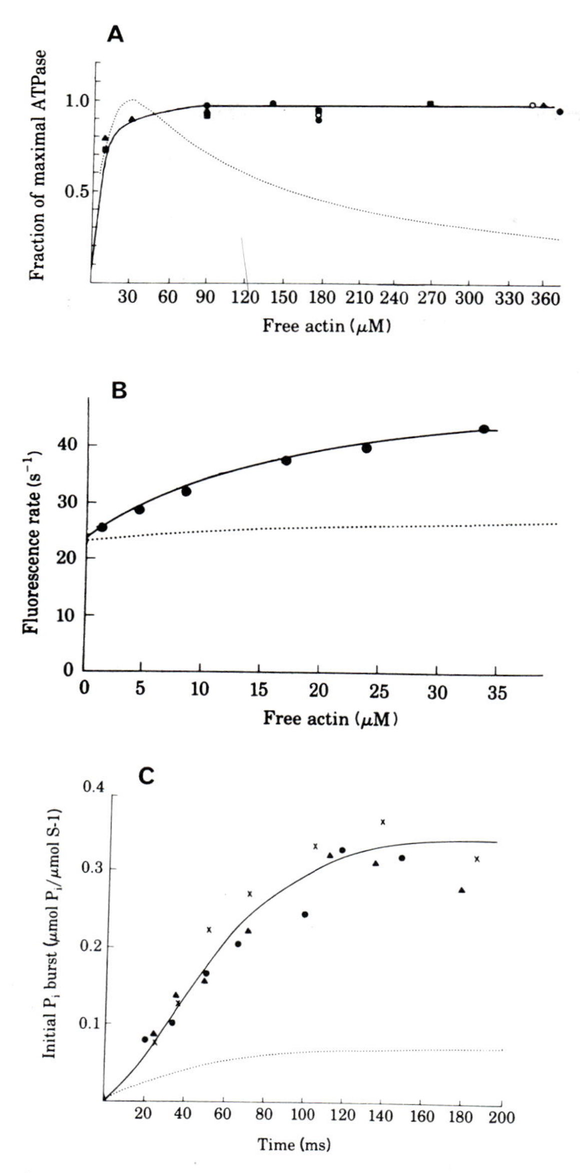

FIG. 3.16. Comparisons of the experimental results (●, ▲, ×) with predictions of the Lymn-Taylor reaction scheme (dotted lines) (from Stein *et al.*, 1981). (A) Steady-state acto-S-1 ATPase rate as a function of actin concentration. (B) Rate of rapid phase of ATP cleavage by acto-S-1 as a function of actin concentration. (C) Determination of P_i burst at high actin concentration (72 μM).

of the Lymn–Taylor reaction scheme, the values of k_{+4}, k_{-4}, and k_6 were measured. The predictions are shown by the dotted lines in the figure and do not conform to the experimental results.

1. The steady-state ATPase (Fig. 3.16A) rises monotonically with actin concentration and shows no inhibition at high actin concentration.
2. The rate of the phosphate burst increases at high actin concentration, instead of decreasing as predicted (Fig. 3.16B).
3. The size of the phsophate burst does not decrease as much as predicted by the Lymn–Taylor reaction mechanism (Fig. 3.16C). The small reduction that is observed may be due to the difficulty in mixing ATP into the concentrated actin/S-1 solution.

What alterations in the Lymn–Taylor mechanism are required to account for these results? In fact they can be explained by introducing a pathway in Eq. (67) in which ATP is hydrolysed by the acto-S-1 complex (a non-dissociating pathway).

$$\begin{array}{ccccccccc} \mathrm{AM + ATP} & \underset{k_{-1}}{\overset{k_{+1}}{\rightleftharpoons}} & \mathrm{AM \cdot ATP} & \xrightarrow{k_{+2}} & \mathrm{AM^* \cdot ATP} & \underset{k_{-7}}{\overset{k_{+7}}{\rightleftharpoons}} & \mathrm{AM^{**} \cdot ADP \cdot P_i} & \xrightarrow{k_{+6}} & \mathrm{AM + ADP + P_i} \\ & & & & k_{+3} \downarrow\uparrow k_{-3} & & k_{+5} \uparrow\downarrow k_{-5} & & \\ & & & & \mathrm{A + M^* \cdot ATP} & \underset{k_{-4}}{\overset{k_{+4}}{\rightleftharpoons}} & \mathrm{M^{**} \cdot ADP \cdot P_i + A} & & \end{array} \qquad (68)$$

In this scheme the rate constants of reaction 5 would be slightly greater than those of reaction 4 to account for the increased rate of the phosphate burst at high actin concentration. The inclusion of a nondissociating pathway is not in conflict with the observation of Lymn and Taylor that addition of ATP to acto-S-1 causes a dissociation prior to ATP cleavage. This is because their observations were made under conditions (moderate ionic strength and fairly low actin concentration) that shift the equilibrium of reaction 3 in favour of M*·ATP. Under conditions that shift the equilibrium toward AM*·ATP (low ionic strength and high actin concentration), however, the nondissociating pathway becomes important.

Thus, there are two mechanisms [Eqs. (67) and (68)] that have been experimentally developed to account for the actomyosin ATPase. The evidence for each seems strong under the experimental conditions for which it was proposed: for the Lymn–Taylor mechanism, moderate ionic strength, low actin concentration and for the Eisenberg–Stein mechanism, low ionic strength, high actin concentration. Which reaction mechanism is followed within the intact muscle thus depends on the conditions that exist there. As these conditions (high ionic strength, very high effective actin concentration) cannot be duplicated *in vitro*, the question cannot be answered conclusively by experiments in solution. However, the question has been approached by experiments using myofibrils (Sleep, 1981; Sleep and Smith, 1981).

4
Heat Production and Chemical Change

4.1 Introduction

In Section 4.2 of this chapter are described the results of measurements of the heat produced by skeletal muscles under different conditions. Why would these results be of interest to anyone trying to understand muscle contraction? First, like any other physical measurements, such as force, shortening, optical changes, volume changes, etc., they provide information which, had we the wit to appreciate it, could improve our understanding of the way muscle works. A second use for heat production measurements is in the energy balance method for establishing the identity of the chemical processes that contribute to the energy output during muscle contraction. This method and its results are described in Section 4.3. Third, in circumstances in which it is considered that ATP splitting and the creatine kinase reaction are the only significant sources of energy output, measurements of the production of heat and work can be used to estimate the rate of ATP splitting. As heat and work can be measured without destruction of the muscle, in contrast to direct measurements of ATP splitting, the influence of various factors on this rate can sometimes be investigated more conveniently in this indirect way. Such studies have been a stimulus to further direct investigations of ATP splitting and to energy balance studies. This stimulation continues.

Many good (and some bad) measurements of muscle heat have been made during the last 50 years. These data are certainly of little use unless they are remembered, and an attempt is made to understand them. In Section 4.2 we summarize what we consider to be the most important aspects of this knowledge of muscle. In Section 4.3 and in Chapter 5 we try to interpret at least a part of it.

Some words used to describe the heat production by muscle have been responsible for a great deal of confusion. For example, the terms *activation heat*, *maintenance heat*, *recovery heat*, *shortening heat*, *the Fenn effect*, and

$$MgATP^{2-} \rightarrow MgADP^{-} + HPO_4^{2-} + H^{+} \qquad (1)$$

$$H^{+} + PCr^{2-} + MgADP^{-} \rightarrow Cr + MgATP^{2-} \qquad (2)$$

$$H^{+} + 2MgADP^{-} \rightarrow AMP^{-} + MgATP^{2-} + Mg^{2+} \qquad (3a)$$

$$H^{+} + AMP^{-} \rightarrow IMP^{-} + NH_4^{+} \qquad (3b)$$

$$\text{Glycogen unit} + 3MgADP^{-} + 3HPO_4^{2-} + H^{+} \rightarrow 2\ \text{lactate}^{-} + 3MgATP^{2-} \qquad (4)$$

$$\text{Glycogen unit} + 6O_2 + 36MgADP^{-} + 36HPO_4^{2-} + 36H^{+} \rightarrow 6CO_2 + 6H_2O + 36MgATP^{2-} \qquad (5)$$

FIG. 4.1 The main reactions occurring during and after contraction can be represented by these equations. As there is no convenient way of showing in such equations the various ionic species that coexist under the conditions in muscle, these equations do not accurately represent the stoichiometry with which hydrogen ions and magnesium ions are involved.

so on are often used in an uncritical and ill-defined manner. In the following description we try to avoid the use of these jargon terms, and to define those few specialized terms we find essential. Other such terms are used with quotation marks, when reference to published work makes their use necessary.

Section 4.3 discusses measurements of the amount of various chemical changes during and after contraction and relates these to the energy output. As an introduction to this section we outline here what chemical changes are expected in contracting and recovering muscle (for a more detailed account, see Kushmerick, 1984).

ATP splitting [reaction (1) in Fig. 4.1] is the reaction by which the mechanical events of contraction are coupled to respiration. ATP is split by actomyosin and also by the Ca^{2+} pump of the SR, and ATP is resynthesized by metabolic processes. There are four such processes to consider.

The creatine phosphokinase (CPK) reaction or Lohmann reaction [reaction (2) in Fig. 4.1]. This process is very effective in maintaining the ATP concentration during muscular activity. The equilibrium constant for this reaction at pH 7 is about 100 (Woledge, 1973); that is,

$$100 = \frac{[ATP] \times [Cr]}{[ADP] \times [PCr]}$$

Writing t for the concentration of ATP as a proportion of total adenosine and p for the concentration of PCr as a proportion of the total creatine,

$$t = \frac{100p}{1 + (99p)}$$

In resting muscle p is about 0.75, so that when the CPK reaction is at equilibrium t is about 0.987. When 90% of the PCr has been used the ATP concentration has fallen only 10% ($p = 0.075$; $t = 0.882$). This behaviour is further illustrated in Fig. 4.2.

Thus, if the CPK reaction is close to equilibrium, the ATP concentration changes very little as the store of PCr is used. The activity of CPK in muscle seems in fact to be sufficient to keep the reaction close to equilibrium. This was first demonstrated by Carlson and Siger (1959, 1960), by measuring the concentration of PCr, Cr, and ATP in muscle (in which other reactions resynthesizing ATP had been blocked) during a series of twitches. Their results are shown in Fig. 4.2B. More recently, "saturation transfer NMR" has been used to measure, in intact muscles, unidirectional fluxes through the CPK reaction. In both perfused cat muscle (Meyer *et al.*, 1982) and superfused frog muscle (Brown *et al.*, 1980) the fluxes in each direction have been found to be high, and equal (within experimental error), showing that the reaction is indeed close to equilibrium.

The CPK reaction can be blocked in muscle by dinitrofluorobenzene (DNFB). After such inhibition the ATP concentration falls during brief contractions (Cain and Davies, 1962; Infante and Davies, 1965).

The adenylate kinase (or myokinase) reaction [reaction (3A) in Fig. 4.1]. The equilibrium constant of this reaction under the conditions in muscle cells is about unity. It does not play any role in resynthesizing ATP except when the other systems described here are exhausted or inhibited. Under these conditions (Infante and Davies, 1965; Dydynska and Wilkie, 1966; Canfield and Marechal, 1973; Canfield *et al.*, 1973; Curtin and Woledge, 1975), the reaction does occur at an appreciable rate and is rendered more effective by the removal by deamination of the AMP formed [reaction (3B) in Fig. 4.1].

Glycogenolysis [reaction (4) in Fig. 4.1]. As muscle often contains a large amount of glycogen, the amount of ATP that can be resynthesized in this way is considerable. Glycogenolysis can be inhibited by IAA (iodoacetate), which has been used in many studies to simplify the metabolic processes. However, in frog muscle, particularly at 0°C, the rate of glycogenolysis is often found to be very low anyway (for example, Curtin and Woledge, 1975) and its inhibition is therefore unnecessary for many experiments.

The lactate formed in glycogenolysis can be oxidized or rebuilt to glycogen in muscle (Bendall and Taylor, 1970), but this latter process is slow. Generally, any lactate that is not oxidized leaves the muscle cells, as it can cross the surface membrane easily. If the muscle is superfused the amount of lactate leaving the muscle can be measured. De Furia and Kushmerick (1977) have used such measurements to estimate the metabolic costs of contraction under anaerobic conditions (see Section 4.3.6). They have also shown that even under aerobic conditions some lactate is lost from the

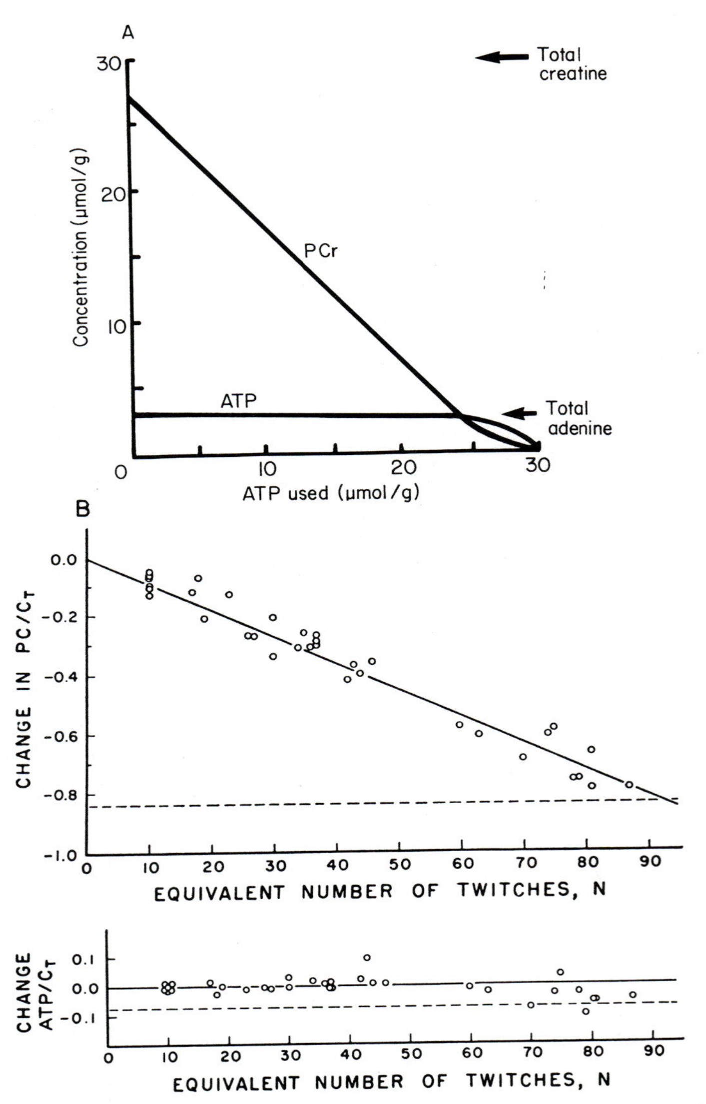

FIG. 4.2. (A) The time course of changes in ATP and PCr concentration that would be expected as ATP is split in a system in which the creatine kinase reaction is in equilibrium. In this example the initial concentration of PCr is taken as 90% of the total creatine concentration (= PCr + free creatine). The total adenine concentration is taken as 10% of the total creatine concentration and the equilibrium constant of the creatine kinase reaction is taken as 100. (B) Observations of the changes in ATP and PCr during a series of twitches of anaerobic, iodoacetate-treated frog muscle. The dotted line shows the maximum change that would occur if all the PCr initially present (85% of the total creatine) were used. Both PCr and ATP have been divided by the total creatine content (= PCr + free creatine) to normalize for muscle size. (From Carlson and Siger, 1960.)

muscle, and account should be taken of this "aerobic glycolysis" when assessing the metabolic costs of contractions under aerobic conditions.

Oxidation of glycogen [reaction (5) of Fig. 4.1]. This reaction produces 12 times more ATP per glycogen unit than does glycogenolysis and is thus capable of producing a very large amount of ATP. The extent of oxidative recovery can be measured from the volume change after absorption of CO_2 (D. K. Hill, 1940a,b) or by using oxygen electrodes (Kushmerick and Paul, 1976a,b; Elzinga *et al.*, 1981; Mahler, 1978a,b). Such measurements have been used by Kushmerick and Paul in assessing the energetic costs of contraction (see Section 4.3.6). In frog muscle, oxidative recovery is slow, full recovery taking at least 1 h at 0°C (D. K. Hill, 1940a). Also in frog muscle, recovery does not start immediately after a brief tetanus but only after a delay of about 10 s (at 25°C) (Elzinga *et al.*, 1984). In mammalian muscle, however, oxidative recovery is much faster and may occur in part *during* a brief tetanic contraction (Crow and Kushmerick, 1982).

Frog sartorius has been the usual preparation for studies of heat production by muscle. In this muscle, the processes of glycolysis and oxidative phosphorylation are rather slow compared to the events during contraction, and they do not occur to any appreciable extent until the contraction is over, even in a rather long tetanus. This observation has led to the idea of separation of the metabolic processes in muscle into those occurring during contraction and relaxation (including ATP splitting and the creatine kinase reaction) and those occurring after relaxation is over. The former have been called "initial processes," the latter "recovery processes." However, we should note that this division is, to some extent, arbitrary; there is no logical difference between these processes. From many points of view the creatine kinase reaction is a "recovery" process and perhaps even ATP splitting should be considered as such. Furthermore, in other muscles the processes occurring during and after a contraction may be different. In particular, glycolysis and oxidation proceed to a significant extent during contraction in some muscles (for example mouse soleus, Crow and Kushmerick, 1982). This is bound to influence the amount of heat produced, even if other factors are the same. The same applies, *a fortiori*, in considering cardiac and smooth muscle. Some caution is therefore appropriate when comparing the heat produced by different tissues.

4.1.1 METHODS

4.1.1.1 *Thermopiles*

In most of the experiments on heat production by muscle the observations have been made with thermopiles designed and built for this purpose. We now give a brief description of these instruments and how they work.

TABLE 4.I
Thermal emf produced by thermocouples often used in muscle thermopiles

Composition of couples	Thermal emf at 0°C (μV/°C per couple)	Reference
Constantan–chromel	57.7	A. V. Hill (1965)
Constantan–silver	40.8	A. V. Hill (1965)
Antimony–bismuth	122	Kaye and Laby (1973)
Palladium gold–platinum iridium	44.5	A. V. Hill (1965)
Constantan–manganin	39.9	A. V. Hill (1965)
Constantan–iron	52.0	A. V. Hill (1965)

Thermopiles measure temperature change and do not measure heat. To determine how much heat has been produced by the muscle requires other information, including heat capacities of the muscle, thermopile, etc., as will be described later in this section. The basis of operation of a thermopile is the thermoelectric effect, which can be stated as follows: when the temperature is different at different locations in an electrically conducting material, an emf is produced. The size of the emf, per degree temperature difference, is a characteristic of the material (the Seebeck coefficient, α; see Table 4.I).

Figure 4.3A is a diagram of one of the repeating elements in a thermopile: a thermocouple consisting of two dissimilar materials, A and B, connected electrically in series. The thermocouple is connected by leads of material, l, to a device for measuring the emf. There are three junctions between different materials in the thermocouple circuit; the temperatures at these points are designated T_1, T_2, and T_3. The emf V produced depends as shown in Eq. (1) on the α values of each material and the temperature difference experienced by it.

$$V = \alpha_1(T_3 - T_1) + \alpha_A(T_R - T_2) + \alpha_B(T_2 - T_3) \tag{1}$$

The couples are usually arranged in the thermopile so that T_2 is controlled by the temperature of the muscle, and T_1 is the same as T_3 and is held at the constant reference temperature T_R. The consequence of $T_1 = T_3$ is that the α value of the leads does not contribute to V.

$$V = \alpha_1(0) + \alpha_A(T_R - T_2) + \alpha_B(T_2 - T_R)$$

Substituting $\Delta T = T_R - T_2$

$$V = \alpha_A(\Delta T) + \alpha_B(-\Delta T)$$

$$V = \Delta T(\alpha_A - \alpha_B)$$

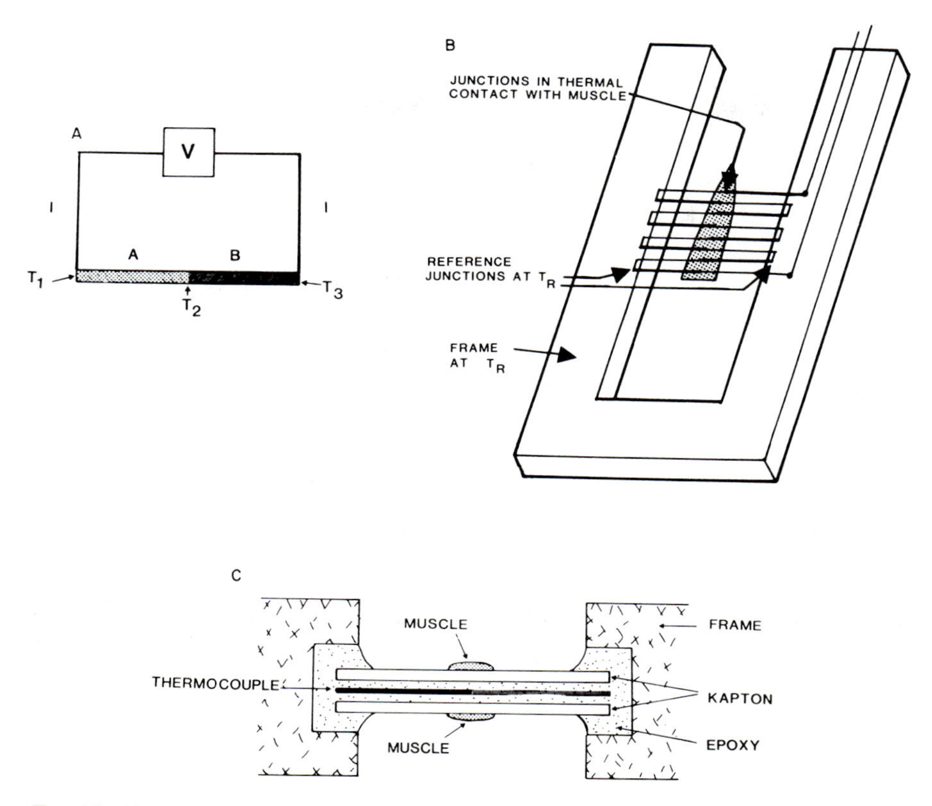

FIG. 4.3. (A) Diagram of a thermocouple of materials A and B connected by leads of material l to a voltage-measuring device. T_1, T_2, and T_3 indicate the different temperatures at junctions between different materials. (B) Diagram of a thermopile consisting of a set of thermocouples on a frame. The position of the muscle is indicated by the shaded area. The junctions in contact with the muscle are at the same temperature as the muscle. The reference junctions and the frame are at the temperature, T_R, that remains constant. (C) Cross section through thermopile and muscle (not drawn to scale). The thickness of the thermopile has been grossly exaggerated compared to its width and the thickness of the muscle, so that the layers of material can be shown.

Thus the size of V depends on the α values for the thermocouple materials (see Table 4.I for examples). As will be clear when specific types of thermopiles are described below, other factors besides the value of α often determine the choice of materials for couples. For example, the choice depends on the ease of actually constructing couples, the size of the couple that it is feasible to make relative to the size of the preparation (this will influence their relative heat capacities), and the resistivity of the materials (this will influence the resistance of the pile, which must be matched to the type of recording system being used).

Figure 4.3B is a diagram of how the thermocouples are usually arranged in a thermopile. In practice, thermopiles usually have many couples rather than the few shown in the diagram. The couples are connected in electrical series and they are arranged so that successive junctions in the electrical circuit are at different temperatures. A thermopile differs from a single couple in the following ways:

1. The size of the signal produced for a particular temperature difference is proportional to the number of couples.

2. The resistance of the thermopile is also proportional to the number of couples. The resistance of each couple depends on the resistivities of the materials from which it is made, its length, and cross-sectional area. The resistance of the thermopile determines the electrical noise it produces and is an important consideration in matching the thermopile to the recording apparatus so as to achieve an optimal signal to noise ratio.

3. Temperature can be observed at more than one location on the muscle by recording the output of different sections of the thermopile. This is particularly useful if heat is produced (or is suspected of being produced) nonuniformly along the length of the muscle, or if the heat capacity of the muscle is not uniformly distributed along its length, or if a temperature gradient exists along the muscle and the muscle is going to move.

It is necessary to electrically insulate the thermocouples from the muscle and the stimulating circuit. This is usually done by fixing thin sheets of insulating material on both sides of the couples (see Fig. 4.3C). For many years sheets of mica have been used, but now plastic film (for example, Kapton, Saran, or Melinex) is preferred when rigid insulation is not needed. Adhesives such as epoxy resins or varnish are used to fill the gap between the couples and to hold the films in place. Another approach that has been successful is the use of Teflon as the adhesive. The couples are placed between sheets of plastic coated with Teflon on the side facing the couples. The sandwich is clamped together, and then heated to about 300°C, which "heat seals" the layers of Teflon to the couples and to each other. These insulating materials contribute, of course, to the total heat capacity of the thermopile. Their heat capacities per unit volume are similar to those of thermocouple materials (see Table 4.II).

A thermopile can be considered as having three main regions as shown in Fig. 4.3C, (1) the part under the muscle, (2) the part between the muscle and the frame, and (3) the frame. It may be useful to briefly outline what actually occurs in each of these locations, and how this influences the design of the various components of the thermopile.

TABLE 4.II

Specific heats of materials

Material	Heat capacity[a]			Notes
	Per unit mass (mJ/°C · mg)		Per unit volume (mJ/°C · mm³)	
Water	4.18	(5)	4.18	
Muscle	3.70	(5)	3.97	
Constantan	0.410	(1)	3.65	60 Cu, 40 Ni[b]
Manganin	0.406	(3)	3.45	86 Cu, 12 Mn, 2 Ni[b]
Silver	0.234	(2)	2.46	
Antimony	0.205	(2)	1.37	
Bismuth	0.121	(2)	1.18	
Epoxy resins	1.0	(1)	1.1–1.40	Araldite[c]
Poly(tetrafluoroethylene)	1.0	(1)	2.14–2.20	Teflon[c]
Polyamide-imide	1.09	(4)	1.49	Kapton[c]
Poly(ethylene terphthalate)	1.25	(1)	1.71–1.73	Melinex, mylar[c]
Poly(vinylidene chloride)	1.34	(1)	2.21–2.30	Saran[c]
Mica	0.88	(3)	2.64	

[a] Numbers in parentheses refer to these references: (1) Kaye and Laby (1973); (2) Wearst (1976–1977); (3) A. V. Hill (1937); (4) Dupont technical information on Kapton, Bulletin HIB; (5) Hill (1965).

[b] Alloy composition.

[c] Trade name.

a. Region of the Thermopile under the Muscle.

Size of the output voltage from the thermopile. For a signal to be produced by a thermopile, the heat produced by the muscle must be shared with the layer of solution around the muscle, the insulating materials, and the thermocouples. This sharing of heat means that the temperature rise is less than if the thermopile were not there. The largest temperature rise (and thus thermopile output) is achieved by minimizing the heat capacities of the solution and thermopile (which act as heat sinks) relative to that of the muscle (which is the heat source).

This relationship is illustrated in Fig. 4.4 and can be described as follows: first, consider a muscle which produces x joules of heat and is not sharing the heat with anything else. The temperature, T_1, will be

$$T_1 = x/(\mathrm{hc_m}) \tag{2}$$

where $\mathrm{hc_m}$ is the heat capacity of muscle. Now consider the situation in which the muscle shares the heat with one other element, the part of the

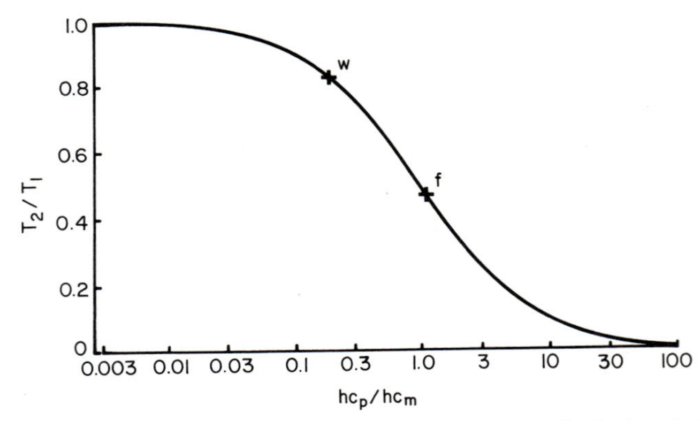

FIG. 4.4. The influence of the relative heat capacity of the thermopile (hc_p) and of the muscle (hc_m) on the thermal signal recorded. T_1 is the temperature rise produced in the muscle alone, and T_2 is the temperature rise produced by the same heat shared between the muscle and the thermopile. The points represent the situation for typical experiments on whole muscle (w) and on a single muscle fibre (f). The heat capacities normalized per millimeter length are

$$\text{w: } hc_m = 12 \text{ mJ/°C} \cdot \text{mm}, \; hc_p = 0.233 \text{ mJ/°C} \cdot \text{mm};$$

$$\text{f: } hc_m = 0.065 \text{ mJ/°C} \cdot \text{mm}, \; hc_p = 0.074 \text{ mJ/°C} \cdot \text{mm}.$$

thermopile in contact with it. In this case the temperature will be less than T_1,

$$T_2 = x/(hc_m + hc_p) \tag{3}$$

where hc_p is the heat capacity of the part of the thermopile with which the muscle shares heat.

Dividing Eq. (3) by Eq. (2) gives the following expression for how the relative heat capacities of muscle and the thermopile influences the temperature (and thus the thermopile output):

$$T_2/T_1 = 1/(1 + hc_p/hc_m) \tag{4}$$

As can be seen in Fig. 4.4, the relation is nonlinear. There are two important consequences:

1. To get signals of adequate size from small preparations it is necessary to use a thermopile with smaller heat capacity, but it need not be proportionally smaller. This is illustrated in the figure by the two points representing a typical whole muscle experiment (w) and a typical experiment with a single muscle fibre (f). The single fibre's heat capacity was about 0.5% of that of the whole muscle, and with the single fibre thermopile, which had a heat capacity 28% of that of the whole muscle thermopile, the signal from the fibre would be more than half as big as that from the whole muscle if the heat produced per unit mass of preparation was the same.

2. For very high or very low values of hc_p/hc_m, changes in this ratio have little effect on thermopile output. So, for example, it may not be worth building a thermopile with a heat capacity less than about 10% of that of the preparation because the technical problems of building it may outweigh the small improvement in the size of the signal.

Speed of response. The speed with which heat is transferred from the muscle to the thermopile determines the accuracy with which the signal from the thermopile reflects the true time course of heat production within the muscle. Best results are achieved (1) by reducing the amount of heat that has to be removed from the muscle (this is another reason for minimizing the heat capacity of the thermopile relative to that of the muscle), and (2) by reducing the distance through which heat must travel to reach the thermocouples. For both reasons thin layers of insulation must be used. When these have been reduced to a minimum, the layers of Ringer's solution and connective tissue on the muscle surface will be the limiting factors. (3) The muscle itself should be thin so that there is the minimum distance through which heat must flow in the muscle itself to replace that transferred to the thermopile, etc.

In thinking about heat flow, muscle is usually considered to be a set of fibres all producing heat in a uniform manner. It is clear, however, that there are a number of fibre types in amphibian skeletal muscle that have different mechanical and structural characteristics (Lannergren *et al.*, 1982; Spurway, 1982, 1984). Often the fibres of a particular type are grouped together in one region of the muscle (see micrographs of frog sartorius muscle in A. V. Hill, 1949c). If different fibre types produce heat at different rates, temperature gradients could be established within the muscle during contraction and the time course of the thermopile output would then reflect not just the production of physiological heat, but also the flow of heat within the muscle. There is some evidence for differences in the time course of heat production by different single fibres (Curtin *et al.*, 1983).

Artefacts and the use of a "protecting region." Many artefacts can affect records of heat production; some of those occurring during muscle shortening can be avoided by using a thermopile that has a "protecting region": a set of thermocouples that are not connected to the voltage-measuring circuit.

An error could be produced if the thermopile is shorter than the muscle, because heat loss is faster from the part of the muscle in contact with the thermopile than it is from the rest of the muscle. Thus, when the muscle shortens a warmer part of the muscle is brought into contact with the thermocouples. The "protecting region" of the thermopile is under the end of the muscle that moves during shortening, and it is long enough that the part of the muscle that lies on the active thermopile after shortening is on the protecting region before shortening. Heat loss through the protecting region is the same as in the rest of the thermocouples and thus no gradients of temperature develop within the muscle (A. V. Hill, 1937). [See also Irving *et al.* (1979) for a description of another, related artefact than can occur when a muscle performs a series of shortenings.]

b. Region between the Muscle and the Frame. When the temperature of the part of the thermopile under the muscle increases, heat flows away from it by various routes. This process of "heat loss" influences the size of the signal produced by the thermopile and its relation to the actual heat production by the muscle. How large and important its effects are depends on the relative rates and characteristics of heat loss and heat production. One of the major routes for heat loss is via the part of the thermopile between the muscle and the frame. Other factors being equal, this heat loss can be reduced by making the couples longer, that is, by increasing the distances between the edges of the muscle and the frame, and by making the couples smaller in cross-sectional area. It is worth noting that the materials used for the couples in general have higher thermal conductivities than the insulating materials and adhesives.

In principle, the size of the electrical signal produced in response to a particular temperature change is larger if the number of couples per unit length of the thermopile is increased (assuming the couples are made from the same materials and have the same cross-sectional area) and the amount of adhesive and insulation material is correspondingly decreased. In practice, however, the signal may not be as much improved as expected because of the more rapid heat loss.

c. Frame. When heat is transferred from the preparation as described above, this heat should be rapidly dispersed in the frame and produce a negligible change in the temperature of the reference junctions. To achieve this, the reference junctions should be in good thermal contact with the frame. The frame itself should be made of a material of high thermal conductivity, and it should have a relatively large heat capacity. If the frame also serves as a mechanical support for the muscle, stimulating electrodes, etc., as well as acting as a heat sink, it will probably be large enough to have a sufficient heat capacity. Frames have been made of anodized aluminum, brass insulated with varnish, or platinum-plated brass.

4.1.1.2 *Types of Thermopiles*

In his book, "Trails and Trials in Physiology," A. V. Hill (1965) gives a full description of the classic Hill–Downing design which had been used in most experiments up to the 1960s. However, since then there have been some major developments that have led to new methods of (1) constructing thermocouples, (2) measuring the sensitivity of thermocouples and heat capacities of thermopiles and muscle, (3) measuring the heat produced along the entire length of the muscle with the "integrating" thermopile, and (4) using Hill–Downing thermopiles for the measurement of heat production by single muscle fibres. Thus, a brief summary of the state of the art of measuring heat production by muscle may be useful (see Table 4.III).

Hill–Downing thermopiles. A main aim of A. V. Hill's efforts was to get an accurate record of the time course of the production of heat during contraction of muscle under various mechanical conditions. To date, the majority of experiments requiring high speed of response and sensitivity have been done on this type of thermopile, and this basic design continues to contribute to our understanding of energetics (Howarth *et al.*, 1975; Gilbert and Matsumoto, 1976; Aubert and Lebacq, 1971; Aubert and Gilbert, 1980; Irving and Woledge, 1981a,b). It is a tribute to Hill and Downing that for the first measurements of the heat produced by a single fibre a similar thermopile (made by J. V. Howarth) was used (Curtin *et al.*, 1983). It is amusing that Hill obviously admired mechanical measurements on single fibres, but thought that "for studying heat production ... no instruments at present conceivable could be sensitive or quick enough" (Hill, 1970, p. 129). In fact, a thermopile essentially the same as his was good enough for the job.

One feature that sets these thermopiles apart from the others is that they are constructed entirely by hand. Individual couples, formed by welding or brazing fine wires together, are flattened to about 8 μm thickness before being assembled into a thermopile. As this requires some unusual skills, thermopiles have not been made in many laboratories. Perhaps for this reason, measurements of heat are not often done in parallel with other types of measurements, even though the information would be complementary. In addition, many quite interesting features of heat production, brought to light in the experiments of A. V. Hill and colleagues, remain to be investigated and understood in terms of recent advances in other aspects of muscle physiology.

Electroplated thermopiles. The idea of forming couples by electroplating a material onto sections of a constantan wire is an old one, and was used by Bozler (1930). He made a silver–constantan thermopile and used it to measure heat produced by a muscle (one rather than a pair of muscles) from the snail. However, it was not used much until N. V. Ricchiuti worked out how to incorporate a very large number of couples, and arranged them so that a muscle could be placed on each of the two faces of the thermopile (Ricchiuti and Mommaerts, 1965). Figure 4.5 illustrates the construction of this type of thermopile.

TABLE 4.III

Some properties of thermopiles of different designs

Name	Type	Composition of couples	Resistance (Ω)	No. of couples[a]	Sensitivity[a] (μV/°C)	Heat capacity[a] (mJ/°C)	Length of thermopile (mm)	Reference
P5	Hill–Downing	Constantan–manganin	68	3.08	119.7	0.53[b]	13.6	Hill (1938)
E5	Electroplated	Constantan–silver	3480	19.26	606.2	1.04[b]	16.2	Rall *et al.* (1976); (E. Homsher, personal communication)
M1	Metal film	Antimony–bismuth	200	4	280	0.25	5.0	Mulieri *et al.* (1977)
VIC	Hill–Downing	Constantan–chromel	223	2.9	168	0.12	6.8	Curtin *et al.* (1983)

[a] Calculated per length of thermopile (mm).

[b] Heat capacities per length for these were calculated from values of the "equivalent half-thickness."

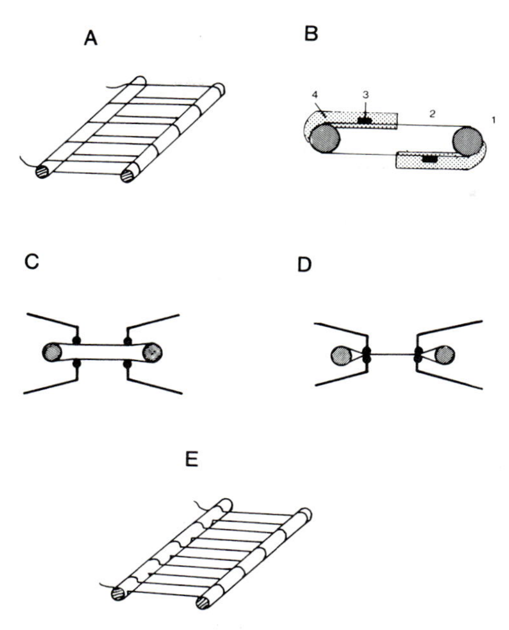

FIG. 4.5. Diagram of some stages in the construction of an electroplated thermopile. (A) Constantan wire is wound around two rods forming a coil. A screw thread on each rod acts as a guide for the wire. (B) A view from the ends of the rods (1). Strips of copper (3) are threaded between the wire to make electrical contact with it. These strips will be connected to the current source during electroplating. Two opposing quarters of the wire coil are coated with wax (4). During plating, areas not covered by wax (2) are plated with silver. After plating the wax and copper strips are removed to expose the constantan. (C and D) A view from the end of the rods. The two sides of the coil of wire are clamped (C) and pushed into a single plane (D). (E) Another view of the thermopile after stage (D). The thermocouple wires are ready to be sandwiched between plastic sheets and held together with epoxy resin. Stimulating electrodes are added and electrical leads connected and the thermopile is then mounted on its frame.

In this type of thermopile, the thermal emf produced at each couple depends on the degree to which the silver, with its high electrical conductivity, is effective in acting as the path for current flow, rather than the alternative path through the constantan, which has a relatively low electrical conductivity. As the thickness of the silver layer increases, the thermal emf approaches its theoretical maximum; however, silver also has a high thermal conductivity, so this results in increased heat loss. In practice, the thermal emf is always less (on the order of 25% less) than the theoretical value of 40 $\mu V/^{\circ}C$ per couple. At one time, the fact that the thermopile sensitivity was not known exactly was a problem that prevented use of these thermopiles in experiments in which the absolute value of the heat production was important. However, as described below, new methods of observing the exact sensitivity of the thermopile have been devised.

Although the output per couple for electroplated silver–constantan is only about half that for constantan–chromel (about 30 vs 60 μV/°C per couple), a large output per unit length of thermopile can be achieved because many couples can be assembled in a unit length of thermopile.

These thermopiles have relatively high resistances compared to Hill–Downing type thermopiles (see Table 4.III).

Thus, the major problems with electroplated thermopiles that were pointed out by Hill (1965) have been solved. The great advantage of these thermopiles is that they are relatively easy to construct. Purpose-made tools are required to construct these thermopiles, but once a set is made, no additional or elaborate apparatus is needed. These thermopiles have been made and used in a number of laboratories (see Homsher *et al.*, 1979; Rall, 1980a,b; Curtin and Woledge, 1981; Gibbs *et al.*, 1966; Yamada and Kometani, 1984.

Metal–film thermopiles. Mulieri *et al.* (1977) have made thermopiles in which antimony–bismuth couples are formed using techniques similar to those used for constructing integrated circuits. A very thin film of material is vacuum-deposited through a mask onto the mica substrate, and then the second material is similarly deposited to form the couples. These materials give a relatively high thermal emf (70 μV/°C per couple), but they are brittle and easily fractured.

The outstanding feature of these thermopiles is their small heat capacity (see Table 4.III), which is an advantage when using small preparations. They have not been widely used, but the first report shows that the heat produced in response to a single stimulation of a preparation weighing less than 2 mg can be measured. It seems likely that this type of design could be developed for use with even smaller preparations.

Integrating thermopiles. The design of the integrating thermopile was devised by Wilkie (1968). One of its main features was a piece of silver (chosen for its high thermal conductivity) placed over the hot junctions. The muscle would then be placed over the silver. The principle is that the heat produced by the muscle (regardless of the distribution of heat production along the length of the muscle) is effectively and rapidly redistributed within the silver so that its temperature is uniform. It is actually the temperature of the silver that is observed.

The second new feature of this design was a resistance wire in thermal contact with the silver. By passing a known amount of current for a known time through the wire, a known amount of heat could be introduced into the system. From this information and observations of the output of the thermopile, the heat capacity of the muscle could be determined. It was essential to have an accurate value of heat capacity, for the absolute value of heat production by the muscle was the point of the experiment.

The major limitation of the integrating thermopiles, which is a consequence of their fairly large heat capacity compared to that of the muscle, is that they respond more slowly than the other types of thermopile discussed here. Thus, they are of limited use for accurately following the time course of heat production (see Wilkie, 1968; Homsher and Rall, 1973; and Rall, 1978).

4.1.1.3 *Working out the Results*

Some calculations must be done to convert records of thermopile output into values of heat production by the muscle. These include (a) subtraction of the signal due to heat produced by the electrical stimulus, (b) "correction" for heat loss and thermopile lag, and (c) conversion of the thermal output signal (μV) into units of heat (mJ).

a. Stimulus Heat. If electrical stimulation of the muscle is used, the stimulus current produces heat as it flows through the muscle and Ringer's solution (which can have a surprisingly high resistance) due to the Joule effect. This produces a temperature rise that is recorded along with that due to contraction of the muscle. The correction for stimulus heat can be based on observations made when the stimulus is applied to an inexcitable muscle. Treatments that cause rigor, such as high temperature or anaesthetics such as Procaine, can be used to make the muscle inexcitable. However, if the muscle swells or otherwise changes its heat capacity as a result of being made inexcitable, the observations will be somewhat in error when applied to normal muscle.

Alternatively, the energy in the stimulus can be calculated from observations of the electrical parameters of the stimulus itself. However, in most cases the amount of heat produced in the muscle by the stimulus is significantly less than the calculated stimulus energy, presumably because a large fraction of the energy is dissipated at or near the electrodes (Woledge, 1968).

b. Heat Loss and Thermopile Lag. Both of these effects cause the temperature change detected by the thermopile (referred to as output) at a particular time to be smaller than that produced by the muscle (input).

Heat loss is the transfer of heat away from the muscle and the part of the thermopile in contact with it. The main route for heat loss is down the thermocouple wires into the frame. Heat loss from a thermopile such as P5 or E5 in Table 4.III with a pair of frog sartorius muscles would occur with a time constant of about 30 s. However, for thermopile VIC and a single fibre the time constant is about 1 s.

Thermopile lag is due to the time required for heat to flow from the muscle into the thermocouples in "contact" with the muscle; the signal cannot be produced until the temperature of these junction changes. Heat must flow through the extracellular space in the muscle, connective tissue, the Ringer's solution layer around the muscle, and the insulation on the thermocouples themselves before reaching the couples. The thermopiles in Table 4.III are all "fast" in that the output lags by only a small percentage behind the input after 1 s of heat production. At very early times (a few milliseconds), however, lag can be quite substantial.

Exponential heat loss. If heat loss is known to follow an exponential time course (which is often the case), then the amount of heat that has been lost up to a time, $r\Delta t$, is

$$\text{heat lost} = (1/\tau)(\tfrac{1}{2})\left(Z_r + 2\sum_{1}^{r-1} Z_r\right)\Delta t$$

where Δt is the time interval of interest and r is the number of the interval, τ is the time constant for the exponential heat loss, and Z_r is the output of the thermopile at the end of r time intervals. (This equation is equivalent to $1/\tau$ multiplied by the area under the record of thermopile output vs time.)

The value of τ is found from separate observations of cooling after the end of a period of heating by, for example, passing current through the thermocouples (the Peltier effect).

A method of correcting records for thermopile lag and heat loss. This method requires that the system behave in a linear way, but the time course of lag and heat loss does not have to be described mathematically.

The relation between heat input from the muscle (F_{time}) and the observed output from the thermopile (Z_{time}) in such a system is

$$Z_n = F_1 b_n + (F_2 - F_1)b_{n-1} + \cdots + (F_n - F_{n-1})b_1$$

where b values are a set of constants and n is an integer multiplied by the duration of the time interval of interest (see Fig. 4.6A). This equation can be expanded to a set of equations as follows:

$$\begin{aligned} Z_1 &= F_1 b_1 \\ Z_2 &= F_1 b_2 + (F_2 - F_1)b_1 \\ Z_3 &= F_1 b_3 + (F_2 - F_1)b_2 + (F_3 - F_2)b_1 \\ &\text{etc.} \end{aligned} \tag{1}$$

To be able to calculate the inputs (the F values) from the observations of the output (Z values), this set of equations must be solved. This can be done by

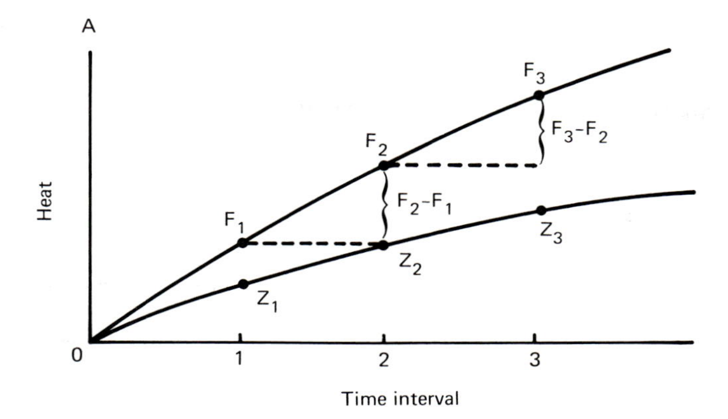

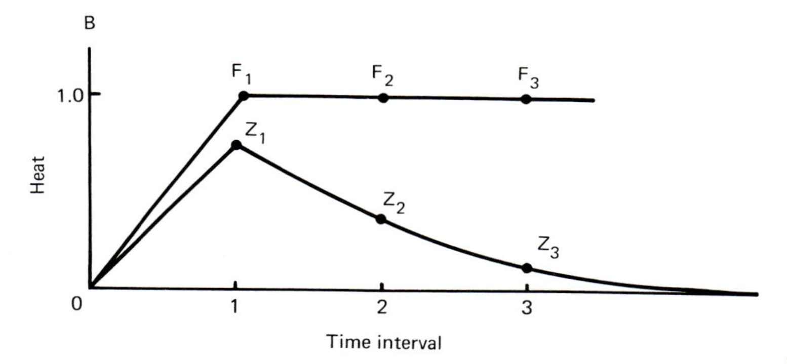

FIG. 4.6. (A) Diagram illustrating the relation between a signal F put into the thermopile (input) and the output signal Z, which is less due to heat loss, for example. (B) Diagram of thermopile response Z to an input F of 1.0 unit during the first time interval. For further explanation, see text.

using the following set of equations in which the a values are functions only of the b values in the above set of equations.

$$\begin{aligned} F_1 &= a_1Z_1 \\ F_2 &= a_1Z_2 + a_2Z_1 \\ F_3 &= a_1Z_3 + a_2Z_2 + a_3Z_1 \\ &\text{etc.} \end{aligned} \tag{2}$$

A control experiment is done in which a unit input ($F = 1.0$) is applied during the first time interval only, so that $F_1 = F_2 = F_3, \ldots = 1.0$ (see Fig. 4.6B). (Note that the duration of the time interval must be the same as that to be used in analysis of the experimental records.) In this special case the relation between the a and b values is simple; the observed values of the thermopile output are the b values. Substituting b for Z in set (2), and $F_n = 1.0$ gives

$$\begin{aligned} 1.0 &= a_1 b_1 \\ 1.0 &= a_1 b_2 + a_2 b_1 \\ 1.0 &= a_1 b_3 + a_2 b_2 + a_3 b_1 \\ &\text{etc.} \end{aligned} \tag{3}$$

which can be solved for the a values

$$\begin{aligned} a_1 &= 1/b_1 \\ a_2 &= (1 - a_1 b_2)/b_1 \\ a_3 &= (1 - a_1 b_1 - a_2 b_2)/b_1 \\ &\text{etc.} \end{aligned} \tag{4}$$

The a values can then be used with other records of thermopile output (Z values) using the set of equations (2).

As mentioned above, the time scale involved in thermopile lag and heat loss are often quite different; lag is important at early times before heat loss begins. If this is the case, the control observations and calculation would be done twice with different time intervals, for example 20 ms for thermopile lag and 1 s for heat loss.

This method has also been discussed by Hill (1937, 1965) and Smith (1973).

c. Calibration of Thermopile Signals by the Peltier Method. In order to convert the signal from a thermopile into an observed amount of heat produced by the muscle, it is necessary to know the value of α (the Seebeck coefficient) for the thermocouples in the pile. This can be used to convert the observed thermopile output into units of temperature change. It is also necessary to know the heat capacity with which the heat produced by the muscle is shared, so that this temperature change can be converted into an amount of heat. Both the heat capacity and α can be obtained by using Peltier heating, that is, the heating produced in the hot junctions of the thermopile by passing current through it. This method was developed by Kretzschmar and Wilkie (1972, 1975).

When current is passed through a thermopile, there are two thermal effects: one is Joule heating, as with any electrical resistance, and is proportional to the square of the current. The other is the Peltier effect, which is directly proportional to the current and is the transport of heat from the cold junctions to the hot junctions, or vice versa. For small currents (for instance less than 100 μA), the Peltier effect is much larger than the Joule effect, with the thermopiles used for muscle research, and Joule heating is negligible. The rate at which heat is transported by the Peltier effect (in watts) is given by $ITn\alpha$, where I is the current flowing (amperes), T is the absolute temperature, n is the number of junctions, and α is the Seebeck coefficient (volts per °C). If α is known the rate of Peltier heating can be calculated from this expression. If α is not known it can be found as explained below.

Transport of heat by the Peltier effect tends to cause temperature changes at both the hot junctions and the cold junctions. The effect at the cold junctions is very small, because these junctions are in good thermal contact with the very large heat capacity of the frame. However, at the hot junctions the heat is distributed only into the heat capacity of the thermopile, and the muscle over the thermopile, and a significant temperature change occurs. This temperature change can be observed by recording from the thermopile and, if α is known, can be expressed in degrees Centigrade. After making allowance for heat loss in the usual way, the rate of change of temperature (dT_0/dt) can be used to find the heat capacity (hc) with which the Peltier heat is being shared.

$$\text{hc} = ITn\alpha/(dT_0/dt)$$

A convenient way to find dT_0/dt would be to take the initial rate of change of temperature, at the start of a period of heating, when heat loss is negligible. However, because the current passed through the thermopile causes a relatively large voltage drop across it, the recording is usually not started until immediately after the heating has stopped. If the period of heating has been long enough for a steady temperature to have been reached, the cooling after it is over, which can be easily observed, has the same time course as the heating at the start, which cannot be observed easily. (This fact is a consequence of the linearity of the system.) Thus, dT_0/dt can be taken as the initial rate of cooling after a sufficiently long period of Peltier heating. If the cooling curve can be described by a single exponential, with time constant τ:

$$T = T_0 e^{-t/\tau}$$

the initial rate of cooling is T_0/τ. Both T_0 and τ can be found easily from a semilogarithmic plot of the cooling curve. If two exponential terms are needed to describe the cooling curve, the initial slope is given by $(T1_0/\tau_1 + T2_0/\tau_2)$.

If the value of α is unknown it can be found with the following method. A series of metal blocks is prepared of known heat capacity (hc_s) which are mounted on the thermopile, one at a time, in place of the muscle. Records of

Peltier heating are made with each of these heat capacity standards, and the initial rate of change of thermopile output (dV_0/dt) is observed. dV_0/dt is related to hc_s as follows:

$$\frac{dV_0}{dt} = \frac{dT_0}{dt} n\alpha = \frac{ITn^2\alpha^2}{(hc_p + hc_s)}$$

$$\frac{1}{(dV_0/dt)} = \frac{1}{ITn^2\alpha^2}(hc_p + hc_s)$$

A graph of $1/(dV_0/dt)$ against hc_s has a gradient of $1/(ITn^2\alpha^2)$ from which α can be found, and from the intercept hc_p can be found.

4.1.1.4 *Chemical Methods*

These procedures are used to determine the extent of various chemical reactions that occur in muscle. As usual, there are a number of methods, and the choice of which to use depends on the purpose of the experiment.

For experiments on recovery processes, the progress of oxidative processes can be followed continuously by measuring the changes in O_2 content of the chamber containing the muscle. Various methods of O_2 measurement have been used. D. K. Hill (1940a) used a differential volumetric method, and the oxygen electrode has been used in a number of different experiments (Kushmerick and Paul, 1976a; Mahler, 1978a; Elzinga *et al.*, 1984). Similarly, the appearance of lactate in the bath can be used to assess the extent of glycolysis (Infante and Davies, 1965; De Furia and Kushmerick, 1977). In these cases, the time resolution is limited by the time for diffusion between the bath and the muscle.

To determine the extent of initial processes such as ATP splitting or the creatine kinase reaction, the procedures are more complicated and usually involve destruction of the muscle. Therefore, two muscles from the same animal are used (for example, the sartorius muscles from the right and left sides of the animal). One muscle acts as a control, for example remaining unstimulated while the other is stimulated, or whatever is required by the experimental design. The difference in the content of substrate (or product) of the reaction of interest in the two muscles gives the extent of the reaction that has occurred as a result of the experimental procedure. A control from the same animal is used because the extent of the reaction that occurs is often quite small compared to the variation between the chemical content of the muscles from different individual animals. This biological variation between animals also means that it is necessary to repeat the experiment on a number of muscle pairs and to use statistics when quantitative information is required.

Three main steps are involved in the chemical experiments on initial processes: (1) stopping the reactions at the appropriate time, (2) extracting the chemicals of interest from the muscle, and (3) measuring the amounts of the chemicals of interest in the extracts.

Stopping the reactions. This is usually done by freezing the muscles rapidly either by immersing them in melting Freon (Cain and Davies, 1964) or isopentane (Mommaerts and Schilling, 1964) cooled with liquid nitrogen, or by flattening them between blocks of metal that have been precooled in liquid nitrogen (Kretzschmar, 1970). (See also the discussion earlier in this chapter about the speed of freezing by these methods.) Incidentally, immersion in liquid nitrogen freezes muscle relatively slowly because the layer of nitrogen around the tissue vaporizes and acts as an insulating layer preventing heat flow out of the muscle.

Extraction. Two main methods of extracting chemicals from muscle have been used extensively (with a number of variations): (1) grinding or pulverizing frozen muscle in acid (such as perchloric or trichloroacetic acid), removing the insoluble muscle residue, and neutralizing the supernatant (for example, Seraydarian *et al.*, 1962; Dydynska and Wilkie, 1966; Kushmerick *et al.*, 1969; Canfield and Marechal, 1973), or (2) soaking the frozen muscle at low temperature (−35°C) for several days in a solution of EDTA and methanol (Cain, 1960; Kushmerick *et al.*, 1969). An ethanol/citrate extraction method has also been used (Seraydarian *et al.*, 1962). These procedures remove small molecules into solution and inactivate most enzymes. However, some enzymes such as adenylate kinase require more drastic treatments to inactivate them (Kushmerick *et al.*, 1969).

Quantitative analysis. A large number of standard biochemical methods have been used. Many are chemical methods or enzymatic analyses for which "kits" are available commercially. In some laboratories these have been adapted for use with an Auto-analyzer system. Various chromatographic methods have also been used. (Papers from laboratories of, for example, Davies, Homsher, Kushmerick, Maréchal, Mommaerts, Wilkie, and Woledge, describe methods routinely and occasionally used.) More recently, high-pressure liquid chromatography (HPLC) (Levy *et al.*, 1976) and isotachophoresis (Woledge and Reilly, 1981) have been used. These have the advantage that a number of compounds can be detected in a single run and since these are less specific than enzymatic methods, unsuspected changes or compounds can sometimes be detected. Undoubtedly, improved methods will be developed.

The time resolution in such studies of initial processes is largely set by the time required for freezing, which is less than 0.1 s. The sensitivity, that is, the minimum extent of reaction that can be detected, depends on the accuracy of the quantitative analyses, the initial difference between the control and experimental muscles, and other errors. Extents of reaction of

0.1 μmol/g wet weight of frog muscle can be detected with these methods (0.1 μmol/g is the SEM typical of the usual size of group of muscles used, 9 or more; see Cain and Davies, 1962; Mommaerts and Wallner, 1967; and Kushmerick and Davies, 1969, for example).

Phosphorus nuclear magnetic resonance (NMR) has been used relatively recently to follow changes in the amounts of phosphorus-containing compounds in muscle (Hoult *et al.*, 1974). An advantage of NMR is that the muscle is not destroyed, so repeated measurements can detect progressive changes in one muscle. In other words, a separate control muscle is not needed and this source of error due to biological variation can be eliminated. Also, measurements can be repeated on the same preparation under the same conditions; the averaged signal to noise ratio can be improved compared to that for a single measurement. However, in practice it can take a long time for a muscle to recover from even a brief contraction. For example, in frog muscle at 0°C, recovery under aerobic conditions takes at least 45 min. NMR is probably better suited to studies of these muscles performing a series of contractions in a steady state in which the extent of recovery between contractions is constant but less than that required for return to the original, precontraction state (Dawson *et al.*, 1977, 1978, 1980).

A property of NMR different from other methods is that the signal to noise ratio depends on the size of the preparation; that is, the size of the signal increases with the size of the muscle more than the noise does. In some experiments on small muscles such as frog sartorius or semitendinosus, it would be possible to use a number of muscles in the chamber being scanned simultaneously, so the signal is essentially the sum from the group of muscle. A problem with this technique is that it gives no information about variation between the different muscles in the group. An alternative is to use larger muscles, perfused to prevent the tissue from becoming anoxic (Meyer *et al.*, 1982), or in studies under anaerobic conditions (Dawson *et al.*, 1978, 1980).

NMR techniques are developing rapidly, but they have not been used to measure small changes, for example those that occur in a single twitch of a frog sartorius. However, clearly NMR is very useful for studies of series of contractions of frog muscle and also contractions of larger preparations, including *in vivo* studies on human subjects (Chance *et al.*, 1980; Ross *et al.*, 1981).

4.2 Heat Production

4.2.1 ISOMETRIC CONTRACTION

We first describe the energy liberation during tetanic contractions at a sarcomere spacing corresponding to maximal overlap of the thick and thin filaments. Figure 4.7 shows the heat production, rate of heat production, and force during an isometric tetanus at 0°C in frog skeletal muscle. During

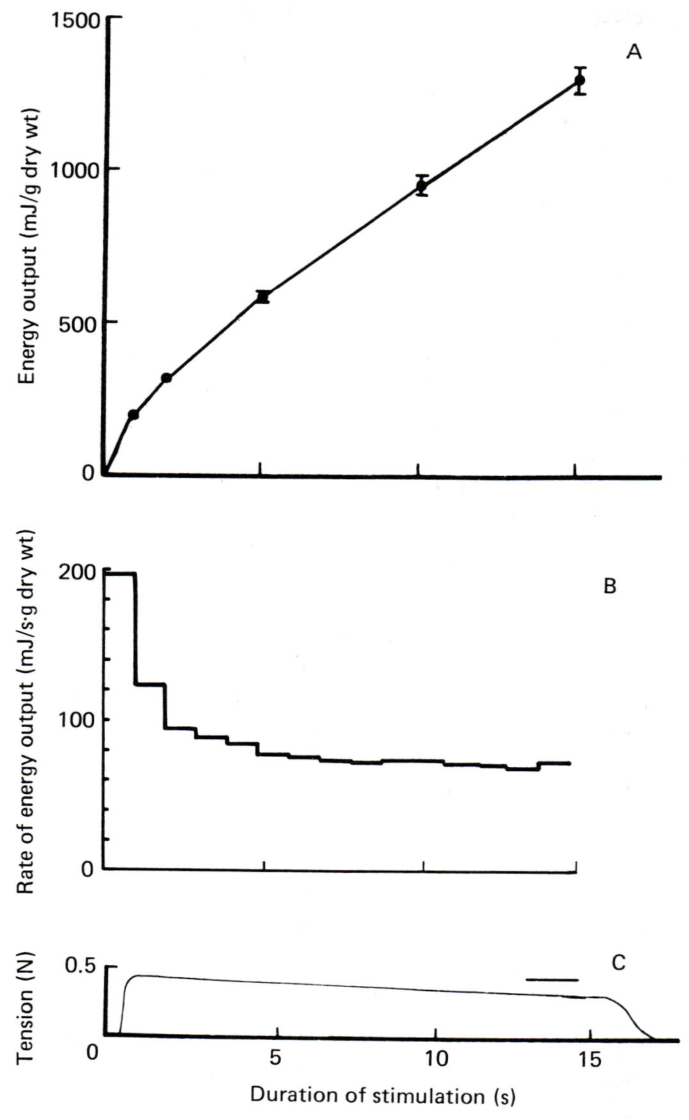

FIG. 4.7. Energy output (A), rate of energy output (B), and tension (C) during an isometric tetanus of frog sartorius muscle at 0°C. These results are from Curtin and Woledge (1979a). (A) and (B) show the mean results for 15 muscles. The bars in (A) show ±1 SE of the mean. (C) shows a representative tension record. The short horizontal line in (C) marks the level of the maximum force exerted. In this muscle, force had fallen by the end of the tetanus to 78% of its maximum value. The mean value for the 15 muscles was 85%. The energy output includes the work done against elasticity in series with the muscle. This quantity is only significant for the time (about 0.5 s) during which tension is being developed (see Fig. 4.18).

the rise of tension, some work is done against series compliance external to the muscle, but once the tension reaches its plateau, work is negligible and heat is the only form of energy produced. Thus during the plateau of tension, one need only measure heat rate to find the rate of energy production. It is striking that although the force exerted by the muscle is relatively constant after 0.5 s, energy is being liberated continuously. In frog muscle at 0°C the rate of heat production is about 14 mW/g. Hill (1949a) called this the "maintenance heat rate." [Although the rate of heat production in Fig. 4.7B appears to be constant after about 5 s of the tetanus, in fact it often continues to decline slowly as does the isometric tension (Fig. 4.7C).] Similar behaviour is found in most muscles that have been investigated, and for this reason it is generally considered that muscles come into a steady state during an isometric tetanic contraction. However, in certain other muscles, for example, rat EDL and chicken posterior latissimus dorsi, there is a sizeable decrease in tension while heat production continues at a constant rate. In these cases the ratio of the rate of heat production to tension is not constant (the value in Table 4.IV is based on maximum isometric tension).

Figure 4.7B also shows that the heat rate, like the tension, does not reach its steady-state value immediately after the start of stimulation. The heat production begins at a very high rate (>40 mW/g) and over 5 s or so falls to a steady rate, whereas the force reaches a steady value within 0.5 s. Aubert (1956) has described the tetanic heat production as the sum of two terms: a time-independent heat rate, which he calls the "stable maintenance heat rate," h_b, and a time-dependent heat rate, h_a, the "labile maintenance heat rate," which is greatest at the beginning of the tetanus and declines to zero in 3–10 s. The time course of the heat rate, h_t, in an isometric tetanus is given by the following equation:

$$h_t = h_a(e^{-\alpha t}) + h_b \qquad (1)$$

where t is time in seconds and α is a rate constant in s^{-1}. Integration of Eq. (1) from the beginning of the tetanus to a time t gives the total heat production H_t as in Eq. (2).

$$H_t = (h_a/\alpha)(1 - e^{-\alpha t}) + h_b t \qquad (2)$$

Aubert's equations are empirical descriptions of the heat production. They do not imply anything about the number or identity of processes contributing to the heat production or their time courses.

Although quantification of the heat production alone provides much useful information, the heat plus work must also be measured for purposes such as energy balance comparisons. While Eq. (2) was devised to describe heat production, it has been found (Curtin and Woledge, 1977; Homsher and Yamada, 1984) that an equation of the same form describes the *energy* liberation. In this case, the value of the $h_a/(\alpha)$ is about 5–10% greater than that which described heat production alone. This is because the rise of force at the beginning of the tetanus has two energetic consequences:

1. As the isometric force rises, the muscle performs *work* against external compliance; this work must be included in the energy production. Estimates of the amount of this work can be made from the force–extension curve of this compliance.
2. Muscles may also possess thermoelastic properties, so that as force rises, some heat is absorbed (see Section 4.2.8). Figure 4.8 shows the heat production in the early part of an isometric tetanus and the effects of making these two corrections.

When the stimulus frequency is increased above the minimum necessary to fully tetanize frog muscle, neither the steady-state tetanic heat rate nor the total amount of "labile heat" production is affected. However, it is reported by Aubert (1956) that the rate of labile heat production may be slightly increased.

When two isometric tetanic contractions are separated by a short time interval, it is found that the steady-state tetanic heat rate in the second tetanus is somewhat reduced compared to the first (Fig. 4.9), but the total labile maintenance heat is reduced to a greater extent. A second tetanus 3 s after the first produces only 35% as much labile maintenance heat as does the first tetanus (Curtin and Woledge, 1977). The greater the duration of time between tetani, the greater is the extent of recovery of the muscle's ability to produce a second burst of labile heat; the time constant for its recovery at 0°C is approximately 2.5 min (Aubert, 1968).

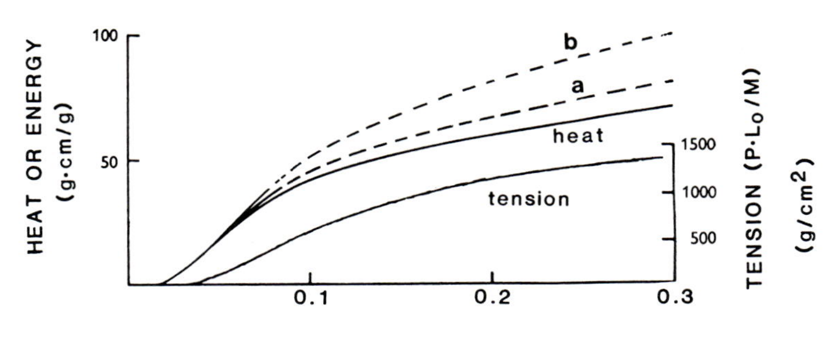

FIG. 4.8. Heat production and tension during the early part of an isometric contraction. The solid lines show the observations made on toad sartorius muscles at 0°C. The dotted lines show the effect of adding, to the heat produced, (a) the thermoelastic heat absorption that may have taken place, and (b) the work done against elasticity in series with the muscle as tension is developed. (From Woledge, 1963.)

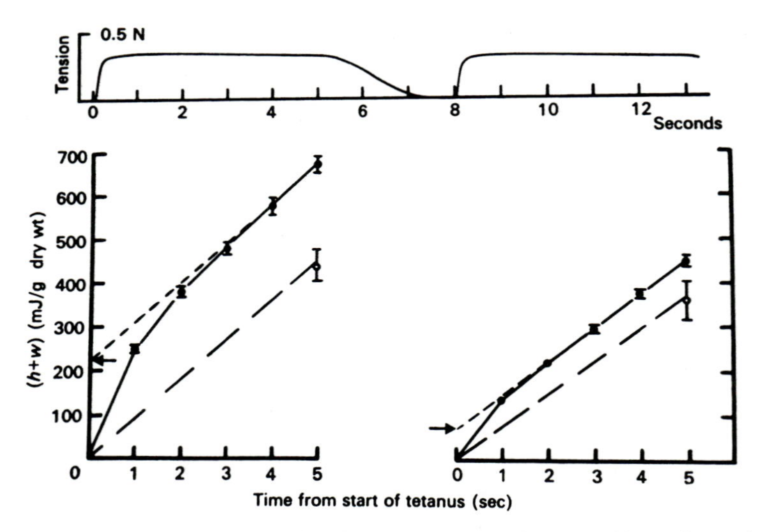

FIG. 4.9. Energy production and tension during two 5-s tetani separated by an interval of 3 s. The results shown are the means of 30 experiments made with sartorius muscles of *R. temporaria* at 0°C. The bars show ±1 SEM. The tension record is a representative example. The mean value of the maximum tension in the second tetanus was 90% of that in the first. The mean rate of stable heat production (shown here by the dotted lines) in the second tetanus was 83% of that in the first, and the amount of labile heat produced (shown by the arrows) was reduced to 35% of the value in the first tetanus. (From Curtin and Woledge, 1977.)

4.2.1.1 *Effects of Changing Extracellular pH*

Aubert (1956) also investigated the effect of extracellular pH on heat production. Bicarbonate/CO_2-buffered Ringer's solutions and also phosphate-buffered Ringer's solutions were used. For conditions more alkaline than neutral (achieved by using low CO_2 partial pressure), the rates of stable heat production (h_b/P) and of labile heat production (α in s^{-1}) were greater than rates at pH values closer to neutral. However, the total amount of labile heat (h_a/α) was not affected by pH. As expected from the fact that CO_2 is highly permeant, the muscles behaved as if the intracellular pH was more acid when they were in bicarbonate/CO_2 buffer than in phosphate buffer at the same pH.

Aubert comments that these changes in heat production were surprisingly small considering that the extracellular pH was being changed by half a pH unit. He concludes that it is unlikely that the reduction in the rate of heat production during a tetanus (as labile heat production diminishes) could be due completely to a change in intracellular pH; the muscle would have to become strongly acid during a tetanus, whereas measurements by

Dubuisson (1939) and D. K. Hill (1940c) suggest that the intracellular pH becomes more alkaline, as expected for net PCr splitting.

4.2.1.2 *Effects of Temperature*

Aubert (1956) also observed the effects of temperature (at 0 and 10°C) on heat and tension production in an isometric tetanus. Heat was, of course, produced much faster at 10°C; the initial rate of labile heat production (h_a) and its rate constant (α) as well as the stable rate were all increased. The total amount of labile heat (h_a/α), however, was increased by only a modest amount ($Q_{10} = 1.5$). The rate of stable heat production was strongly temperature dependent, even when normalized for the tension; the Q_{10} values for h_b and h_b/P were 5.1 and 3.1, respectively. Both of these parameters have a greater Q_{10} than that of maximum isometric tension (about 1.3; see Table 2.III). On the other hand, the value for maximum velocity of shortening [2.67 (Edman, 1979); see Table 2.III] is similar to the value 3.1 for h_b/P.

Table 4.IV shows the results of some other observations of isometric heat production at various temperatures. These seem to agree in general with the observations made by Aubert.

4.2.1.3 *Dependence of Isometric Heat Production on Muscle Length*

As first described in detail by Hill (1925) and by Fenn and Latchford (1933), heat production during isometric contraction, like tension, is greatest at a length close to the length *in vivo*, and is less at lengths smaller or greater than this. Even at a muscle length sufficient to eliminate nearly all overlap between thick and thin filaments, a considerable amount of heat production still occurs during isometric contraction (Fig. 4.10). This shows that processes other than actin–myosin interaction are producing this heat. These processes most probably also contribute to the heat produced at shorter muscle lengths. As Fig. 4.10 suggests, the heat produced at both long and short lengths can be described by Aubert's equation (see above) as the sum of labile and stable components. Therefore, in order to give a more detailed description of the length dependence of isometric heat production we discuss the behaviour of these two components separately.

Stable heat rate. Figure 4.11A compares the results of five investigations of the length dependence of stable heat rate in frog muscle. The different investigators did not all define the "reference length" of the muscle in the same way (see legend, Fig. 4.11), but nevertheless the results are in close agreement with one another. They show that at lengths longer than the reference length the stable heat rate declines linearly to reach a value of about 28% at a length such that filament overlap would be eliminated in

TABLE 4.IV

A. Heat production during a tetanus of muscle from various species at various temperatures

Reference	Species	Muscle	T (°C)	Steady-state rate of energy liberation (mW/g)	Steady-state rate of energy liberation $\left(\frac{mW/g}{N/mm^2}\right)^a$	Labile heat production (mJ/g)
Hill and Woledge (1962)	*R. temporaria*	Sartorius	0	13.5	0.08	46
Hill and Woledge (1962)	*R. temporaria*	Sartorius	17	117	0.54	69
Homsher and Kean (1978)	*R. pipiens*	Sartorius	0	11	0.05	26
Homsher and Kean (1978)	*R. pipiens*	Sartorius	20	170	0.56	—
Curtin and Woledge (1981)	*R. temporaria*	Semitendinosus	0	6.4	0.07	28
Homsher *et al.* (1972)	*R. pipiens*	Semitendinosus	0	7.0	0.05	14
Floyd and Smith (1971)	*R. temporaria*	Slow fibres in iliofibularis	20	6.0	0.023	0
Gibbs and Gibson (1972)	Rat	Soleus	27	21	0.10	0
Wendt and Gibbs (1973)	Rat	EDL	27	13.6	0.60[b]	0
Rall and Schottelius (1973); Canfield (1971); Bridge (1976)	Chicken	Anterior latissimus dorsi (tonic)	20	2–9	0.06–0.1	0
Rall and Schotellius (1973); Canfield (1971); Bridge (1976)	Chicken	Posterior latissimus dorsi (phasic)	20	15–58	0.4–0.8[b]	4
Woledge (1968)	Tortoise (*Testudo* sp.)	Rectus fermoris	0	0.35	0.002	0

Abbott (1951)	Toad (*B. bufo*)	Sartorius	0	4.0	—	10
Woledge (1982)	Terrapin (*Pseudemys elegans*)	Retractor capitis	0	0.96	0.01	9

B. Heat production during twitch of muscle from various species at various temperatures

Reference	Species	Muscle	T (°C)	Twitch heat $(H)^c$ (mJ/g wet wt)	$Pl/H^{c,d}$
Rall and Schottelius (1973)	Chicken	ALD	20	0.79 ± 0.04 (20)	14.5 ± 0.3
Rall and Schottelius (1973)	Chicken	PLD	20	2.05 ± 0.12 (10)	9.4 ± 0.3
Hill (1958)	*R. temporaria*	Sartorius	0	13.8	10.5
Hill and Woledge (1962)	*R. temporaria*	Sartorius	0	15.5	—
Hill and Woledge (1962)	*R. temporaria*	Sartorius	17	14.6	9
Homsher *et al.* (1972)	*R. pipiens*	Semitendinosus	0	9.6–14.0	16.3
Homsher *et al.* (1972)	*R. pipiens*	Semitendinosus	20	5.6–14.8	10.8
Hill (1950a)	*Testudo* sp.	Rectus femoris	0	9.0	10.6

[a] Rate of energy liberation divided by force production.

[b] Calculated from peak tetanic force.

[c] In some cases a mean value ±SEM is given; the number of values is given in parentheses. In other cases, a single value or range is given.

[d] P is the maximum tension produced and l is muscle length.

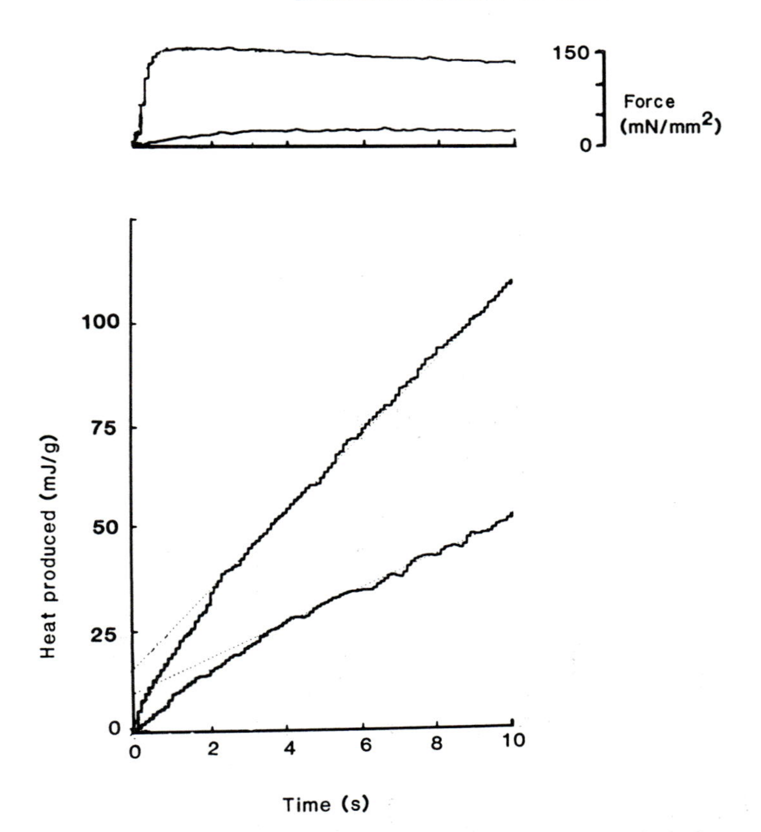

FIG. 4.10. Records of tension (upper traces) and heat produced (lower) during a 5-s tetanus of frog semitendinosus muscle at 0°C. Two sets of records made at different muscle lengths are superimposed. The muscle was near L_0 when it produced the larger tension and heat. It had been stretched before stimulation to a longer length before the tetanus that gave the smaller amount of tension and heat. (From Smith, 1972b.)

most of the muscle. At lengths less than the reference length, the heat rate also falls but less steeply than it does at long lengths. These results are compatible with there being a small range of lengths at which the heat rate is independent of length. However, the results do not demonstrate a plateau region clearly. Abbott (1951) reports a set of observations of toad (*Bufo bufo*) muscle and points out that the plateau in the heat–length relation is more marked in this species than in frog muscle.

The length dependence of the stable heat rate resembles that of tension in that it declines from its optimum more steeply at long lengths than at short lengths, and possibly shows a short plateau region. However, at lengths

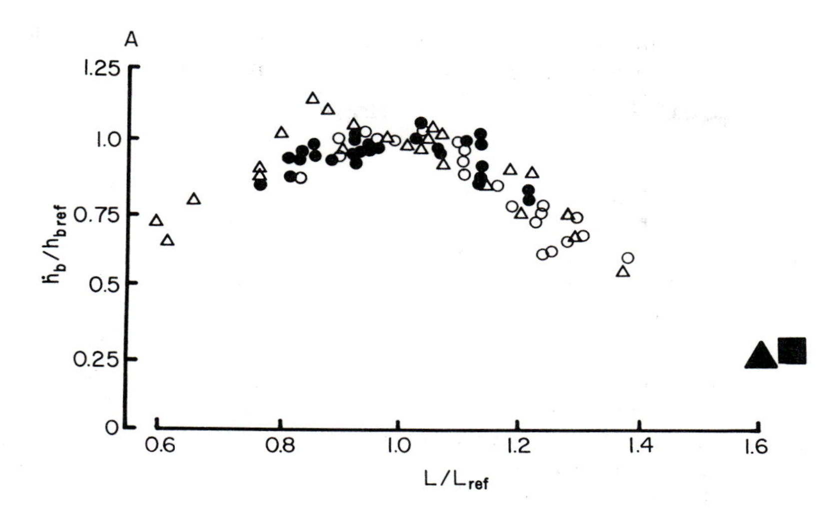

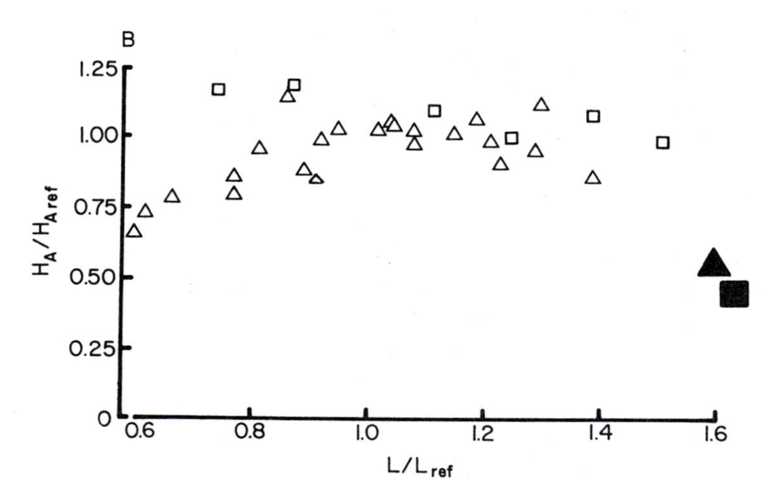

FIG. 4.11. The dependence of heat production on muscle length in an isometric tetanus of frog muscle. (A) shows normalized values of the rate of production of "stable" heat, and (B) shows normalized values of the total quantity of "labile" heat. The reference values of the heat used for normalization are the heat produced at the reference length below.

(□) Smith (1972b), L_{ref} = length for maximum force, semitendinosus, *R. temporaria*. (●) Abbott (1951), L_{ref} = *in situ* length, sartorius, *R. temporaria*. (△) Aubert (1956), L_{ref} = *in situ* length, sartorius, *R. temporaria*. (○) Aubert and Gilbert (1980), L_{ref} = 0.9 of the length giving maximum tension, sartorius, *R. temporaria*. (■) Curtin and Woledge (1981), L_{ref} = length at which resting sarcomere length was 2.2 μm, semitendinosus, *R. temporaria*. (▲) Homsher *et al.* (1972), L_{ref} = *in situ* length, semitendinosus, *R. pipiens*.

below the optimum, the relation of heat rate to length forms a single line, whereas that of tension and length does not. The tension–length relation is steeper at lengths less than 0.8 of the reference length than at longer lengths. In addition, unlike tension, the heat rate does not approach zero at very short and very long lengths.

Is the length optimal for the heat rate the same as that optimal for tension development? The answer to this question is not clear. In a number of experiments the optimal length has been found to be the same for heat rate and force [Aubert (1956) using sartorius, Homsher *et al.* (1972) and Smith (1972a,b) using semitendinosus, Elzinga *et al.* (1982) using toe muscle]. However, in experiments on frog sartorius, Aubert and Gilbert (1980) found that the optimal length for the rate of stable heat production was shorter than that for tension development as had been reported by Hill (1925). This apparent conflict should be resolved by additional experiments on sartorius muscles. It may be relevant to note that Aubert and Gilbert (1980) report not the maximum tension but the final tension at the end of the 20-s tetani. Furthermore, they noted that the ratio of the maximum to final tension in the tetanus did depend on the muscle length.

As the stable heat at optimum muscle length is known to be accounted for by ATP splitting (see Section 4.3.2.1 of this chapter), it is interesting to enquire whether the rate of ATP splitting in a long tetanus shows the same dependence on muscle length as the heat rate does. The answer is that it does. The published studies on this point agree in finding that when muscle are stretched to a length sufficient to prevent most actin–myosin interaction the rate of ATP splitting is, like the heat rate, about 20–30% of its maximum value (Sandberg and Carlson, 1966; Infante *et al.*, 1967; Cain *et al.*, 1962). At muscle lengths too short for tension development the rate of ATP splitting is reduced to 30–40% of its maximum value.

Interpretations. The simple interpretation of the results discussed above is that the stable rate of heat production is due to ATP splitting by two processes: actin–myosin interaction, which is proportional to filament overlap, and calcium pumping by the SR. The rate of this second process is not strongly influenced by muscle length.

The reason that the heat production rate and the rate of ATPase activity should decline at short muscle lengths is not obvious. It could be that some crossbridge sites are prevented from acting because of the distortion of the filament arrangement at short sacromere length caused by the double overlap of filaments. Another possibility is that activation is not maximal at short muscle lengths. Taylor (1974) has shown, by microscopic examination of contracting single fibres allowed to shorten in tetanic contractions, that some myofibrils become "wavy," because activation is not maximal in tetanic contractions at lengths less than 1.7 μm.

Labile heat. The length dependence of the labile heat is different from that of the stable heat rate. The results of four investigations are shown in Fig. 4.11B and although they are rather more scattered than those for stable heat they agree in showing that there is a broad plateau in the relationship extending perhaps from 0.9 to $1.3L_{ref}$. However, at lengths greater than $1.4L_{ref}$ (corresponding roughly to a sarcomere length of 3.1 μm), the labile heat is reduced.

The fact that, over quite a wide range, labile heat is independent of length strongly suggests that it is not derived from actomyosin activity, but comes from some activation process. We suggest in Sections 4.3.2.5 and 4.3.2.6 of this chapter that labile heat may be derived from the binding of calcium ions to parvalbumin. If this is so, the fact that it is reduced at long muscle lengths could be because the internal free calcium ion concentration is raised by stretching the resting muscle to these lengths, so that parvalbumin is already partially saturated with calcium, leaving fewer sites to take up calcium on stimulation.

4.2.1.4 *Effects of Species and Muscle Type*

As seen in Table 4.IV, muscles of different species have quite different steady-state rates of heat production ranging from a low of 0.35 mW/g in the tortoise to 134 mW/g in rat "fast" muscles (EDL). Differences appear even within the same genus; for example, the steady-state heat rate in *R. temporaria* is about 40% greater than that of *R. pipiens.* The same applies to functionally different muscles within the same animal; "slow" muscles have steady-state heat rates about six times less than muscles that produce rapid movements. This difference can be compared with the difference in rate of ATP splitting (Crow and Kushmerick, 1982). In general, the greater V_{max} for shortening, the greater the steady-state tetanic heat rate. For example, tortoise muscle has the lowest value of both parameters and extensor digitorum longus (EDL) of rat has the highest values of both.

Finally, Table 4.IV shows that fast muscle fibres of frogs, toads, and *Pseudemys elegans scripta* are the only types of muscles thus far reported that produce an appreciable amount of "labile heat."

4.2.2 HEAT PRODUCTION DURING AND AFTER SHORTENING

Figure 4.12 is a record of the rate of heat production during a tetanic contraction that includes a period of shortening. During the shortening the heat rate is higher than in the periods when the muscle is isometric. This phenomenon was first noted by Hill (1938), who measured the heat rate (h)

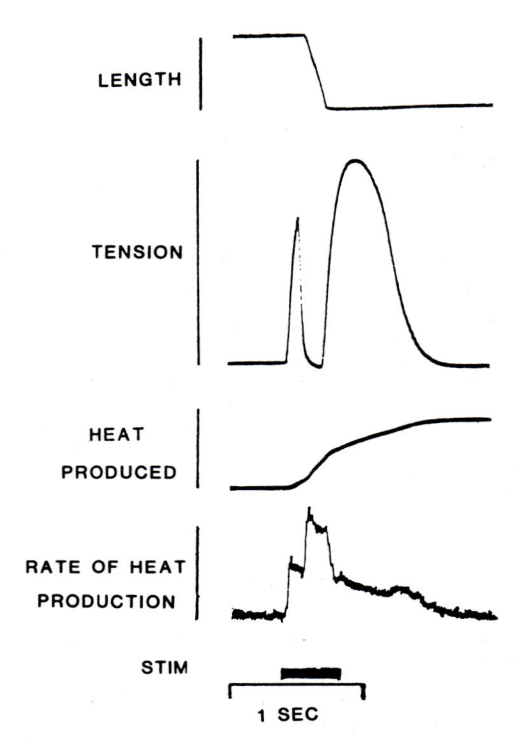

FIG. 4.12. A record of heat production during shortening. Records of a 0.4-s tetanus at 0°C of a pair of frog sartorius muscles (L_0 = 30 mm) during which the muscles shortened a distance of 6 mm at a constant velocity near to V_{max}. Note the large increases in the rate of heat production during the period of shortening. (Unpublished recordings made by R. C. Woledge and V. A. Dickinson in 1971.)

during shortening at different velocities (V). His results and those from some later investigations are shown in Fig. 4.13. Hill originally suggested that h might be a linear function of V, but it now seems that the relation of h to V becomes less steep at high velocities. The shortening heat coefficient α (the symbol a was used by Hill in 1938) can be defined from results plotted as in Fig. 4.13; α is the slope of the chords (two examples are shown in Fig. 4.13B by dotted lines). The shortening heat produced in a time interval t is $\int_0^t \alpha V\, dt$. For an alternative definition of the shortening heat see Homsher and Rall (1973) and Homsher *et al.* (1973).

There are rather few published observations of heat rates during shortening at high velocity because it is difficult to measure the rate of heat production during the necessarily brief period of rapid shortening. For example,

when shortening at V_{max} a frog muscle at 0°C takes only about 0.15 s to traverse the plateau of the length–tension curve. Because of this difficulty, α has been determined more often in a less direct way, which we shall call the "difference method." Measurements are made of the total heat produced over a time interval that includes a period of shortening and also tension redevelopment after the end of shortening. From the total heat is subtracted the heat produced over the same time interval in an isometric contraction. This "isometric control" is usually calculated as a mean of observations at different muscle lengths weighted in proportion to the time the muscle has spent at each length during the contraction with shortening.

There are a number of reasons for supposing that the α values from these experiments would be different from those obtained as described above from the heat rates. For example, with the difference method, the observed heat includes the thermal effects of the drop in tension at the start of the shortening and of the rise in tension after the end of shortening. These events, the fall and redevelopment of tension, are complex; they include internal shortening, possibly thermoelastic changes, and any after-effects of the shortening itself. For this reason the heat rate during redevelopment of tension may be greater than in the isometric control contraction (Hill, 1938, 1964a). In spite of these complexities, the values for α obtained by the difference method are about the same as those obtained from heat rates during shortening (Hill, 1938; Woledge, 1968). A number of useful results have been obtained with the difference method and these are described next.

The shortening heat coefficient and sarcomere length. The value of α depends on muscle length; it decreases with filament overlap at sarcomere lengths greater than 2.2 μm (Lebacq, 1980; Homsher *et al.*, 1983; Yamada and Kometani, 1984). On this particular point the best experiments have been done on frog semitendinosus muscle, which is more extensible than satorius muscle. This is relevant because when unstimulated muscle shortens a significant heat absorption occurs. In more extensible muscles such as semitendinosus this heat effect is smaller, but still not negligible (Fig. 4.14A). It seems likely that this energy change occurs in parallel elastic structures in the muscle and that it is the same in stimulated and unstimulated muscle; however, this point has not been demonstrated experimentally. Assuming that it is the same, the heat changes observed during shortening of unstimulated muscle must be subtracted from those for stimulated muscle before calculating the shortening heat. This procedure leads to the rather simple result that α is proportional to the filament overlap. The ratio of α to the isometric force at the same sarcomere length is constant.

Shortening heat and distance shortened. Hill (1938) concluded that the shortening heat was proportional to the distance shortened (dl). This view was based on a comparison of a series of contractions in which the muscle

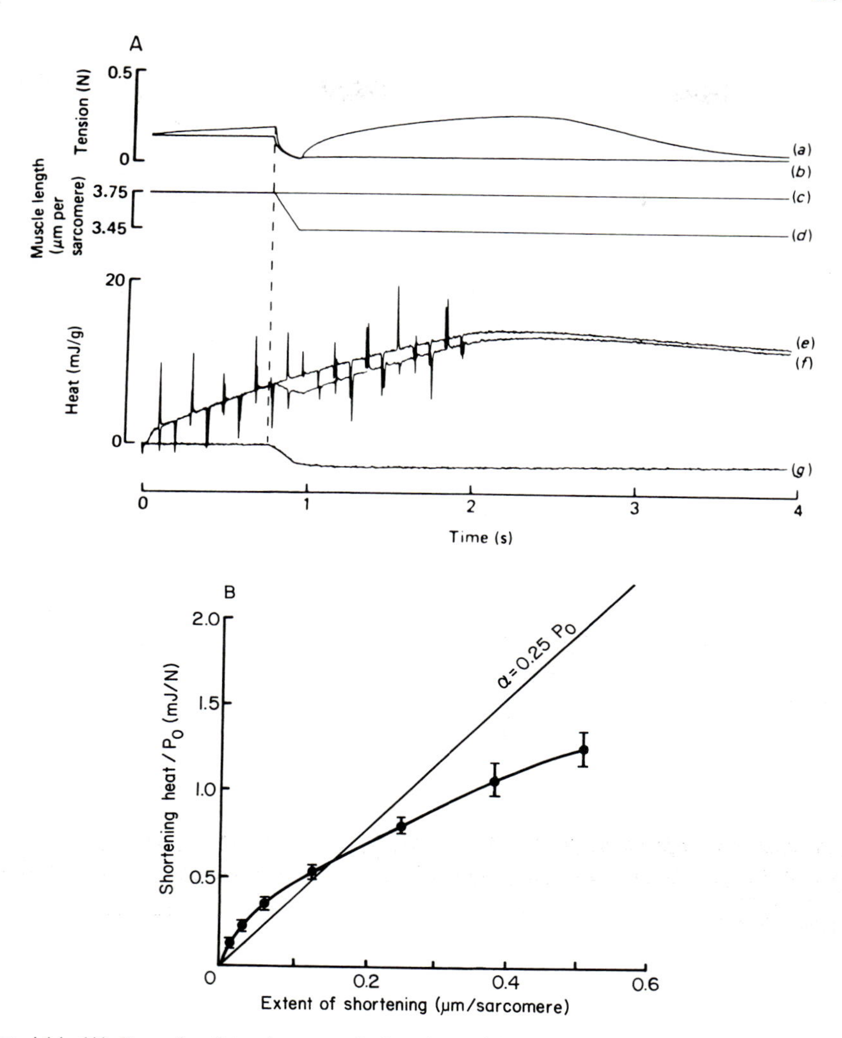

FIG. 4.14. (A) Records of tension, muscle length, and heat in an experiment on frog semitendinosus muscles at 0°C. Record (e) shows the heat produced during a 2-s tetanus under isometric conditions and constant muscle length (c); (f) shows the heat produced during a 2-s tetanus with shortening (d); (g) shows the heat absorption during release (d) of the resting muscle. Records (a) and (b) show the tension during release of tetanized and resting muscle, respectively. (From Homsher *et al.*, 1983.) (B) Relation between shortening heat (normalized by the maximum isometric force) and distance shortened during isovelocity releases. Each point is the mean ($\pm$SEM, n = 4–10) for a particular distance shortened. Frog sartorius (*R. temporaria*) at 0°C. The straight line is the shortening heat predicted if the shortening heat coefficient is assumed to be constant and equal to $0.25P_0$. (Based on results of Irving and Woledge, 1981a.)

shortened different distances from the same initial length, which was probably a sarcomere length of 2.5 and 3.0 μm. Irving and Woledge (1981a) studied a series of shortenings ending at the same final length. They found the relationship between the shortening heat and *dl* to be nonlinear (Fig. 4.14B). The nonlinearity was shown not to be due to the variation of α with sarcomere length. Why are the results different than Hill's? There are at least two possibilities. In Hill's experiments, (1) the shortening heat for small values of *dl* was less than for larger values of *dl* because the filament overlap was less and (2) the shortening heat for small *dl* was underestimated because account was not taken of the heat absorption during length change in unstimulated muscle, which, as mentioned above, probably also takes place in stimulated muscle.

As shown in Fig. 4.14B, in the experiments of Irving and Woledge (1981a), a significant amount of shortening heat was produced when the distance shortened was small. This contrasts with the results of Gilbert and Matsumoto (1976), who could not detect any shortening heat with similar, small releases. The reason for this discrepancy is not known.

The nonlinearity in the shortening heat curve in Fig. 4.14B can be described by stating that in a contraction containing two contiguous periods of shortening, the shortening heat associated with the second is reduced, due to the occurrence of the first. The reduction is found to be less if there is an interval of isometric contraction between the two period of shortening (Irving and Woledge, 1981b). After an isometric interval of 0.5 s, the recovery is almost complete. [An earlier, brief report of much longer recovery time (Dickinson and Woledge, 1973) was found by Irving *et al.* (1979) to be due to a technical error.]

Shortening heat and velocity of shortening. As already mentioned, Hill originally suggested that shortening heat is independent of velocity (and therefore force) but, in fact, his mean values (Hill, 1938; his Table III) suggest that shortening heat might be less at high velocities. In later experiments Hill (1964a) confirmed this dependence on velocity, and the results are shown in Fig. 4.13F and G, where they can be compared with those obtained directly from the heat rates (Fig. 13A–E).

The nonlinear relation of shortening heat to velocity is probably related to the nonlinear dependence of shortening heat to distance shortened. It has been shown that shortening heat recovers during an interval between two periods of rapid shortening, so it might also recover to some extent during shortening itself. If so, the reason that α is less at high velocities might be that there is less time for this continuous recovery. This idea suggests that the results of experiments on the relation between α and v might depend on the value of *dl* used; and the result of experiments on the relation of shortening heat to *dl* might depend on the value of v. These points should be investigated.

Shortening heat and duration of stimulation. An experiment described by Aubert (1956) suggests that the shortening heat is the same whether the shortening occurs early, after 1 s of stimulation, when the labile heat rate is high, or after 5 s, when labile heat production has stopped.

Shortening heat at different temperatures. Most observations of shortening heat have been made at 0°C. Hill (1938) reports some observations at temperatures between 9 and 19°C which show that the value for α is increased by up to 50% at the higher temperature. The isometric tension P_0 is also greater at higher temperature so that the ratio of α/P_0 is independent of temperature.

Shortening heat in different species. Most work has been done on frog muscle in which shortening heat is a conspicuous phenomenon. The value of the α is about $0.25P_0$ in frog muscle. In tortoise muscle the value of the shortening heat constant is much less, $0.04P_0$. However, shortening heat is still a conspicuous feature of heat production by tortoise muscle because the isometric heat rate is also much less than in frog muscle. It has been reported that shortening heat is not conspicuous in chicken muscle (Rall and Schottelius, 1973) nor in rat soleus and EDL muscles (Gibbs and Gibson, 1972). Detailed studies of the amount of shortening heat in these and other muscles should be made. It would be interesting to have a well-documented example of a substantial amount of shortening without shortening heat.

Energy output during shortening. When muscle is shortening, at velocity V under tension P, external work is being done at a rate equal to PV. The rate of energy output by the muscle is obtained by adding this power output to the observed rate of heat production. Measurements of the rate of energy output obtained in this way are shown in Fig. 4.15 as a function of P and also V. Hill (1938) suggested on the basis of calculated (rather than measured) shortening heat and observations of power output (his Fig. 11) that the rate of energy output is a linear function of P, as shown by the straight line on Fig. 4.15A. The calculation of the shortening heat was based on the assumption that α was independent of load. In fact, Hill's 1938 results suggest that α increases with load, as was later confirmed (Hill, 1964a). As can be seen in Fig. 4.15A, the measurements of heat and work at low forces (high velocities) all fall below the expected straight line, and so it is clear that energy is not a linear function of force.

It seems from the results in Fig. 4.15 that the maximum rate of energy liberation occurs in the region of $0.3P_0$ ($0.4V_{max}$); the rate observed at zero load is only slightly less. The maximum rate of energy liberation is about five times the rate of energy liberation in isometric contraction of frog muscle (in tortoise muscle it is eight times greater).

The form of the relation between energy output and load may depend on the extent of shortening that has already occurred at the time the measurements are made. This point needs further investigation.

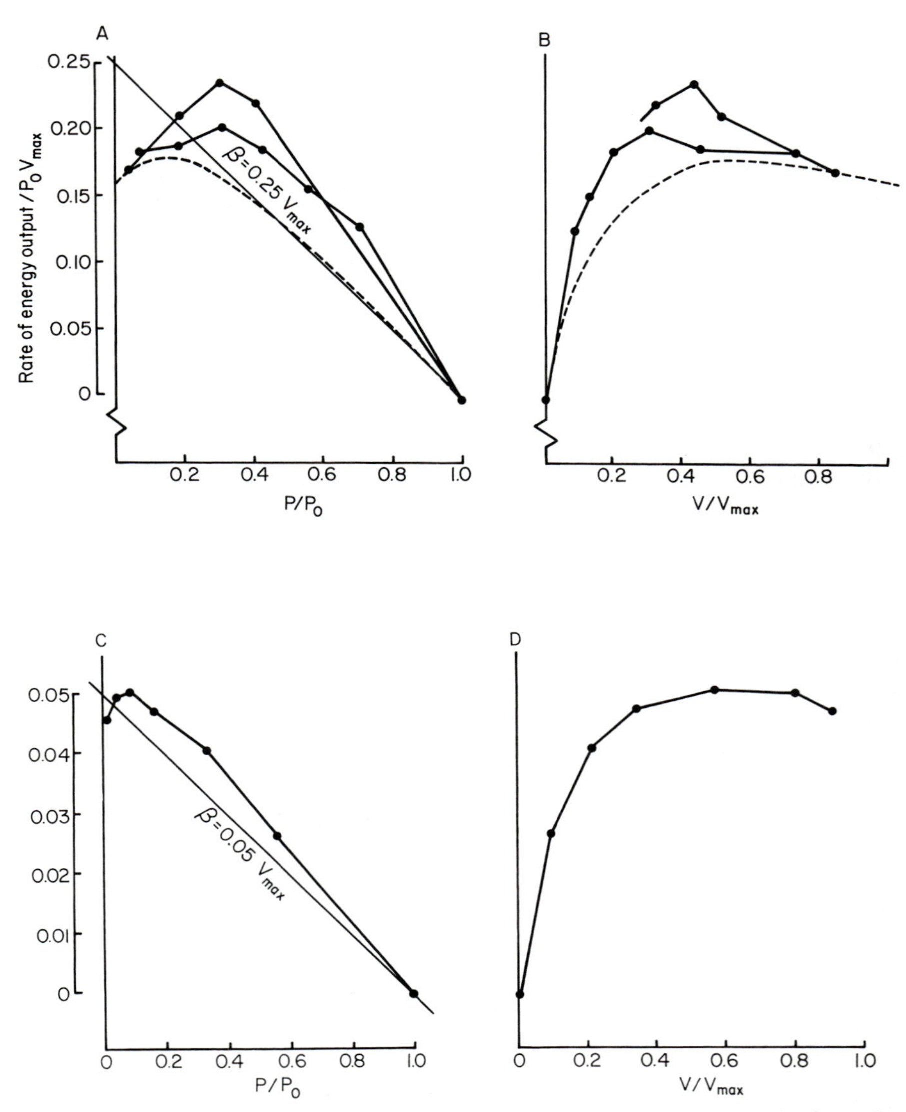

FIG. 4.15. The rate of energy output (normalized by $P_0 V_{max}$) as a function of relative force (A and C) and of relative velocity (B and D). (A) and (B) were recalculated from the tabulated results and figures for frog sartorius, 0°C, in Hill (1964a). (C) and (D) are for rectus femoris of *Testudo* at 0°C (Woledge, 1968).

The rate of energy output probably has a proportional relation to the filament overlap at sarcomere lengths beyond 2.2 μm because this is the case for the isometric force and heat rate and for the shortening heat constant, but there are no direct observations on this point.

Work output as a proportion of total energy output. The ratio of work rate to rate of energy output is shown in Fig. 4.16 as a function of load. In the region of $P = 0.5P_0$ the ratio has a broad optimum value of 0.45 for frog muscle and 0.77 for tortoise muscle. Measurements have also been made of the ratios of the *total* amounts of work to the total energy output in short tetanic contractions; the maximum value for frog muscle is 0.39 (Hill, 1939) and 0.70 for tortoise muscle (Woledge, 1968).

The ratio of work to energy output was called "efficiency" or "mechanical efficiency" by Hill. It should be clearly distinguished from the thermodynamic efficiency (work/free energy output, see Section 4.3.4 of this chapter). The basis of this distinction, which was clearly set out by Wilkie (1960), is that the energy output includes the entropy change ($T\Delta S$) as well as the free energy output.

4.2.3 HEAT PRODUCTION WHEN ACTIVE MUSCLE IS STRETCHED

When an active muscle is stretched, its mechanical and energetic behaviour is strikingly different from that during shortening or in isometric contraction and, after the end of the period of stretching, differences persist. This is perhaps not surprising because, in all of the experiments discussed here the extent of filament sliding during the stretch is large compared to the distance over which it is thought crossbridges can remain attached. Thus, during the stretches, crossbridges are probably being broken off forcibly by a process different from the mechanism of detachment during shortening. During and after the stretch, the behaviour of the muscle can be expected to reflect the properties of bridges broken off in this way. The mechanical behaviour of active muscle during and after stretches has been considered in Chapter 2, Section 2.2.2, and the chemical events are considered in Section 4.3.5 of this chapter. Here we review measurements of the heat produced by, and work done on, active muscle during stretches. Probably the most interesting observations are of the change of internal energy in the muscle, that is, the sum of heat produced (h) by the muscle and the work done by it (w). During stretches a quantity of work (w_{ap}) is done *on* the muscle by the apparatus; this is equivalent to a negative amount of work done *by* the muscle. The change in internal energy is thus given by $h - w_{ap}$. (We are following our usual sign convention according to which a decrease in internal energy of the system is a positive quantity.)

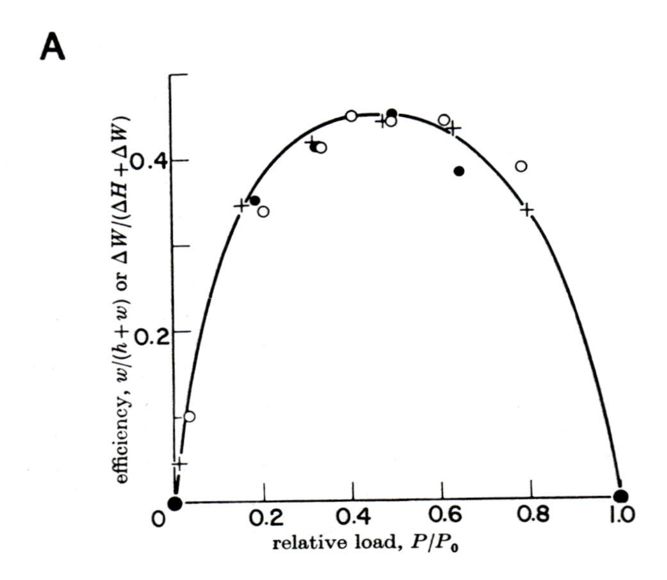

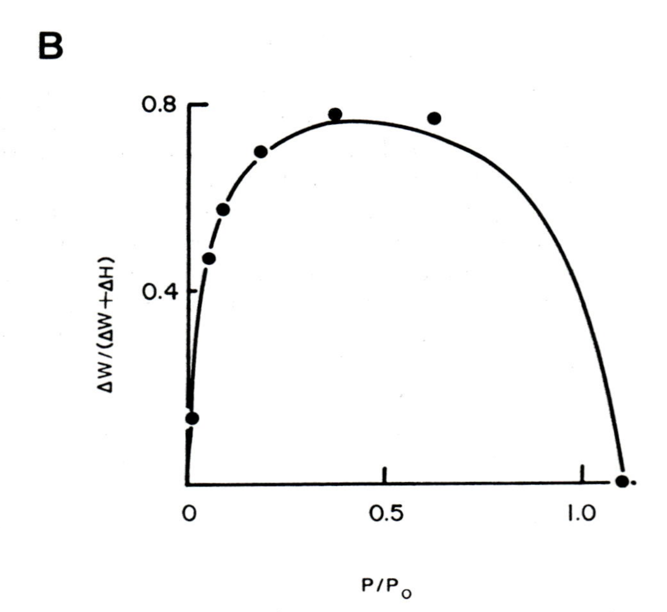

FIG. 4.16. Work/(heat + work) as a function of load in tetanic contractions at 0°C. The work and heat were measured during a brief interval while the muscle was shortening isotonically. (A) Hill (1964b), results of three experiments are shown by different symbols, sartorii, *R. temporaria*. (B) Woledge (1968), rectus femoris, tortoise (*Testudo*).

Before considering the experimental results it seems worthwhile pointing out some errors that are more likely to arise in these experiments than in others. In at least some cases the published data do not include enough detail to make clear whether or not the results have been influenced by these errors. In any attempt to repeat and extend these measurements special attention should be given to these pitfalls:

1. Inaccurate measurements can result from heat being produced nonuniformly within the muscle. Two separate errors can occur.

A. Heat may be produced nonuniformly along the length of the muscle. As temperature change is usually measured from only part of the muscle length, the measured change may not be representative of the muscle as a whole. This error can occur in any type of contraction, but is likely to be worse during stretch, because the muscle may "give" (suddenly lengthen by a relatively large amount) in one small region causing the release of a large amount of energy there. For an example of this behaviour, see the discussion of relaxation heat by Hill and Howarth (1957b). (Muscle is in fact stretched during relaxation from "isometric" contractions.) This error can be reduced by measuring heat production from a longer region of the muscle.

B. Nonuniform heat production through the thickness of the muscle can produce temperature gradients which persist for up to 1 or 2 s in a thick muscle and makes measurements of the time course of heat production uncertain. At longer times thermal equilibrium has taken place and accurate measurements can be achieved. For this reason studies of total heat production in a contraction–relaxation cycle are likely to be unaffected. Examples of this type of error, again studied in relaxation, are given by Hill (1961a,b). The error can be reduced by using thinner muscles; the time required for thermal equilibration is reduced in proportion to the square of the muscle thickness.

2. Another problem is the accuracy of the calibration for heat production. In studies of stretches, the change in internal energy in the muscle is found by subtracting the work done on the muscle from the observed heat production. As the difference between these two quantities is usually small compared to the absolute values, it is critical that the calibration for both types of measurement be accurate. The accurate calibration of the work measurements is easy, but, as Hill and Woledge (1962) have pointed out, incorrect heat calibrations have in the past affected a number of studies, including those on stretches by Fenn (1924). The "Peltier method" of Kretzschmar and Wilkie (1975) has now become established as the easiest way to obtain accurate calibration and should be used in any future work on stretches.

Four experimental studies have been published of the energy changes during stretches of tetanically stimulated muscle.

Hill (1938) and Abbott and Aubert (1951) used isotonic stretches in which the velocity was initially fairly high (this is the "give" studied by Katz, 1939; see Chapter 2, Section 2.2.2), but became progressively slower, eventually falling to less than 1% of V_{max}. During this slow lengthening the rate of heat production is found to be less than in isometric contraction. An example is shown in Fig. 4.17A. Of course the rate of production of energy ($h - w_{ap}$) is reduced even more (line 3 in c), but it does not become less than zero. As can also be seen in Fig. 4.17A, during the initial rapid lengthening at

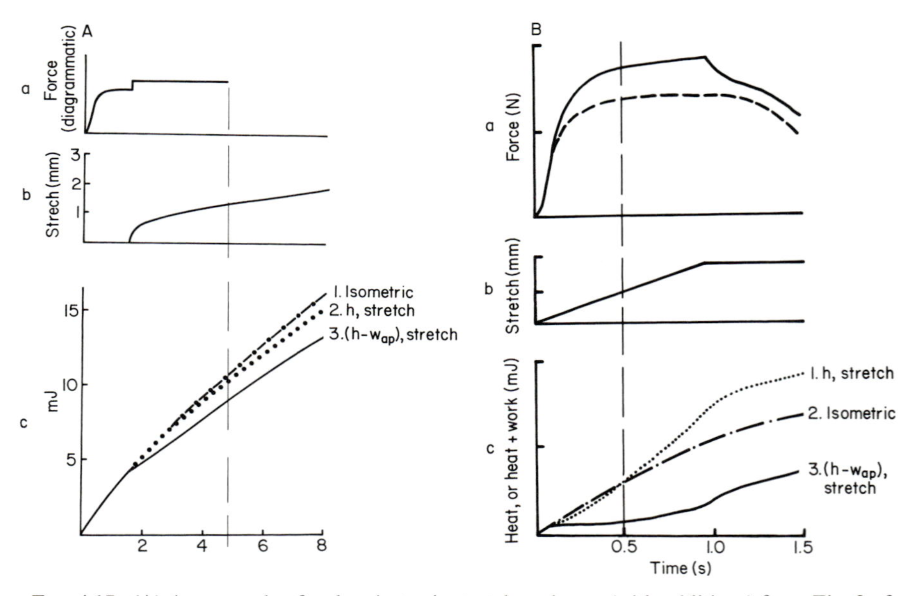

FIG. 4.17. (A) An example of a slow isotonic stretch, redrawn (with additions) from Fig. 2 of Abbott and Aubert (1951). The muscles, frog sartorii at 0°C, were stimulated tetanically for 8 s. After 2 s of isometric contraction an isotonic load greater than the isometric force was applied. Force was not recorded in these experiments, but it must have followed a time course similar to (a). When the isotonic load was applied the muscle began to lengthen (b), first rapidly then more slowly. The heat production recorded during this contraction is shown by line 2 in (c). The heat + work produced by the muscle is shown by line 3. These lines can be compared with the isometric heat production (line 1). (There is of course no work done in the isometric contraction, except during tension development.) (B) An example of isovelocity stretch at moderate speed, about 0.1 V_{max}, redrawn (with additions) from Fig. 5 of Abbott *et al.* (1951). The muscles, frog sartorii at 0°C, were stimulated tetanically for 1 s, and stretched by 3 mm as shown in (b). The force recorded is shown in (a) compared to an isometric contraction. (c) shows the heat (line 1) recorded and ($h - w_{ap}$) (line 3) calculated from the heat and the work done in the stretch. These results can be compared with the heat recorded in the isometric contraction (line 2).

the start of the stretch, the rate of heat production is higher than it is later in the tetanus (it is equal to the isometric rate in the example shown).

Abbott *et al.* (1951) used isovelocity stretches to study the effects of stretching at higher rates (about 10% V_{max}). An example of their results is shown in Fig. 4.17B. For most of the stretch the rate of heat production is greater than in an isometric contraction although at the start of the stretch it is less. The rate of production of $h - w_{ap}$ is very low, almost zero, throughout the stretch. In contrast to these results it was found that stretches applied during relaxation did not reduce the rate of production of $h - w_{ap}$ below that during relaxation after a tetanus without stretch.

Hill and Howarth (1959) studied the heat and work produced during isovelocity stretches at about 50% V_{max}. Figure 4.18A shows an example of their results. The heat production was considerably greater than in an isometric contraction. The most striking feature of the results is that the rate of production of $h - w_{ap}$ (slope of the line in the Fig. 4.18A) was actually negative during stretch, that is, the internal energy of the muscle was increasing. The interpretation of results such as those shown in Figs. 4.17A, 17B, and 18A is complicated by the fact that (1) some of the work done by the apparatus was used to stretch compliance in the tendon and the apparatus itself, which are in series with the muscles, and (2) there may be a thermoelastic heat absorption (see Section 4.2.8 of this chapter). These complications are largely avoided by considering the rates of heat and work production at times when tension is almost constant, as has been done in constructing Fig. 4.18B, which shows the rates of work, of heat, and of energy production as a function of the rate of stretch. Only with very slow stretches is the heat production less than the isometric rate. However, the rate of production of $h - w_{ap}$ decreases as the rate of stretch increases and can be strongly negative. Surprisingly, there is no published experimental study of the influence of speed of stretch on heat and work production and thus the real form of the lines sketched in Fig. 4.18B is not known.

Another way in which the complications due to external elasticity and thermoelasticity can be avoided is to measure the total heat produced in a contraction–relaxation cycle, and to investigate the effects of stretch on this quantity. This has the further advantage that the results are less likely to be affected by errors due to the nonuniformity of heat production discussed above. Experiments of this type have been reported by Fenn (1924), Abbott *et al.* (1951), and Hill and Howarth (1959). It is clear from all these results that the total energy can be reduced considerably below the isometric level by stretching the muscle. However, it cannot be reduced below zero, even though in these contractions, usually twitches or short tetani, the *rate* of production of $h - w_{ap}$ does fall below zero, as in the longer tetani discussed above and shown in Fig. 4.18A. It could be that the quantity is always

positive for contraction–relaxation cycles because an endothermic process occurring during the stretch is reversed quickly when the stretch is over, perhaps during relaxation.

This suggestion is supported by the observation that after the end of a stretch the rate of production of energy is greater than in an isometric contraction. This can be seen, for example, in the results shown in Figs. 4.17B and 4.18A, and was also observed by Curtin and Woledge (1979b). In that experiment, illustrated in Fig. 4.19, the stretch occurred between 1.0 and 1.08 s after stimulation began. In the interval from 1.5 to 2.5 s, the heat production was higher than in the isometric control. The extra heat production was too large to be due to the degradation of the work done on the muscle by the elasticity in series with it. One possibility is that this heat is due to the reversal of an endothermic process caused by the stretch. This experiment, in which some measurements of ATP splitting were also made, is discussed further in Section 4.3.5 of this chapter.

What is happening during stretch to cause the decreases in the rate of production of $h - w_{ap}$? Three suggestions have been made.

1. The process(es) that produce the heat during isometric contraction are prevented. It seems very plausible, for instance, that the ATP splitting by myofibrils, one of the major contributions to the isometric heat production, might be prevented by stretch. However, this prevention could not explain, in a simple way, how the rate of production of $h - w_{ap}$ becomes negative.

2. These processes are not just prevented, but actually reversed. There are other examples of the reversibility of ATP-driven energy conversion processes, so this also should be considered a plausible suggestion.

3. An extra endothermic process occurs during stretch. An example of such a process would be the storage of work in a mechanically strained structure.

FIG. 4.18. (A) An example of isovelocity stretch at high speed, about $0.5V_{max}$, redrawn (with additions) from Fig. 6 of Hill and Howarth (1959). The muscles, toad (*Bufo bufo*) sartorii at 0°C, were stimulated tetanically for the whole of the period shown and were stretched by 15% L_0 at about $0.5V_{max}$ as shown in (c). The tension recorded is shown in (a). (c) shows that heat recorded (line 1) (not given in the original figure, but reconstructed from the other information there) and the $(h - w_{ap})$ (line 3) calculated by Hill and Howarth from the heat and the work done by the apparatus. These lines can be compared with the heat recorded in the isometric contraction (line 2). (B) A possible relation between speed of lengthening and rate of heat and $(h - w_{ap})$ production. As no systematic study of this relation has been published, this figure was constructed from the examples in Figs. 4.17A, 17B, and 18A, by measuring the appropriate rates at the times shown by the vertical broken lines in those figures. The rates have been normalized by dividing by the isometric heat rate.

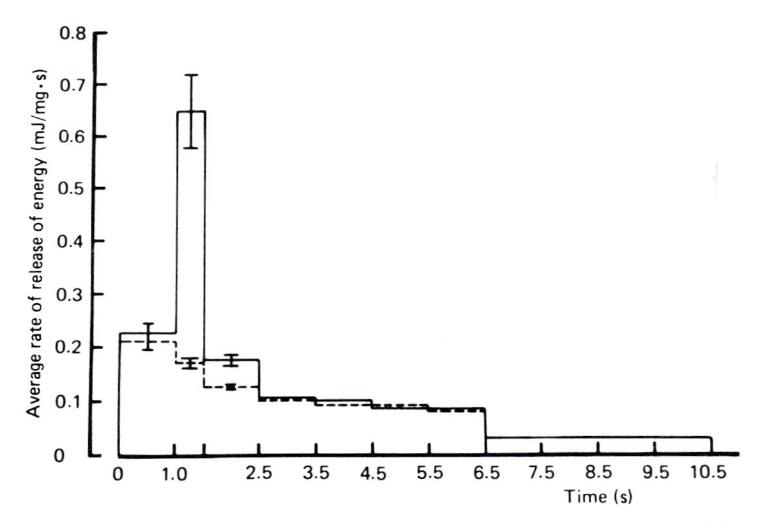

FIG. 4.19. The average rate of release of energy during time intervals during a 6.5-s tetanus and relaxation of frog sartorius muscle at 0°C. The dotted line shows the results for a tetanus that was isometric throughout; the solid line shows results for a tetanus in which the muscle was stretched by 3 mm during part of the interval from 1.0 to 1.5 s. (From Curtin and Woledge, 1979b.) Means of 15 observations; bars show ± 1 SE where this is large enough.

As we describe below there is clear experimental evidence that (1) occurs but that (2) does not occur. This leaves (3) as the most likely explanation of the negative rate of production of $h - w_{ap}$. The observation that this process seems to be reversed before or during relaxation is compatible with the idea that it is the storage of work in some stressed component in the muscle. Cavagna and Citterio (1974) and Edman *et al.* (1978) have suggested, on the basis of mechanical observations, that work can be stored in muscle in this way.

If it is correct that the effect of stretch in a complete cycle of contraction and relaxation is to prevent (and not reverse) ATP splitting by the contractile apparatus, it would be expected that the part of the heat production derived from activation processes, about 30% of the total in a twitch (see Section 4.2.7), would remain unaltered by a stretch. The results of Hill and Howarth (1959), however, suggest that the $h - w_{ap}$ in a twitch can be reduced to less than 30% (see Fig. 4.20). Either this result is in error, perhaps because of one of the difficulties listed above, or the interpretation we are suggesting here is wrong. Hill writes (1965, p. 362) that "Howarth and I are obstinately convinced that the experimental results are in general correct." The experiments should be repeated.

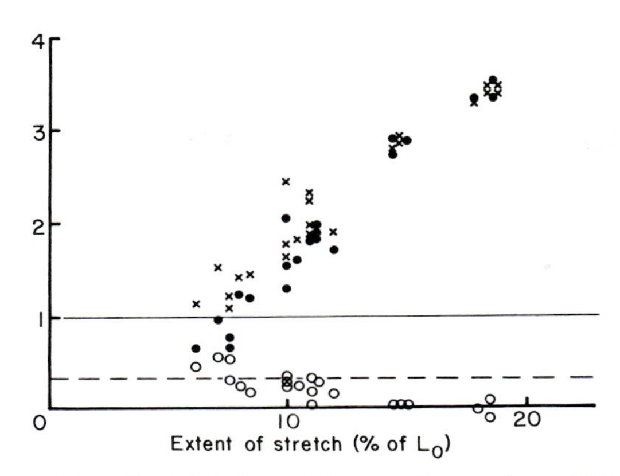

FIG. 4.20. Total quantities of heat h (×), work done on the muscle w_{ap} (●), and the difference $h - w_{ap}$(○). From Tables 1, 2, and 3 of Hill and Howarth (1959). The results are for twitches and have been divided by the isometric heat. The dotted line marks a value of 30% of the heat in an isometric twitch (see text).

As regards interpretations of the results our ideas, with the benefit of hindsight and the results of later chemical experiments, are different from those of Hill and his colleagues. They clearly favoured possibility (2), the reversal of ATP splitting during stretch. Two of the principal reasons behind their thinking were, first, the isometric heat production was considered at that time to be the result of "activation processes" and not directly related to the contractile machinery itself. The possibility that this heat production could be slowed or stopped by a stretch was therefore considered very unlikely. Second, it was considered that the series elastic component of muscle was the only component that could store mechanical energy, and that its properties could not change during contraction.

4.2.4 RECOVERY HEAT

At the end of relaxation after a tetanic contraction, the condition of a muscle is different in a number of ways than before the contraction. PCr has been hydrolysed and other changes have occurred, some probably related to the distribution of calcium within the muscle. If the muscle is immediately subjected to further stimulation it behaves differently than in the initial contraction; for instance, it produces less heat and relaxes less quickly. If it remains unstimulated the muscle will return to a state very similar to that before the initial tetanus; oxygen is consumed, PCr is resynthesized, and the ability to respond to stimulation returns to its original level. In aerobic conditions, muscles can undergo many such cycles of stimulation and recovery with little change in their properties. In frog muscle this process of recovery takes about an hour at 0°C, and during this time the rate of heat production is greater than the resting rate. This extra heat production is known as "recovery heat." The study of this heat production in relation to the known chemical changes occurring at the same time ought to give information about the recovery processes. In particular we might expect that the reversal of the unidentified process that produces part of the heat during contraction would be detectable in this way.

4.2.4.1 *Observations of Recovery Heat*

A. V. Hill (1965) has reviewed much of the earlier literature concerning recovery heat. The main observation is that the amount of recovery heat in frog muscle under aerobic conditions is between 1.2 and 1.5 times the amount of the initial heat, that is, the amount of heat produced during stimulation. Under anaerobic conditions the recovery heat (R) is much less, whereas the initial heat (I) is almost unchanged. The ratio r ($=R/I$, often called the "recovery ratio") is reduced to about 0.3. However, it seems that under anaerobic conditions the recovery process is incomplete, because the muscle's ability to respond to stimulation diminishes and it fails to respond at all after a few cycles of contraction and "recovery." More recently, Godfraind-de Becker (1972, 1973) has made a very careful study of the ratio r in frog and toad muscles. Table 4.V summarizes her results and compares them with the earlier observations. She finds r to be smaller when the muscle is in phosphate-buffered solution than when it is in one buffered with bicarbonate/CO_2 at the same pH. The intracellular pH would

TABLE 4.V
Summary of observations of r, ratio of recovery heat to initial heat.

Animal	T (°C)	Buffer[a]	r (mean ± SEM)	n[b]	Duration (s)	Reference
Frog	20	—	1.46 ± 0.03	20	0.03–0.5	Hartree and Hill (1922)
Frog	18	—	1.17 ± 0.080	10	0.10–0.30	Hartree (1928)
Frog	22.5	—	1.52 ± 0.032	10	—	Cattell (1934)
Frog	20	P	1.42 ± 0.05	12	0.5	Godfraind-de Becker (1973)
Toad	20	P	1.65 ± 0.07	9	0.5	
		P	1.48 ± 0.02	2	1.0	
		P	1.32 ± 0.15	3	1.5	
		B	2.02 ± 0.08	10	0.5	
		B	1.81 ± 0.09	3	1.5	

[a] Buffer in the Ringer's solution: P, phosphate; B, bicarbonate and CO_2.
[b] Number of observations.

be lower in bicarbonate/CO_2 buffered solution (Bolton and Vaughan-Jones, 1977). Increasing the duration of the tetanus causes a fall in the value of *r*.

Much information about the time course of aerobic and anaerobic recovery heat has been published (Hill, 1965; D. K. Hill, 1940b; Godfraind-de Becker, 1972, 1973). The time course of O_2 consumption (at 0°C) was found by D. K. Hill (1940a) to be very similar to that of the heat production. The time course of PCr resynthesis (at 0°C) appears (Fig. 4.21A) to be slower than either heat production or oxygen consumption (Dydynska and Wilkie, 1966; Dawson *et al.*, 1977). Unfortunately, similar information about these variables at the higher temperature at which most observations of recovery heat have been made is not available. One can conclude from the similarity of the time course of these three variables that, as expected, oxidative phosphorylation coupled to the resynthesis of PCr is responsible for a major part of the recovery heat.

4.2.4.2 *Interpretation of Recovery Heat*

A rigorous interpretation of recovery heat requires additional information about the changes in the concentration of PCr and the amount of O_2 consumption and/or lactate produced; in particular all the measurements must be made under conditions that are more comparable than in Fig. 4.21A. However, we will consider some results about recovery in relation to a specific hypothesis about the processes that are occurring. The tentative conclusions that can be drawn from this analysis may stimulate a proper energy balance study of the recovery process.

Consider first the simple case in which the initial heat (I) is produced solely by the splitting of PCr and the recovery heat (R) by the oxidative resynthesis of this PCr. The value of r ($=R/I$) will be the ratio of the molar enthalpy change (ΔH) for these two processes. ΔH for PCr splitting (ΔH_{PCr}) is usually taken as -34 kJ/mol (Curtin and Woledge, 1978). If the heat produced by uncoupled oxidation is ΔH_{O_2} per mole of O_2 used, and if the P:O_2 ratio is p, then the heat produced by the coupled oxidative resynthesis of PCr, expressed per mole of PCr, is given by

$$\Delta H_{res} = (\Delta H_{O_2}/p) - \Delta H_{PCr} \quad (1)$$

For oxidation of glycogen the value of p would be expected to be 6.5 (Woledge, 1971) and ΔH_{O_2}, -470 kJ/mol (Carpenter, 1939). ΔH_{res} is then 38 kJ/mol. The recovery ratio of this case, which we shall call r_0 should thus be given by

$$r_0 = \Delta H_{res}/\Delta H_{PCr} = -38/-34 = 1.13 \quad (2)$$

In a more complicated, but a more realistic case, a proportion (z) of the initial heat will be produced by an unidentified process and the remainder,

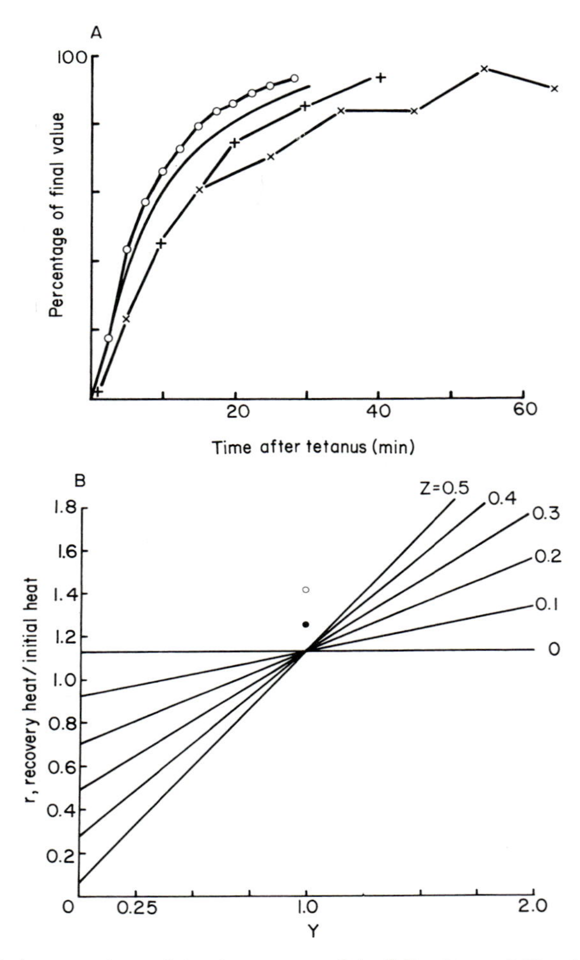

FIG. 4.21. (A) A comparison of the time course of the following variables during recovery of frog muscle after a tetanus. (○) O_2 consumption after a 12-s isometric tetanus at 0°C (D. K. Hill, 1940a.) Solid line, Heat production. Average of 10 observations with tetanus durations between 2 and 24 s at 0°C (D. K. Hill, 1940b). (+) Recovery of PCr content after a 30-s tetanus at 0°C (Dydynska and Wilkie, 1966). (×) Recovery of PCr content after a 25-s tetanus at 4°C (Dawson *et al.*, 1977). For comparison, the points (×) have been plotted on a time scale expanded by 1.428, corresponding to an assumed Q_{10} of about 2.3 for the recovery process. For (○) and solid line the amount of heat and oxygen consumption required for *complete* recovery were not measured. It has been assumed here that these processes had gone to 90% completion at the time the last observations were made. (B) The relation between the recovery ratio (r = recovery heat/initial heat) and Y, the ratio of the heat of reaction of the unidentified process and the heat from PCr splitting during recovery required to reverse it (see text). The relation has been calculated for a number of different values of z, the proportion of the initial heat that is due to the unidentified process. In each case the value of ΔH_m for PCr splitting has been assumed to be -34 kJ/mol. (○) and (●) Values of r for values of ΔH_m for PCr splitting equal to -30 and -32 kJ/mol, respectively, and $Y = 1$ and $z = 0$.

$(1 - z)I$, by splitting of PCr. In this case one would expect the following four processes to contribute to the recovery heat:

1. The oxidative resynthesis of the PCr split during the contraction. The heat produced by this process will be $r_0(1 - z)I$.
2. The reversal of the unidentified process, which will absorb the same amount of heat as that originally produced, $-zI$.
3. The PCr splitting required to drive (2). (We are assuming that the reversal of the unidentified process is coupled to PCr splitting. The stoichiometry of the coupling of PCr splitting to the reversal of the unidentified process is unknown.) We denote by Y the ratio of the heat produced by this PCr splitting to that absorbed by (2). If $Y = 1$, the coupled process, (2) plus (3), is thermally neutral; for $Y < 1$ it is endothermic; for $Y > 1$ it is exothermic. The heat produced by (3) is YzI.
4. The resynthesis by oxidative phosphorylation of the PCr splitting (3), which occurs after contraction to reverse the unidentified process. This will produce heat amounting to $r_0 YzI$.

The total recovery heat R is thus the sum of these four terms:

$$R = r_0(1 - z)I - zI + YzI + r_0 YzI \tag{3}$$

and the ratio r can be found

$$r = R/I = r_0(1 - z) - z + Yz + r_0 Yz \tag{4}$$

$$= r_0 - z(1 + r_0)(1 - Y) \tag{5}$$

This relation of r, Y, and z is illustrated in Fig. 4.21B. The value of r is equal to r_0 (1.13) if there is no unexplained heat ($z = 0$) and also if the unexplained heat is reversed by PCr splitting in a thermally neutral process ($Y = 1$).

As shown in Table 4.V, the value of r observed for short (0.5 s) tetani of frog muscle at 20°C is about 1.4, which is certainly greater than the expected value of r_0 (1.13). The proportion of unexplained energy (z) in a 0.5 s contraction is unknown, but we might expect it to be similar to that in a 5-sec tetanus at 0°C, which was found by Curtin and Woledge (1979a) to be 0.29. Thus, Eq. (5) or Fig. 4.21B shows that Y is about 1.4. A value of $Y > 1$ indicates that the coupled process, reversal of the unidentified reaction by PCr splitting, is exothermic.

In longer contractions, the proportion of the initial heat that is unexplained would be expected to be less (Homsher *et al.*, 1979; Curtin and Woledge, 1979a). If we accept that the reversal of the unidentified process during recovery is exothermic, the value of the recovery ratio r should be smaller for tetani of longer duration. As shown in Table 4.V, Godfraind-de Becker found that r was indeed smaller (1.32) for a 1.5-s tetanus of toad muscle at 20°C than it was (1.65) for a 0.5-s tetanus.

The increase in r she observed on changing to a bicarbonate/CO_2 buffer can also be understood in a qualitative way. This change of solution would cause a fall in intracellular pH that would reduce ΔH_{PCr} (Woledge, 1973). This would cause r_0 to increase [see Eq. (2) above and Fig. 4.21B].

In conclusion, the known facts about the recovery ratio can be plausibly explained if it is assumed that the unidentified process that produces part of the initial heat is reversed during recovery by being coupled to PCr splitting, with the coupled process being exothermic.

4.2.4.3 *Negative Delayed Heat*

It has been reported by several authors (Hartree, 1932; Furusawa and Hartree, 1926; Bugnard, 1934; Hill, 1961a, 1965) that muscles absorb heat for a short time after the end of a short tetanus, or a twitch. Some examples of this phenomenon are shown in Fig. 4.22. The effect is not very large, no more than 5–10% of the heat produced during the contraction. The heat absorbed is larger under anaerobic conditions, presumably because it is not truncated so soon by the onset of positive recovery heat. Inhibition of glycolysis by IAA does not abolish the effect. The possibility exists that its appearance is an artefact, caused by a small, but systematic error in heat loss correction. If such an error did exist it would produce an apparent heat absorption with a time course similar to that observed. If the effect is

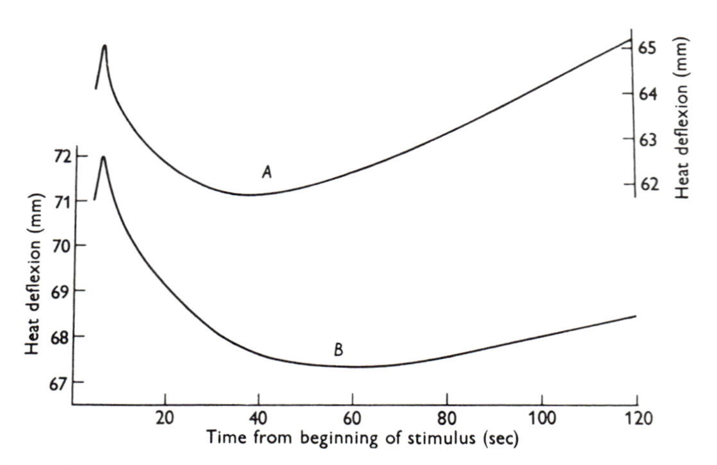

FIG. 4.22. Final stage of initial heat production at 0°C, followed by negative heat. Toad sartorii, in O_2 (A) and in N_2 for 57 min (B). 3-s tetanus. (A) Scale of deflexion on right, 1 unit = 0.67 mJ/g. Initial heat maximum = 43.5 mJ/g; negative heat to minimum = 2.22 mJ/g. (B) Scale of deflexion on left, 1 unit = 0.62 mJ/g. Initial heat maximum = 44.6 mJ/g; negative heat to minimum = 2.85 mJ/g. (From Hill, 1961a.)

not artefactual, it must represent the heat change produced by some recovery process in the muscle that can occur even in anaerobic, IAA-treated muscle. The nature of such a process is mysterious, unless it represents the coupled reversal of the process that produces the unexplained part of the initial heat. This suggestion is of course incompatible with that made above that this process is exothermic. In conclusion, there seems to be a number of poorly understood phenomena occurring during recovery, which will therefore probably prove to be a fruitful field for further investigation.

4.2.5 HEAT PRODUCTION BY RESTING MUSCLE

The amount of resting heat production is of interest in relation to the chemical processes that occur in the resting muscle. The contribution of specific reactions to the resting metabolism can be investigated. This is done by measuring the heat production in the absence and presence of an inhibitor of the reaction being investigated. Studies have also been done to try to elucidate the mechanisms by which resting heat production is affected by changes in the experimental conditions, such as stretch of the muscle (Feng effect), increase in extracellular potassium concentration (Solandt effect), or increase in osmolarity.

It would be useful to have a value for the ratio of the maximum heat production to that under resting conditions, which could be compared with, for example, the corresponding ratio for the ATPase rate of isolated actomyosin. This ratio can be obtained from results from different experiments (Table 4.VI), but the appropriate measurements do not seem to have been done in the same experiment in which it could be certain that the conditions were the same.

The sodium/potassium pump. The contribution of this pump to the resting heat production was investigated by Chinet *et al.* (1977) using a specially designed calorimeter. They measured the rate of heat production of resting soleus muscle of the rat and other tissues in the presence and absence of ouabain, a specific inhibitor of the sodium/potassium pump. Addition of the inhibitor resulted in a transient decrease of about 5% in the heat rate, which was attributed to block of the pump; parallel studies of efflux of sodium supported this idea. Subsequently the heat rate increased; it is not clear what reactions were responsible for this change.

The Feng effect: stretched muscle. Feng's (1932) original observation was that resting frog sartorius muscle produces more heat after it is stretched to a longer length by a load on an isotonic lever, than it does when at slack length. The effect was reversible. In a more quantitative study Clinch (1968) showed that the heat rate was constant at lengths between L_0 (measured *in situ* with the legs at right angles to the midline of the body) and $1.2L_0$, and

TABLE 4.VI
Rates of heat production during rest and isometric contraction

Resting

Preparation	Heat rate (mJ/g wet wt · min)	T (°C)	Reference
Sartorius, *R. temporaria*	16	18	Solandt (1936)
	10	20	Hill and Howarth (1957b)
	10.9 ± 2.5 (SD) (n = 30)	20	Clinch (1968)
	3.5 ± 0.59 (SEM) (n = 8)	17–20	Yamada (1970)
Soleus, rat	158 ± 3.8 (SEM) (n = 23)	30	Chinet *et al.* (1977)

Isometric tetanus

Preparation	Duration of Tetanus (s)	Mean heat rate[a] (mJ/g · min)	T (°C)	Reference
Sartorius, *R. temporaria*	2	7668	20	Canfield *et al.* (1973)
	5	1128	0	Curtin and Woledge (1979a)
	6	1331	0	Homsher *et al.* (1979)
Soleus, rat	2	1260	27	Gibbs and Gibson (1972)

[a] Mean rate during isometric contraction lasting for the duration indicated; the rate is in fact not constant during such contractions. See Section 4.2.1 for more detail on this point.

increased at longer lengths. When the muscle was at $1.3L_0$, the heat rate was between three and five times the basal level.

Feng (1932) was interested in the chemical events accompanying this change in heat production and found that muscle treated with iodoacetate or fatigued by stimulation did not increase its heat production so much as normal muscle does when stretched. He measured oxygen consumption and found that it increased along with heat production. If the conditions were anaerobic during the stretch about half as much heat was produced as in aerobic conditions, and an oxygen debt was established; that is, when oxygen was reintroduced extra oxygen was used (Feng, 1932; Euler, 1935). In terms of modern ideas about chemical reactions in muscle, these findings are consistent with the idea that stretch increases, for example, the basal rate of ATP hydrolysis by the sodium or calcium pump and/or myosin, and that oxygen, when available, is used to resynthesize the ATP. ATP hydrolysis and resynthesis would each produce about an equal amount of heat. The idea that calcium may be involved is supported by Clinch's (1968) finding that a number of twitch potentiators enhanced the Feng effect.

All these studies showed that pH influenced the increase in resting heat with stretch; alkaline conditions enhanced it and acid conditions reduced it.

It is also interesting that the Feng effect does not occur in all species. It is reliably present in sartorius muscles of *R. temporaria*, but absent or very small in the sartorius of *R. pipiens* and *R. esculenta* and the toad, *Bufo bufo*. Feng (1932) suggests that the effect is only present in more extensible muscles.

The Solandt effect: increased extracellular potassium. Solandt (1936) discovered that the heat production by resting muscles was enhanced by an increase in the concentration of potassium in the extracellular solution. The effect was quite large; heat rate could be increased to more than 20 times its basal value. Hill and Howarth (1957a) investigated this effect more extensively, and established that heat rate increased with potassium concentration in the range 10–18 m*M*. The membrane potential would be expected to be depolarized from the usual value of about −90 mV to between −65 and −50 mV under these conditions. They confirmed that the increased heat production occurs in the absence of mechanical activity. Although the muscle may twitch when first exposed to high potassium solution, it soon becomes inactive and continues to produce heat at a high rate for a long time. Hill and Howarth (1957a) also examined the effects of other ions in combination with potassium and as replacements for it. They concluded that the Solandt effect is mediated by the depolarization of the membrane potential. But, as the authors say, the real question was simply pushed on a stage by this finding; exactly how a change in membrane potential leads to increased heat production remained unclear.

The results of experiments on the oxygen requirements of the Solandt effect are relevant to this question. Hill and Howarth (1957a) found that under anaerobic conditions the effect was very small, only about one-tenth as large as in oxygen. They conclude from this that the heat production induced by high potassium is not simply due to an increase in the rate of the reactions that occur in muscle in normal Ringer's solution. If this were the case, then about half as much heat would be produced in the absence of oxygen as in its presence.

Increased osmolarity of the extracellular solution. Yamada (1970) investigated the increase in resting heat production that occurs when a muscle is placed in hypertonic Ringer's solution (in most experiments sodium and potassium were increased in the same proportion, to two or three times their normal concentrations). The heat rate increased to 10–20 times its basal rate. This effect appeared to be unrelated to the membrane potential, because the effect occurred both when the fibres were completely depolarized and when the membrane potential was close to its normal value. Thus, the effect would appear to be due to intracellular changes. Procaine, which is believed to block calcium release from the SR, reduces heat production in hypertonic solution. On the basis of this fact, Yamada (1970) favoured the idea that the increased heat was due to an increase in myosin ATPase, triggered by an increase in intracellular free calcium. However, this explanation does not account for his finding that the effect on heat production is almost completely absent when oxygen is replaced with nitrogen. If myosin were hydrolysing ATP at a higher than usual rate, this reaction would still produce heat in nitrogen even if the heat production due to ATP resynthesis by oxidative reactions were prevented. A direct effect on mitochondria due to the hypertonic extracellular solution seems a likely explanation of the results and of the Solandt effect.

4.2.6 HEAT PRODUCTION DURING RELAXATION

The heat produced by a frog sartorius muscle after the end of a tetanus, during the time when tension is declining to zero, was studied in detail by Hill (1961b). An example of one of his records is shown in Fig. 4.23. The rate of heat production begins to fall very soon after the end of the tetanus, at a time when little or no fall of tension has occurred. At the time when appreciable fall of tension begins, the heat rate rises again, passes through a maximum, and declines to zero at about the same time as tension. The secondary rise in the rate of heat production is thought to be due to the following processes: (1) the dissipation in the muscle of work done on it by external series elasticity; (2) the dissipation of work done on one region of the muscle as it is extended by shortening in another region—it is well known that relative movements of this kind occur in muscle during relaxation (see Chapter 2, Section 2.2.5); and (3) the thermoelastic effect of the fall in tension.

Hill suggests that, after removal of these effects, the remainder of the heat production might follow a time course like the dotted line in Fig. 4.23. It is hard, however, to establish quantitatively that this is so. One of the reasons for this is that the heat production due to process (2) cannot easily be estimated. Another reason is that the heat production during relaxation is very nonuniform both along the length of the muscle and through the thickness of the muscle (Hill, 1961b).

4.2.7 TWITCHES

A. V. Hill stated (1949a, p. 196) that one of the reasons for investigating the energetics of the twitch is that "... the single twitch is the elementary unit of all muscular responses." However 16 years later he had come to a different point of view when he wrote "unless special reasons exist ... employ tetanic contractions instead of twitches ... a lot of confusion can be caused

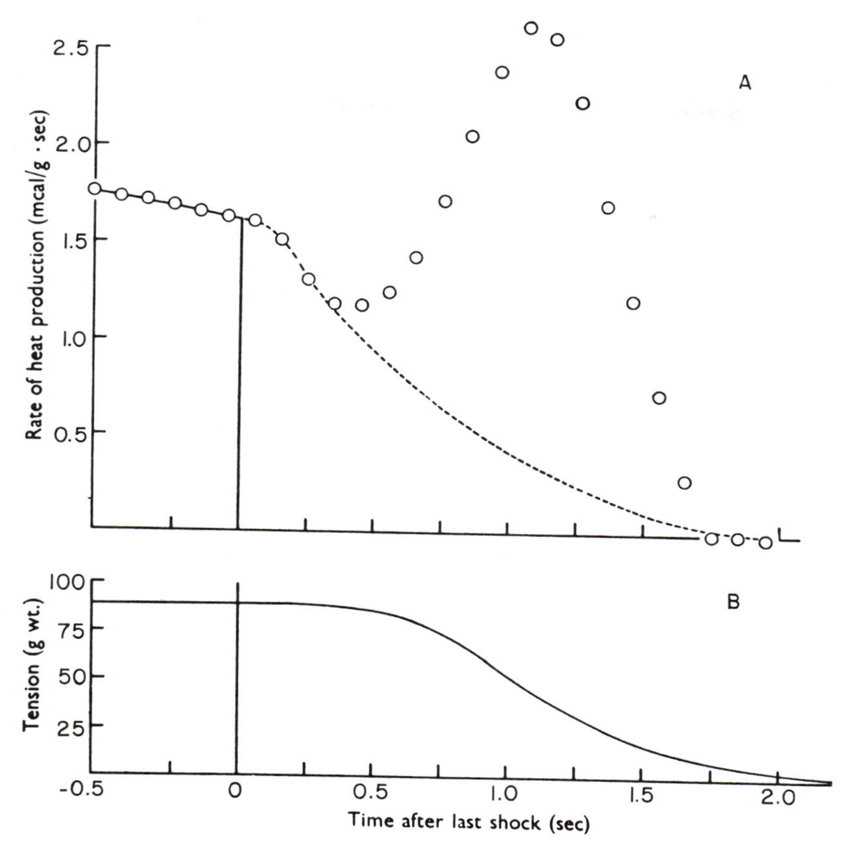

FIG. 4.23. (A) Heat production (○) before and after the last shock (at time 0) of a 3.3-s isometric tetanus of toad sartorii at 0°C. Corrected for lag in 0.1-s time units. The dotted line indicates the hypothetical heat production that would be detected if dissipation of work could be prevented. The tension is shown in the lower record (B) (Hill, 1961b.)

by the use of twitches" (Hill, 1965, p. 140). The isometric twitch is not, as it once seemed, a good reference for comparison of the energy output by shortening and lengthening muscle. The reasons are the following considerations:

1. There may be considerable redistributions of length during an isometric twitch. As tension is developing, there is internal shortening because of the compliance of the tendons and the apparatus. During relaxation, there is both shortening and lengthening due to this compliance and internal sarcomere length inequalities (Cleworth and Edman, 1969, see also

Chapter 2, Section 2.2.7). The amount of tension development is influenced by the amount of compliance external to and in series with the fibres.

2. The muscle is probably not maximally activated (that is, the troponin sites are not fully saturated by calcium) at any time during a twitch, and in any case the degree of activation is continually varying. The evidence for this idea is that treatment of the muscle with a variety of potentiators can increase the twitch force (Sandow, 1965).

3. The extent and duration of mechanical activity are affected by (a) the temperature (Jewell and Wilkie, 1960); (b) the initial muscle length (Edman and Kiessling, 1971), (c) the time since the last activation (Connolly *et al.*, 1971); (d) the speed and extent of shortening (Briden and Alpert, 1971; Dickinson and Woledge, 1973; Edman, 1980).

4. The initial sarcomere length of a muscle cannot be set to values less than 2.0 μm. An attempt to use shorter sarcomere lengths simply increases the amount of internal shortening at the start of the twitch (Hill, 1949a).

Thus, the various factors that would be expected to alter the energy liberation (shortening, lengthening, time for which the muscle is activated, amount of calcium bound to the troponin, and muscle length) interact with one another in a complicated way. It is therefore hard to design an experiment that isolates the effect on the energy liberation of a single variable. In spite of these difficulties, it has been possible to observe behaviour that is qualitatively similar to that seen during tetanic contractions, and some useful experiments have been made.

4.2.7.1 *Isometric twitches*

a. Near Optimum Length. Some observation of the amount of heat produced in a twitch of various kinds of muscle are summarized in Table 4.IV. In these experiments care was taken to minimize the compliance in series with the muscle. There are considerable variations in the amount of heat produced, but when the results are normalized as force:heat ratios (PL_0/H), much of this variation is eliminated. The use of this particular dimensionless ratio is traditional, but has no fundamental significance. If it were true that a twitch was an "elementary unit of muscular response," in that each crossbridge went through just one cycle, the number would be of considerable interest. However, it is unlikely that this simple view of the events during a twitch is correct.

b. Beyond Optimum Length. As muscle length is increased beyond that optimal for tension development (L_0), the amount of heat produced in a twitch declines (Fig. 4.24). At a sarcomere length of 3.65 μm, at which, in

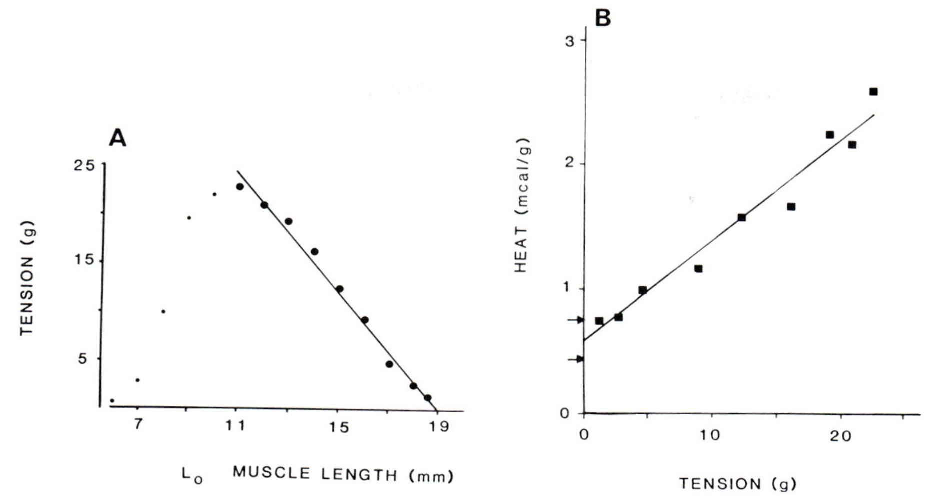

FIG. 4.24. Results of an experiment on semitendinosus of frog (*R. temporaria*) at 0°C. (A) The relation between twitch tension and muscle length. The line is the least-squares best fit to the results at and beyond L_0 (large dots). (B) The relation between heat and tension produced in twitches at muscle lengths at and beyond L_0. The line is the least-squares best fit assuming the errors to be associated with the heat measurements. The arrows indicate the calculated error limits of the best estimate of the zero tension intercept. All the results are from the same experiment. (From Smith, 1972b.)

frog muscle, there is no overlap between thick and thin filaments, the twitch heat is about 20–55% of its value at L_0. This heat production presumably comes from activation processes and not from crossbridge cycling. (For the results of similar experiments in tetanic contractions, see Section 4.2.1.) Twitches have an advantage over tetani for investigating the heat produced when there is no crossbridge activity because at these lengths very little (<2%) of the maximum twitch force is exerted, whereas much larger forces may be produced gradually in a tetanus, presumably by the processes responsible for "creep" at L_0 (see Chapter 2, Section 2.2.1). It is not certain to what extent the heat from activation processes is dependent on sarcomere length, but four lines of argument suggest that the dependence is small:

1. If muscles are stretched to sarcomere lengths beyond 3.6 μm there is little further decline in heat produced (J. A. Rall, personal communication).
2. Twitch force is proportional to filament overlap at sarcomere lengths beyond 2.2 μm at 0°C (Close, 1972b). Such behaviour would not be observed if there were substantial variations in the amount of activation. In the experiments by Close, similar behaviour is seen in about half the muscles at 20°C; however in the remainder twitch force increases between 2.2 and

2.8 μm. This suggests that activation *is* length dependent in some muscles at high temperature.

3. If twitch force is varied by varying temperature, the relation between force and heat produced falls on the same line as that found when force is varied by using different muscle lengths (>2.2 μm) at a fixed temperature. This is the expected result if the activation processes produce the same heat at all lengths and temperatures (Rall, 1979).

4. If muscles are placed in a solution in which 99% of the water is replaced by D_2O, no force is developed in response to a single stimulus. The heat produced under these conditions is not dependent on length (>2.2 μm) and is about half that produced in normal solution, at a sarcomere length of 3.65 μm (Rall, 1980a).

For these reasons, it seems that heat produced in a twitch at a sarcomere length of 3.65 μm (h_{act}) may be a reasonable estimate of the heat produced by processes other than actomyosin interaction at other sarcomere lengths. Thus, h_{act} should be proportional to the amount of calcium released. It is found that agents that are thought to change the amount of calcium released by the sarcoplasmic reticulum produce the expected change in h_{act} (Rall, 1982a).

Changes in h_{act} have been used to study the changes induced in the activation processes by previous activity (Rall, 1980b). As shown in Fig. 4.25A, h_{act} is depressed to 20% by a single stimulus and recovers with a time constant of about 1 s. After a tetanus, the recovery follows a biphasic time course as shown in Fig. 4.25B.

4.2.7.2 *Twitches with Shortening*

a. The Fenn Effect. In an extremely influential energetic study, Fenn (1923, 1924) investigated the energy liberation in frog skeletal muscle contracting against various afterloads. [For a discussion of the Fenn effect and review of recent studies of it, see Rall (1982b).] The temperature ranged between 0°C and room temperature [i.e., 7°C in the cellar of the Hill's home near Manchester (Fenn, 1923)] and the initial muscle length was near L_0. The results of his experiment, corrected for calibration error (Hill and Woledge, 1962), are shown in Fig. 4.26A. The major findings can be summarized as follows: (1) When a muscle shortens and does work, additional energy is liberated above that appearing in the isometric contraction. (2) The additional energy liberated is approximately equal to the work done. Fenn concluded that "the energy liberated by the contraction . . . is not dependent solely upon the initial mechanical and physiological condition of the muscle, but *can be modified by the nature of the load which the muscle discovers it must lift after the stimulus*." This conclusion invalidated the visco-elastic

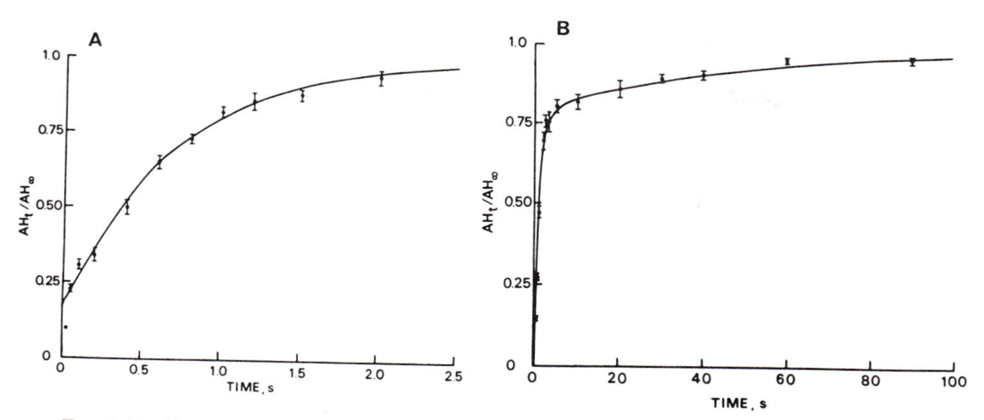

FIG. 4.25. (A) Repriming of the energy liberation at long muscle lengths in response to a stimulus after a twitch. The energy liberation (h_{act}, or AH in this figure) in response to a stimulus at some time t after a twitch of a fully rested muscle divided by energy (AH_∞) produced during a twitch of a fully rested muscle, vs the duration of the time interval between the twitches. Average results $\pm$SEM from posttwitch experiments ($n = 7$). The line is derived from a least-squares fit for a single exponential using the data from 0.05 to 2 s. The data at 0.01 s is approximately that expected from stimulus heat. (B) Repriming of the energy liberation at long muscle lengths in response to a stimulus after a tetanus of 3–80 s duration. The energy liberation (h_{act}) in response to a stimulus at some time t after a tetanus, (h_{act}), divided by AH_∞, is plotted vs the time interval between the tetanus and the twitch. Averages of 50 observations from 13 muscles pairs for tetani of 3, 40, and 80 s duration. The line is derived from a least-squares fit of the sum of two exponentials to the data. The rate constants differ by 48-fold. (From Rall, 1980b.)

theory of contraction which held that the amount of energy liberated in any twitch was a constant. Fenn's conclusion introduced a completely new, and still valid, idea to muscle energetics, i.e., that the driving reactions are regulated by the force and/or shortening occurring in contraction. The significance of his conclusion, rather than the details of the results described in (1) and (2) above, is what makes Fenn's papers genuine classics. A better appreciation of this idea can be gained from the results in Fig. 4.26B, which are for different initial muscle lengths and temperatures. Clearly the results are different from Fenn's in that the energy liberation is not equal to a constant plus the work done (see above). Hill's (1930) experiment at a higher temperature differs from Fenn's also in that the energy production does not exceed the isometric values. But, these experiments do show the generality of Fenn's conclusion that the energy liberation is dependent on the mechanical conditions prevailing during contraction.

The observed variation in the total energy output in these experiments (Fig. 4.26B) can be explained as follows. At $L > L_0$ the isometric heat

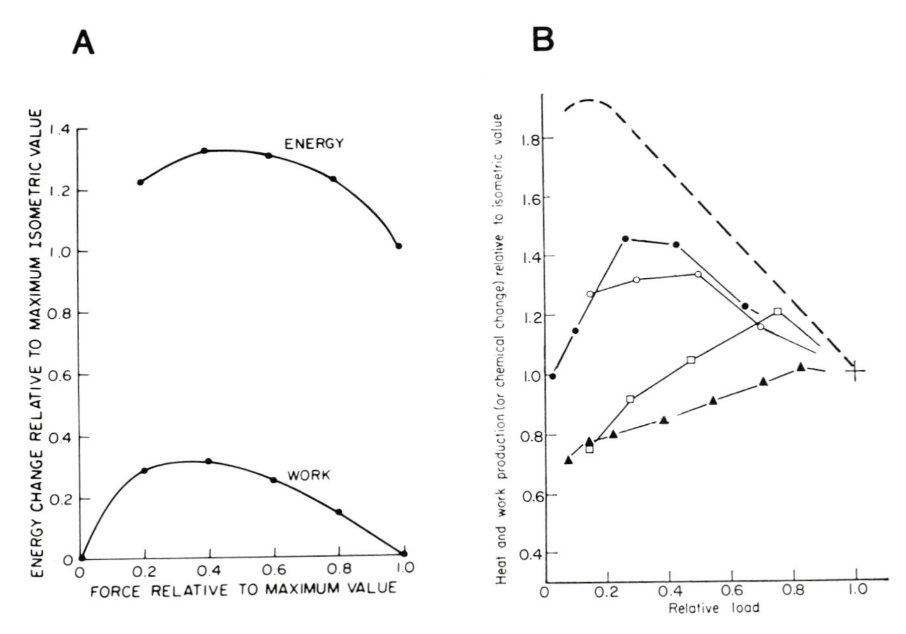

FIG. 4.26. (A) Energy liberation and work as a function of fractional load in afterloaded isotonic contractions of frog skeletal muscle. Energy liberated relative to the maximum isometric value in a twitch. (From Rall, 1982b.) (B) Results of different investigators for the variation in total heat (or chemical change) with load, in afterloaded isotonic twitches. All these experiments were made with sartorius muscles. Where heat was measured it includes the work done. (○) Fenn (1923), *R. temporaria*, 8°C: heat production. (●) Carlson *et al.* (1963), *R. pipiens*, 0°C: heat production. (□) Fischer (1931), *R. pipiens*, 12°C: oxygen consumption. (▲) Hill (1930), *R. esculenta*, 16°C: heat production. The dotted line gives a rough idea of the variation implied by the equation $E = A + ax + W$. (Based on Woledge, 1971.)

production is less but the distance a muscle can shorten under a given relative force is greater than at L_0. At $L < L_0$, the main factor is the smaller distance that can be shortened because of the proximity of the initial length to the steep left-hand portion of the length–force curve. The main reason that the experiments at 20°C differ from those near 0°C is the smaller distance the muscle can shorten at the higher temperature. This result is obtained because the duration of the twitch is reduced more by the temperature rise that the velocity of shortening is increased.

b. Hill's 1949 Twitch Studies. In these experiments (1949a–c) Hill sought to extend to twitches the observations he had made in tetani (Hill, 1938) (see Chapter 2, Section 2.2.2, and Section 4.2.2 of this chapter). If the energy production in a twitch were like that produced during an interval in a tetanus, it would be described by the following relation:

$$\begin{aligned} e &= e_0 + ax + Px \\ &= \text{constant} + ax + \text{work} \end{aligned} \tag{1}$$

where e is the energy produced in a twitch with shortening, e_0 is that produced in an isometric (no shortening) twitch, x is the distance shortened, a is a constant, and P is the force. The equation can be rearranged

$$e - Px = e_0 + ax \tag{2}$$

Where $e - Px$ is the heat production in a twitch with shortening. As shown in Fig. 4.27, Hill (1949a) observed the rate of heat production *during* shortening. The initial length was varied so the distance shortened was different. He found that the longer the initial muscle length, the greater the distance shortened and the greater the heat produced. These results and others were described by the relation similar to Eq. (2),

$$\text{heat} = A + ax \tag{3}$$

that is, the heat was equal to a constant A, plus a term proportional to the distance shortened, x. While this statement does describe the variation in heat produced during shortening with distance shortened, it is not as universally applicable as Hill claimed (1949a). For example, Hill himself recognized (1949b) that the constant A, determined in experiments like those in Fig. 4.27, was considerably less than the heat produced in an isometric twitch [e_0 in Eq. (2)]. In the experiments in Fig. 4.27 the load was the same for all distances shortened. If, however, the load is varied over a wide range. the results do not conform to Eq. (3). The value of A *increases* as the load increases; it eventually is equal to the isometric heat when the load is sufficient to prevent any shortening.

These limitations of Eq. (3) were not appreciated and it came to be regarded as a description of heat production of twitches under all conditions. Consequently, in a number of papers it is clear that the authors expected the heat + work to depend on load as shown by the dashed line in Fig. 4.26B [calculated from Eq. (3)]. The observations, however, show that less heat is produced during a twitch with light load than this unfortunate prediction.

4.2.7.3 *Twitches with Stretches*

As described in Section 4.2.3, when tetanized muscle is lengthened, the net rate of energy produced by the muscle ($h_s - w$, where h_s is the heat produced and w the work done on the muscle) is less than the energy produced in a comparable period of isometric contraction, h_i. Hill and Howarth (1959)

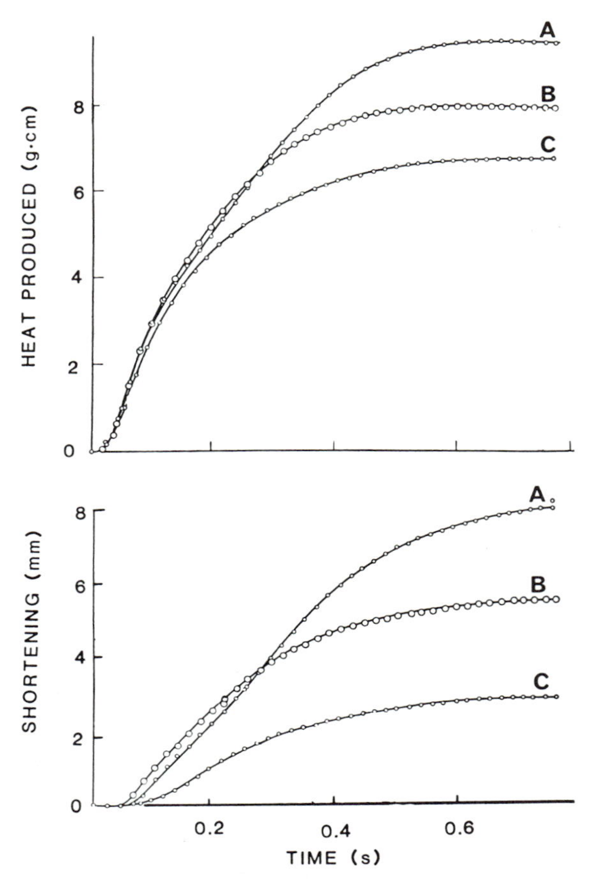

FIG. 4.27. The effect of shortening on the heat production in a twitch. Superimposed records of heat and shortening in three twitches. Records from two semimembranosi of toad, 58 mg, 0°C, isotonic load 0.17 g. The initial muscle lengths were different in each case: A, 18.5 mm; B, 15.5 mm; C, 12.5 mm. (From Hill, 1949a.)

extended these experiments to twitches and found that $h_s - w$ was clearly less than h_i (see Fig. 4.20). Thus, the behaviour in the twitch was again similar to that in a tetanus.

A striking feature of Hill and Howarth's results was that when the stretch began shortly after the stimulus was applied, $h_s - w$ was zero (within experimental error) as though the chemical reactions in the muscle had been totally prevented by the stretch. (They never found $h_s - w$ to be significantly below zero, however.) These results are quite surprising because 30% of the heat production in an isometric twitch is thought to come from the processes involved in calcium release and transport, which presumably cannot be prevented by stretching the muscle after the stimulus. It is hard to decide whether Hill and Howarth's experiments really are, as they claim, accurate enough to distinguish between a value of zero and 0.3 h_i for $h_s - w$. Various systematic errors (described in Section 4.2.3) might have influenced their results. These experiments are intrinsically difficult because $(h_s - w)$ is obtained as the difference between two relatively large quantities measured in different ways. It is very desirable to repeat Hill and Howarth's experiments with special attention to the possible systematic errors, in particular, those due to longitudinal inhomogeneities. The heat should be measured from as much of the muscle length as possible and the work done should be measured for the same part of the muscle by recording the length change of this region.

4.2.8 THERMOELASTIC HEAT

When an active muscle is subjected to a small and fairly rapid shortening, the tension drops and there is an increase in the rate of heat production (Fig. 4.28A), which is, as far as has been determined, simultaneous with the length change. Small rapid stretches have a converse effect (dotted line in Fig. 4.28A); that is, the tension rises and the rate of heat production decreases and may become negative. It has been suggested that these changes in heat rate are due to the occurrence of a "thermoelastic effect" in muscle. We will discuss the meaning of, and the evidence for this suggestion, but first we describe the observed heat production in more detail.

4.2.8.1 *Observations on Muscle*

We assume, for the purpose of this argument, that the processes producing heat during isometric contraction continue during and after a release or stretch. If this is so we can subtract the isometric heat production from the other records, as shown in Fig. 4.28B. However, it should be kept in mind that these length changes may in fact alter the rates of the isometric heat-producing reactions. After subtraction of the isometric control, it can be seen that the extra heat produced during and after the release diminishes as tension redevelops; the converse effect is seen after a stretch. Thus, the muscle behaves as though each element of tension change ΔP caused a proportional heat change ΔQ.[1] This proportionality has usually been expressed in the form

$$\Delta Q = -RL_0 \Delta P$$

[1] Note that ΔQ is positive for a heat production, unlike q_{rev} on pp. 237–241, which follows the thermodynamic convention that heat production is negative.

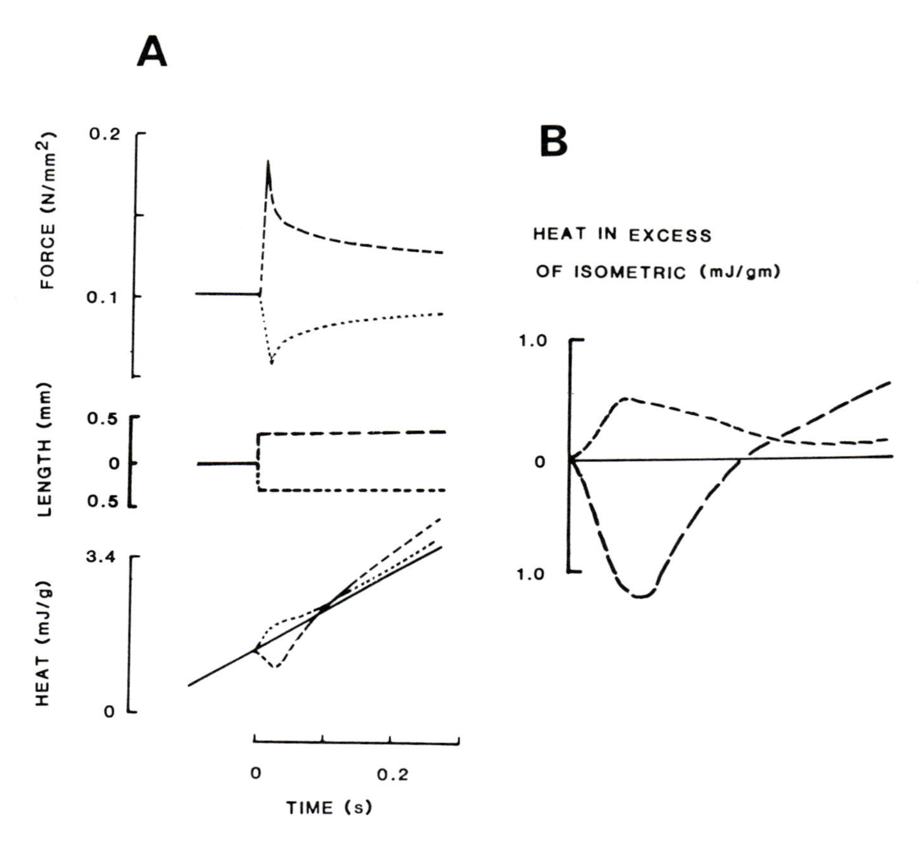

FIG. 4.28. (A) The force and heat production in part of a tetanic contraction which is either isometric (solid line) or contains a small stretch (dashed) or release (dotted). (B) Result of subtracting heat production under isometric conditions from that produced during stretch (dashed) or during release (dotted). [Both (A) and (B) from Gilbert and Matsumoto (1976).]

where RL_0 is a constant, R is a dimensionless quantity, equal to about 0.01 (see the open symbols in Fig. 4.29B) and L_0 is the muscle length.

The amount of heat produced is somewhat dependent on the speed of the length change (Woledge, 1961). This is, at least in part, due to friction between the muscle and thermopile. Artefacts due to this friction would be likely to make experiments with very rapid length changes hard to interpret (Gilbert, 1982). In all the experiments that have been reported to date the length changes that have been used are slow enough that the process that restores tension from T_1 to T_2 (see Chapter 2, Section 2.2.4) would be expected to be almost completed during the release; thus any heat production from this process would be included in the observed excess heat production caused by the length changes.

Attempts have been made to detect heat absorption accompanying rise of tension in active muscle in two other types of experiments: (1) applying a stretch during an isotonic shortening to cause a fairly abrupt rise of tension, and comparing the heat production with that when the contraction continued isotonic (Woledge, 1961); (2) comparing the heat production in an afterloaded isotonic contraction in which rise of tension ceased abruptly as shortening starts, with that in an isometric contraction in which tension continues to rise (Woledge, 1961). In both of these experiments the difference in heat production may be partly due to shortening heat and partly due to "thermoelastic" effects of the tension change. A division of the heat production into these two parts depends on assumptions about the time course of the shortening heat. The value of R was found to be 0.006 in these experiments, which is less than that found in experiments involving small length change imposed during the plateau of an isometric tetanus.

Similar experiments made with muscles in rigor are simpler to interpret for the following reasons: there is no heat production in the absence of a length change, and there is very little redevelopment of force after the length change. An example of heat change during release and stretch of rigor muscle is shown in Fig. 4.29A. It can be seen that there is heat produced when tension drops, and heat absorbed when tension rises, and these effects can be seen to be almost equal in size. However, the amount of heat change is only about half of that seen in active muscle (Fig. 3.29B).

4.2.8.2 *Definition of Thermoelastic Heat*

In this section we discuss the meaning of the term *thermoelastic heat* and consider whether the effects observed in active and rigor muscle can properly be given this label.

In any system undergoing a reversible change (that is, one for which the work done on the system can be completely recovered by reversing the change) the flow of heat out of the system (q_{rev}) is related to the work done (w_{rev}) by

$$q_{rev} = -T(dw_{rev}/dT) \quad (1)$$

The proof of this relation can be found in thermodynamics textbooks (for example, McGlashan, 1979) and is shown for the special case of interest here in Table 4.VII. We will apply this relation to a change in length of an elastic body. If the work done by volume change against the ambient pressure is negligible, the work done by the body during the length change is $\bar{P}\Delta L$, where $\bar{P}$ is the average force during the length change, ΔL.

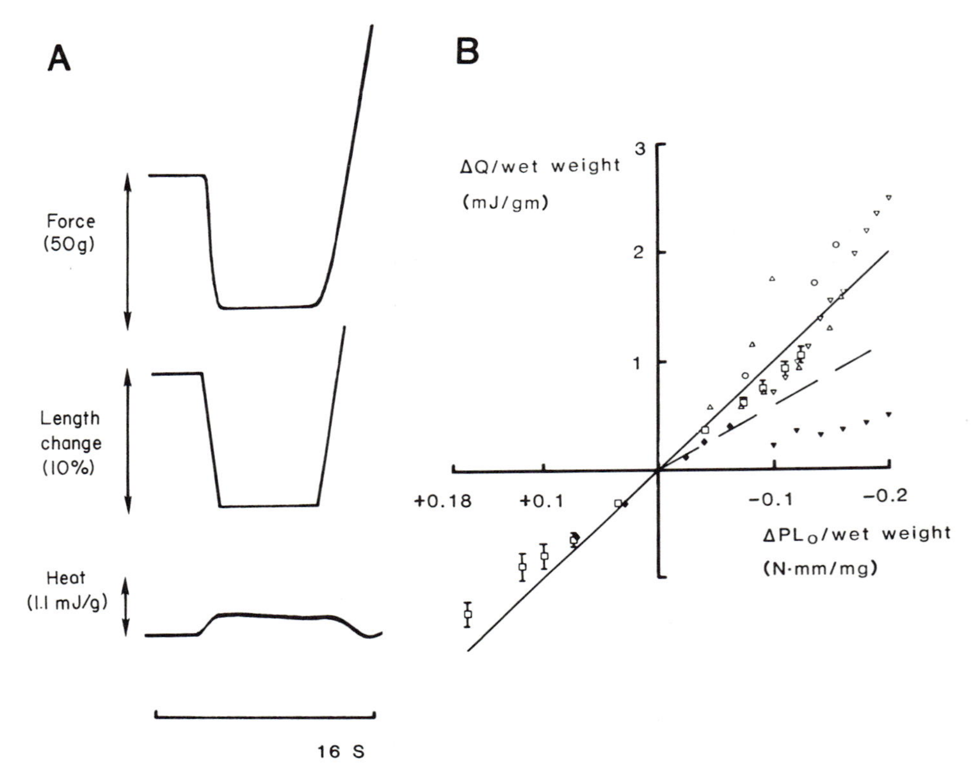

FIG. 4.29. (A) Records of force and heat production during release of frog sartorius muscle in rigor. (From Aubert, 1956, his Fig. 49, p. 272.) (B) Summary of thermoelastic heat observations. ΔQ is the heat produced in the muscle when tension changes by ΔP. Open symbols are for active muscle; closed symbols are for muscle in rigor. Results are from (□) Gilbert and Matsumoto (1976), bars indicate ± 1 SEM; (○) Woledge (1961); (△) Hill (1953); (▼, ▽) Chaplain (1971); (◆) Aubert (1956). All these results are for 0°C except Aubert's, which are at 18°C. The solid line represents the relation $\Delta Q = -0.01 L_0 \Delta P$ and the dashed line $\Delta Q = -0.006 L_0 \Delta P$.

Bodies that expand when heated produce less force (P) at a particular length at higher temperature (see Fig. 4.30). Therefore, at the higher temperature they produce less work for the same change in length. The difference between work done at the higher temperature and that done at the first temperature is a negative quantity, dw_{rev}/dT (see Fig. 4.30). As Eq. (1) shows, heat would be given out when the body is released; when the body is stretched an equal amount of heat would be absorbed. This is thermoelastic heat. It is produced when an elastic body changes length. In this description we have been considering bodies that expand when heated; these are said to have normal thermoelasticity. Bodies that shrink when heated, such as rubber,

TABLE 4.VII

Proof of the relation $q_{rev} = -T(dw_{rev}/dT)$

Consider an elastic body of length L exerting a force P. The internal energy, free energy, and entropy of the body are represented by U, F, and S, respectively.

$$F = U - TS \tag{T-1}$$

For any change

$$\Delta F = \Delta U - T\Delta S - S\Delta T \tag{T-2}$$

If the work done on the system during the change is w and the heat flow into it is q

$$\Delta U = q + w \tag{T-3}$$

$$\Delta F = q + w - T\Delta S - S\Delta T \tag{T-4}$$

For a reversible change

$$q_{rev} = T\Delta S \tag{T-5}$$

$$\Delta F = w_{rev} - S\Delta T \tag{T-6}$$

If work done by volume change is negligible, $w = P\Delta L$

$$\Delta F = P\Delta L - S\Delta T \tag{T-7}$$

At constant T

$$\Delta F = P\Delta L \tag{T-8}$$

$$dF/dL = P \tag{T-9}$$

At constant L

$$\Delta F = -S\Delta T \tag{T-10}$$

$$\partial F/\partial T = -S \tag{T-11}$$

From Eq. (T-9) $\partial P/\partial T = \partial^2 F/\partial L\, \partial T$ and from Eq. (T-11) $-\partial s/\partial L = \partial^2 F/T\, \partial L$. Thus

$$\partial P/\partial T = -\partial S/\partial L \tag{T-12}$$

$$q_{rev} = T\Delta S = T\Delta L \frac{\partial S}{\partial L} = -T\Delta L \frac{\partial P}{\partial T} = -T \frac{\partial w_{rev}}{\partial T}$$

have rubber-like thermoelasticity. The same arguments apply, and consequently heat is absorbed when they are released to shorter lengths.

In the special case in which the elasticity is linear (Fig. 4.30) and the stiffness independent of temperature, q_{rev} is proportional to the change of force (ΔP) during the change of length: the constant of proportionality (R) is usually defined by

$$q_{rev} = RL_0 \Delta P \tag{2}$$

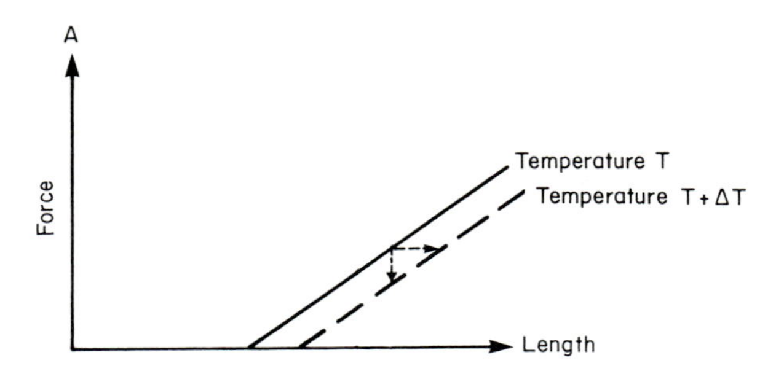

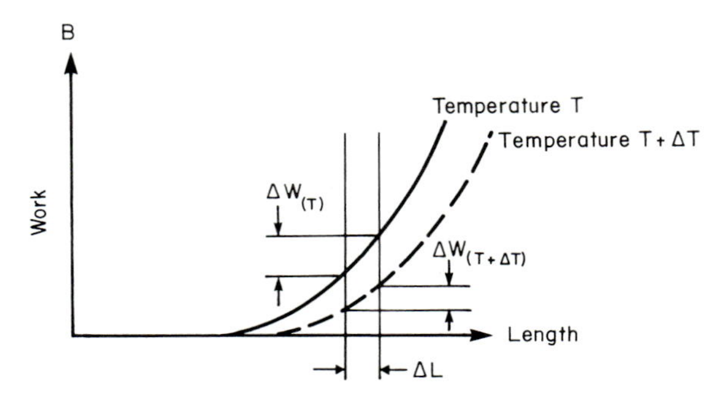

FIG. 4.30. Diagram illustrating the force (A) and the work (B) done in extension of an elastic body at two different temperatures. In (A) the dotted arrows show the change in length that occurs on heating at constant force, and the change in force that occurs on heating at constant length. (B) shows the work done (Δw) during a length change (ΔL) at the two temperatures. Note that Δw is less at the higher temperature.

in which L_0 is included so that R can be dimensionless and have the following simple relation to the coefficient of linear expansion, α.

$$R = \alpha T \tag{3}$$

$$q_{\text{rev}} = \alpha T L_0 \Delta P \tag{4}$$

The derivation of this relation is given in Table 4.VIII.

In systems in which there is no dissipation of energy (heat for release $=$ $-$heat for stretch), the only heat change during a change of length is the thermoelastic heat. Using the relation shown in Eq. (4), α can be determined from observations of q_{rev}, or vice versa. In active muscle, however, metabolism results in considerable energy dissipation, and changes in muscle length can be expected to produce heat changes by influencing those processes as

TABLE 4.VIII

Derivation of the relation $q_{rev} = RL_0 \Delta P$

In any elastic body the work done w_{rev} during a change of length ΔL is $P\Delta L$. Thus

$$\partial w_{rev}/\partial T = \Delta L(\partial P/\partial T)$$

If the elasticity is linear $(\partial P/\partial l)$ is a constant. The tension change ΔP accompanying a length change ΔL is $\Delta L(\partial P/\partial l)$

$$\frac{\partial w_{rev}}{\partial T} = \Delta L\left(\frac{\partial P}{\partial l}\right)\left(\frac{\partial l}{\partial T}\right) = \Delta P \frac{\partial l}{\partial T}$$

As shown in Table 4.VII the heat change q_{rev} is equal to $-T(\partial w_{rev}/\partial T)$

$$q_{rev} = -T(\partial w_{rev}/\partial T) = -T\Delta P(\partial l/\partial T)$$

The coefficient of linear expansion α is defined by

$$\alpha = -\frac{1}{L_0}\frac{\partial l}{\partial T}$$

where L_0 is the length of the body when $P = 0$. Thus

$$q_{rev} = -T\Delta P\, L_0 \alpha$$

Defining R as equal to $-T\alpha$

$$q_{rev} = R\, L_0 \Delta P$$

well as by the thermoelastic process. In principle, the thermoelastic contribution to the observed changes could be found from mechanical measurements at different temperatures. If muscle is released sufficiently rapidly the tension (T) falls linearly along the T_1 curve (see Chapter 2, Section 2.2.4). The process responsible for the T_1 force can be considered reversible because the T_1 force is independent of the speed of release. If we knew how the T_1 curve was shifted by an abrupt temperature change we could find the size of the corresponding thermoelastic heat produced (or absorbed), but this information is not available. Alternatively, knowing that the T_1 curve is linear we could use Eq. (3) to find the thermoelastic heat from the coefficient of expansion of the whole active muscle, or of the various components that carry the force: the actin filament, the crossbridges, and the myosin backbone. Unfortunately, these coefficients of expansion are also unknown. Thus at present we do not have the experimental evidence from which the thermoelastic heat of active muscle can be determined directly.

In rigor muscle, the coefficient of expansion α has been found to be 2×10^{-5} (Wohlish and Clamann, 1931). R $(= T\alpha)$ is thus 0.006 and in good agreement with the observed heat change in rigor muscle. Therefore, there is little doubt that the heat observed during length change in rigor muscle is thermoelastic in nature.

If we assume that the structures bearing tension in active muscle are the same as those in rigor muscle, then the same value of R would be expected. However, as we have seen, the value of R, at least as assessed by release as stretch experiments summarized in Fig. 4.29B, is clearly higher than that found in rigor muscle. This discrepancy suggests that there is an additional process occurring in the active muscle, which is responsible for about half of the heat change.

Huxley and Simmons (1973) have suggested that the process by which tension redevelops from T_1 to T_2 is exothermic. Could it be responsible for the large heat changes in active muscle that we have been considering? Since the recovery of tension from T_1 to T_2 does not occur in rigor muscle, it seems a good candidate. However, the amount of tension recovery ($T_2 - T_1$) during this phase is not a linear function of the size of the release; the value of $T_2 - T_1$ reaches its maximum for a release that results in T_2 being about 80% of T_0 (the initial force). For larger releases $T_2 - T_1$ is constant. In contrast, the heat measurements show that extra heat continues to increase as the distance released increases and the heat remains proportional to the change in tension (before release—after release (Fig. 4.29B)). Therefore, the process responsible for recovery of force from T_1 to T_2 is not responsible for the heat production after a release.

In conclusion, the full amount of the extra heat produced during a release of an active muscle cannot be explained by the thermoelastic effect of change of tension and remains puzzling.

4.3 Relation of Chemical Changes to Energy Output

4.3.1 ENERGY BALANCE

4.3.1.1 *Introduction*

One approach to the study of energetic aspects of contraction which has been actively pursued is the energy balance technique. At the time of writing, the most recent reviews on it are those by Curtin and Woledge (1978) and Homsher and Kean (1978), both of which concentrate on experiments involving measurements of heat production, and the one by Kushmerick (1984), which gives a very stimulating account of metabolic studies. Here, we only outline the basis of the approach here and summarize current thinking and recent results. For a more complete account, the above reviews should be consulted.

The goal of the energy balance approach is to measure the energy produced as heat and work during a contraction, and to compare this with the amount of energy that can be explained by chemical reactions. To do this one makes use of the following relationship:

$$\text{heat} + \text{work} = \sum^{i} \Delta H_{m} \xi_{i} \tag{1}$$

where the left side of the equation is the energy produced as heat and work. The right side of the equation is the sum for all the reactions (i) that occur of the extent of each reaction ξ multiplied by its molar enthalpy change ΔH_m. When a particular chemical reaction occurs under specific conditions of pH, temperature, and ionic strength, etc., an energy change occurs (ΔH_m in J/mol reaction). In this context the useful feature of this relationship is that the ΔH_m (energy per mole of reaction) is the same when the reaction occurs *in vitro* as when it occurs *in vivo. In vitro* all the energy appears as heat, whereas *in vivo* some of the energy may be converted into mechanical work, the rest appearing as heat.

How is an energy balance comparison actually made? The first step is to make a hypothesis about what chemical reactions occur and produce the energy; these reactions are the ones that will be included in the right side of the equation. In all the energy balance comparisons that have been done, only metabolic reactions have been taken into account (that is, ATP splitting and the reactions that resynthesize it). This is certainly the simplest idea, and should be explored fully before attempting to develop the techniques that would be necessary to make quantitative measurements of the extents of, for example, protein reactions during contraction of intact muscle. A choice must also be made about the time period during which chemical reactions are expected to occur and balance $h + w$ production. In most energy balance studies the hypothesis has been that the heat and work is *immediately* "paid for" by ATP hydrolysis. In other words, it has been assumed that the energy from ATP hydrolysis is not stored in the muscle for a significant period of time. An alternative hypothesis that could be tested in an energy balance study is that $h + w$ represents energy previously stored in the muscle, and that ATP splitting occurs after $h + w$ production replenishes the store.

Having defined the hypothesis to be tested, the four terms in Eq. (1) must be evaluated. The following quantities are found from observations made during contraction of the muscle:

1. Heat production, which has usually been evaluated from the increase in temperature of the muscle as measured with a thermopile.
2. Mechanical work done by the muscle on any external load, against the compliance of the apparatus to which it is attached, and the compliance of its tendons. This can be determined from the tension produced by the muscle,

observations of distance moved, and the compliance of the apparatus and tendons.

3. The extent of the chemical reactions that occur during the contraction. This has been done by rapidly freezing the muscle to stop the chemical reactions, removing the metabolites from the frozen muscle by extraction techniques, and then measuring the amounts of the relevant compounds in the extract by specific analytical methods. From comparisons of the levels of metabolites in experimental and control muscles, the extent of reactions can be estimated from the disappearance of substrate and/or the appearance of product.

4. ΔH_m values must be available from *in vitro* calorimetric studies of reactions of purified chemicals under known conditions. These values must also be appropriate for the conditions that exist in the muscle during contraction. As described in more detail below, this term *cannot* be established by experiments on muscle.

With this information, the energy produced by the chemical reactions that occurred can be calculated from the observed extent of reaction and the appropriate ΔH_m values. This calculated quantity is usually referred to as "explained" energy. It is then compared with the amount of energy the muscle actually produced as heat and work during the contraction, the "observed" heat + work. If the energy from the chemical reactions is equal to the heat + work, that is, if the two sides of Eq. (1) balance, then the energy balance description of the events in contraction is complete. If, on the other hand, the energy from the chemical reactions is insufficient to account for the heat + work, then the original hypothesis describing the events during the contraction is incomplete. Some other reaction occurs during contraction, and as we shall describe, this is in many cases found to be so.

4.3.1.2 *A Numerical Example of an Energy Balance Comparison*

Table 4.IX is an example of the data from an energy balance experiment (Curtin and Woledge, 1979a) on a 5-s isometric tetanus of sartorius (*R. temporaria*) at 0°C. The sum of the heat and work produced during the contraction is shown along with the extents of the ATP cleavage reaction and the creatine kinase reaction. The ΔH_m values of these reactions, which were measured in separate *in vitro* experiments, are −48 and +14 kJ/mol, respectively. The energy that can be explained by each reaction is the product of extent of the reaction and its ΔH_m. The table also compares the observed energy (heat + work) and the total energy that can be explained by the ATP cleavage and creatine kinase reaction. The observed energy is greater than the amount of energy that can be explained by these reactions; this

TABLE 4.IX

Energy balance[a]

Observed energy, heat + work (mJ/g dry wt) = 587.9 ± 20.2 (14)

Evaluation of explained energy

Reaction	ξ, extent of reaction (mol/g dry wt)	ΔH_m (kJ/mol)	$-(\xi \times \Delta H_m)$, explained energy (mJ/g dry wt)
ATP cleavage	12.17 ± 0.65	−48	584.2 ± 31.2
Creatine kinase reaction	11.78 ± 0.61	+14	−164.9 ± 8.5
		Total	419.3 ± 32.2

Energy balance

	(mJ/g dry wt)
Observed energy	587.9 ± 20.2 (14)
Explained energy	419.3 ± 32.2 (30.9)
Difference	168.7 ± 38.1 (44.7) = Unexplained energy

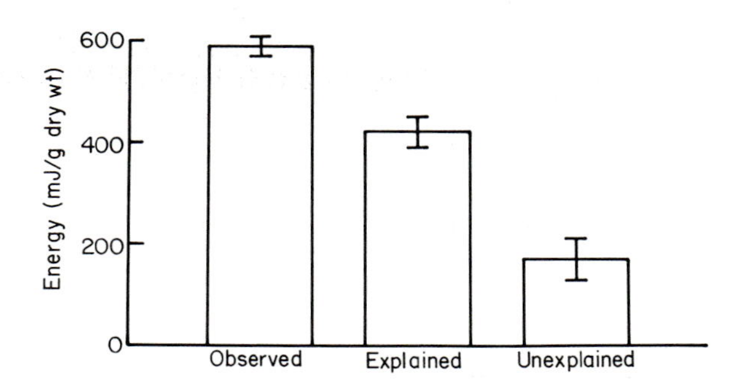

[a] Energy balance comparison for 5-s isometric tetanus of frog sartorius at L_0 and 0°C. Energy values are ±SE. The numbers in parentheses are the degrees of freedom. The sign convention used here is as follows: ξ is positive when the reaction proceeds in the direction ATP → ADP + P_i and PCr + ADP → ATP + Cr; AH_m is negative when heat is produced by the reaction proceeding in the ξ positive direction; the observed, explained, and unexplained energy is positive when energy is released by the muscle. (Results are from Curtin and Woledge, 1979a.)

imbalance suggests that some additional chemical reaction must occur and contribute about 30% of the total heat + work.

4.3.1.3 *Points of Clarification about Energy Balance*

There have been a number of problems and doubts (A. V. Hill would have called them "false trails") that have arisen about energy balance during the past. It seems worthwhile to point out, as in aid to understanding the literature, the ones that have been resolved.

a. What Are the Appropriate ΔH_m Values? These must be determined *in vitro* under defined conditions using purified chemicals so that it is certain that the whole of the observed heat is due to the specific chemical reaction being studied. Values used in energy balance calculations must be appropriate for the conditions of temperature, pH, ionic strength, etc., which exist in the muscle used in the energy balance study.

In several papers (Carlson *et al.*, 1963, 1967; Wilkie, 1968; Gilbert *et al.*, 1971; Homsher *et al.*, 1975), a value of −48 kJ/mol was used for the ΔH_m of PCr splitting, the major net reaction during brief contractions of normal frog muscle at 0°C. This value is too large; its use overestimates the energy that can be explained by this reaction, and correspondingly underestimates the unexplained energy. Using this value, it was incorrectly concluded in some of these studies that there was a balance between the energy explained by metabolic reactions and the observed heat + work. Because the incorrect value for the ΔH_m was used, the realization that there is an important energy gap was delayed for several years. The energy balance calculations for some of the results of these studies have been revised in the review by Curtin and Woledge (1978).

How did this error arise? Carlson and Siger (1960) made a comparison of the "*in vivo* ΔH" for PCr splitting, which ranged from −40 to −48 kJ/mol and was based on heat production by muscle, with values calculated as described below from results of *in vitro* experiments ("*in vivo* ΔH" is observed (heat + work)/ΔPCr; see next section). PCr splitting can be considered as the sum of three reactions that can be written schematically:

$$PCr^{-2} \longrightarrow Cr + HPO_4^{-2} \quad (1)$$

$$HPO_4^{-2} + H^+ \longrightarrow H_2PO_4^- \quad (2)$$

$$BH \longrightarrow B^- + H^+ \quad (3)$$

For reaction (1) they used the value −37.7 kJ/mol from Gellert and Sturtevant (1960). For the protonation of phosphate, reaction (2), they used Bernhardt's (1956) value, −31.8 kJ/mol. Carlson and Siger referred to the mixture of substances involved in buffering, reaction (3), as "physiological buffers" and took the "aggregate heat" of their reactions to be in the range +26.8 to +10.8 kJ/mol. Taking account of the fact that reactions (2) and (3) occur to an extent of 0.5 mol per mole of reaction (1), these values give a total of −40 to −48 kJ/mol of PCr hydrolysis, matching the "*in vivo*" value. (The −48 kJ/mol value is the one that was used in a number of subsequent energy balance studies.)

Carlson and Siger rightly pointed out that little was known about the intracellular buffers and their heats of reaction; our knowledge is still inadequate. However, the main error in their calculation is Bernhardt's value for the heat produced by the protonation of phosphate; it was substantially too large and was, in fact, not the best value available at the time. A much more accurate and extensive study of the protonation of phosphate had already been done by Bates and Acree (1945). From their results the ΔH_m of protonation of phosphate under appropriate conditions was only −10 kJ/mol.

In the experiments of Bernhardt (1956) and Bates and Acree (1945) it was actually the temperature dependence of the equilibrium constant that was observed. The ΔH_m can be calculated from these results and the van't Hoff equation

$$\Delta H_m = \frac{1}{RT^2} \frac{d \ln K}{dT} \tag{4}$$

where R is the gas constant and T is the absolute temperature. Figure 4.31 shows the values of the pK measured for similar conditions in the two experiments. A large number of other measurements were made by Bates and Acree to investigate the effect of concentration of phosphate (range 0–0.1m) and salt (range, 0–0.1m NaCl) and the effect of temperature (range, 0–60°C in 5°C steps). As can be seen from the van't Hoff equation; it is the *slope* of the line at 0°C in the figure that determines the ΔH_m at 0°C. Clearly, the ΔH_m based on Bates and Acree's results is more accurate because there are more data points from which the slope can be determined. Indeed, it is reasonable to suppose that in the temperature range 0–25°C Bernhardt's data are scattered around a line of the same slope as that of the results of Bates and Acree. At higher temperatures, the slopes of the two sets of results are different. The fact that the absolute values of the pK in the two studies are different is not relevant here; possibly it is due to the difference in ionic strength in the two experiments (KCl, 0.7 M, Bernhardt; NaCl, 0.1 M, Bates and Acree; it should be noted that the ionic strength in muscle is probably approximately 0.2 M).

As described above, Carlson and Siger evaluated ΔH_m for PCr splitting to be in the range −40 to −48 kJ/mol using Bernhardt's value for the heat of

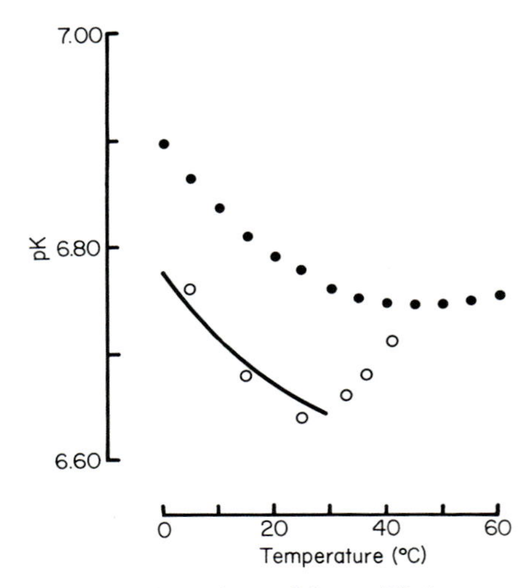

FIG. 4.31. Data on the temperature dependence of the equilibrium constant for the protonation of phosphate [reaction (2) in Table 4.X]. (●) Results of Bates and Acree (1945); (○) data from Bernhardt (1956). The solid line was drawn to have the same slope as a line through the values of Bates and Acree, but shifted vertically for comparison with the results of Bernhardt.

protonation of phosphate. If the more accurate value of -10 kJ/mol is used for the protonation, the ΔH_m for PCr splitting is in the range -29 to -37 kJ/mol (see Table 4.X). This range includes the value of -34 kJ/mol which, as a result of more recent experiments, is now considered appropriate (Curtin and Woledge, 1978).

b. In Vivo ΔH_m. This phrase has been used to describe the ratio of observed heat + work produced by a muscle to the amount of PCr split by the muscle (Carlson *et al.*, 1967; Wilkie, 1968). There are two points to be made about "*in vivo* ΔH" values. First, in such an experiment it has often been assumed that PCr splitting is the only reaction occurring and contributing to the production of heat + work. Unless it can be proven that PCr splitting is the only reaction, it is circular reasoning to use such an "*in vivo* ΔH" to predict the $h + w$ that can be explained by a particular amount of PCr splitting. This logical flaw was even less likely to be recognized after some "*in vivo* ΔH" values turned out to be numerically equal to the -48 kJ/mol incorrectly calculated from *in vitro* data as described in the previous section. This agreement seems to have lent support to the erroneous notion that the "*in vivo* ΔH" was the best value because it was determined under genuine physiological conditions.

TABLE 4.X

Summary of ΔH values[a]

Reaction	ΔH_m (kJ/mol)	ξ (mol)	heat = $\xi \times \Delta H_m$ (kJ)
(1) $PCr \rightarrow Cr + HPO_4^{2-}$	-37.7	1.0	-37.7
(2) $HPO_4^{2-} + H \rightarrow H_2PO_4^-$	-31.8 (a)	0.5	-15.9 (a)
	-10.0 (b)		-5.0 (b)
(3) $BH \rightarrow B^- + H^+$	$+26.8$ to $+10.8$	0.5	$+13.8$ to 5.4
		Total	-40.2 to -48.2 (c)
			-29.3 to -37.3 (d)

[a] The first column shows schematically the three reactions involved in the splitting of phosphocreatine. PCr is phosphocreatine, Cr is creatine, and B represents the intracellular buffers. The second column shows the values of the molar enthalpy change for the reaction. The third column shows the extent of reaction per mole of reaction (1) and the last column shows the heat produced when 1 mol of PCr is split. (a) indicates the value from Bernhardt (1956) and (b) that from Bates and Acree (1945). (c) is the total heat calculated using (a), and (d) the total using (b).

A second point to be made about "*in vivo* ΔH" value is that it can quite validly be compared with the correct ΔH value for PCr splitting, -34 kJ/mol, which is based on *in vitro* experiments. A discrepancy between the two values would be evidence for an unidentified reaction. In addition, comparisons between "*in vivo* ΔH" values can be useful; for example, when contractions under two different conditions are being compared. The hypothesis being tested is that the same set of reactions occur, and the ΔH values are the same under the two sets of conditions. If the hypothesis is correct, the "*in vivo* ΔH" values must be the same. If they are found to be different, the hypothesis must be rejected.

In practice, "*in vivo* ΔH" values present considerable statistical problems which arise when one attempts (1) to test whether a given ratio is different from zero or another ratio, and (2) to evaluate the error of a ratio that is calculated from two mean values both of which have errors.

c. Time and Speed of Freezing. The energy balance literature contains a number of changes in attitude about the stage during or after stimulation at which an energy imbalance can be detected. It is clear now that the time at which freezing begins can be delayed for a relatively long time and the unexplained energy is still evident. For example, an energy gap is observed when freezing occurs after relaxation is complete (at 0°C 3 s after the end of tetanic stimulation, Curtin and Woledge, 1974), and also when freezing is delayed for as much as 30 s after the end of stimulation (Carlson *et al.*, 1967; see Curtin and Woledge, 1978). (It should be noted that Carlson *et al.* (1967) did not come to this conclusion, because they used -48 kJ/mol for the ΔH_m for PCr splitting.) As discussed below, one of the outstanding questions in this field is "when exactly is the energy imbalance reversed?"

Another point about freezing that has been controversial is the importance of the speed of freezing. The most rapid method is probably the so-called "hammer method" developed by Kretzschmar (1970), in which the muscles are flattened between precooled metal blocks. The feature of the method that makes it rapid is that flattening reduces the distance through which heat must flow during cooling. When it was first used (Gilbert *et al.*, 1971), doubts were raised about the effectiveness of rapid immersion of muscles (sartorius, less than 1 mm thick) in precooled Freon or isopentane. This method of freezing was used in other laboratories at that time. However, Homsher *et al.* (1975) made a direct comparison of "hammer" and "immersion" freezing and showed that both methods are rapid enough for the energy imbalance in a tetanus to be detected at 0°C.

4.3.2 ISOMETRIC CONTRACTIONS

4.3.2.1 *Time Course of Unexplained Energy in Isometric Contractions*

The results of two studies of the time course of production of unexplained energy are presented in Fig. 4.32. Both studies were on sartorius muscle of *R. temporaria* at 0°C tetanized at L_0 for up to 15 s (Curtin and Woledge, 1979a) or 20 s (Homsher *et al.*, 1979). Under these conditions the creatine kinase reaction was shown to be the only reaction resynthesizing ATP; glycolytic reactions and oxidation did not occur. The results agree well in showing that the unexplained energy is not produced continuously during an isometric tetanus. It is produced during the first 5–8 s of stimulation, but not later, in the last several seconds of a 20-s tetanus. This resolves one of the major uncertainties about energy balance. On the basis of earlier work (Gilbert *et al.*, 1971), it seemed quite possible that there were two phases in the production of unexplained energy in an isometric tetanus. It was thought that during the first phase, which lasted about 2 s, unexplained energy was produced rapidly, and then following this there seemed to be a slower phase which continued through the entire period of stimulation.

These new results provide further evidence that unexplained energy is not produced by a reaction immediately coupled to ATP splitting or the creatine kinase reaction, because they proceed unabated even during the time when the unexplained energy is not being produced. The question of whether there is some energy from reactions directly and immediately involved with

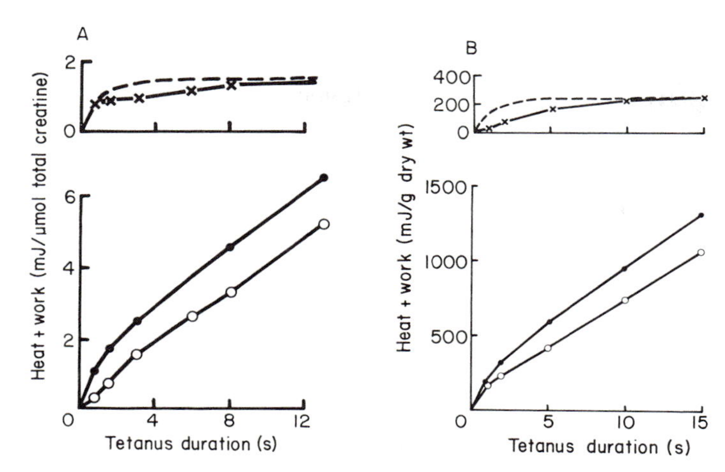

FIG. 4.32. Two investigations of the time course of the production of unexplained energy. Both studies were made with isometric tetanic contractions of sartorius muscle of *R. temporaria* at 0°C. (●) Observed energy output; (○) Explained energy output. (×) Unexplained energy (observed − explained), which is compared with the labile heat + work production shown by the dotted line. (A) is from Homsher *et al.* (1979). (B) is from Curtin and Woledge (1979a).

ATP splitting and the creatine kinase reaction (such as a buffer reaction, etc.) has been raised repeatedly and underlies some of the misuse of "*in vivo* ΔH" values referred to earlier. The evidence from Homsher *et al.* (1979), that there is no unexplained energy produced from 8 to 13 and from 15 to 20 s of a tetanus even though the metabolic reactions are occurring, is the most direct evidence on this question.

These results (Curtin and Woledge, 1979a; Homsher *et al.*, 1979) are also relevant to the question of when reversal of the unidentified process occurs. During stimulation of the muscle under isometric conditions, there is never any excess ATP splitting, so if ATP splitting is responsible for reversing the unidentified reaction, this reversal must start after isometric contraction is complete. (See Sections 4.2.4 and 4.3.2.4 about recovery.)

Another important implication of the time course results is that the unexplained energy seems unlikely to come from a transition of actin and myosin from their resting states to the state characteristic of contracting muscle. This transition and any energy changes associated with it would presumably be complete by the time tetanic tension has reached its plateau less than 0.5 s after stimulation starts, whereas the production of unexplained energy continues beyond this time.

A number of energy balance studies have been done on muscle from *R. pipiens*. It should be noted that sartorius muscles of *R. pipiens* produce a

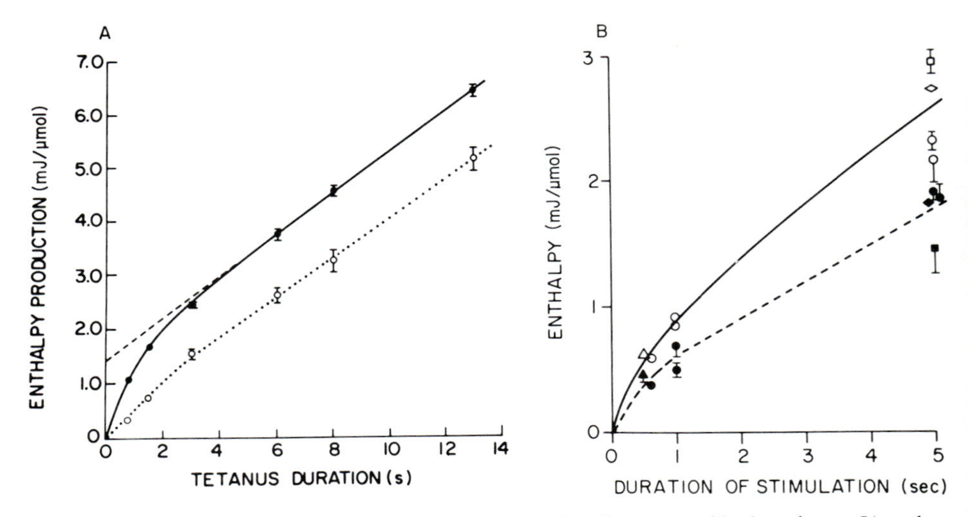

FIG. 4.33. The solid curve shows the observed enthalpy (heat + work) plotted as mJ/μmol of total creatine (ordinate) against tetanus duration in seconds (abscissa). The dotted curve shows the net chemical change (PCr split). This has been multiplied by the molar enthalpy change (−34 kJ/mol for PCr splitting). The bars show ±1 SEM. The difference between the solid and broken line is the unexplained energy. Results in (A) are for sartorius of *R. temporaria* (Homsher *et al.*, 1979) and those in (B) are for sartorius of *R. pipiens* (Homsher and Kean, 1978). Note the difference in scales.

smaller amount of unexplained energy than sartorius from *R. temporaria* under similar conditions in an isometric tetanus (see Fig. 4.33).

4.3.2.2 *Long Sarcomere Length*

Studies have been done on muscles at long lengths to determine whether reduced filament overlap, and thus fewer actin–myosin interactions, would influence the energy balance. Kean *et al.* (1976, 1977) have done experiments on *R. pipiens*, and Curtin and Woledge (1981) on *R. temporaria*, and in both cases there is a significant amount of unexplained energy produced at muscle lengths which are long enough to have prevented most actin–myosin interactions (Fig. 4.34C). This provides further support for the idea that the unexplained energy may not be involved directly with the interaction of actin and myosin.

Although the results agree in this respect, they differ in that Kean *et al.* (1977) found that the amount of unexplained energy at the two muscle lengths was the same, and Curtin and Woledge (1981) found that significantly less unexplained energy was produced at the long muscle length. There are some differences in the experimental details that may explain this discrepancy: the species of frog and the sarcomere length at the long muscle length were different (Kean *et al.*, *R. pipiens*, 3.6 μm; and Curtin and Woledge, *R. temporaria*, 3.8 μm). There may be a genuine species difference such that the source of the unexplained energy is influenced by muscle length in *R. temporaria*, but not in *R. pipiens*. Alternatively, the source of this energy may only be affected by muscle length at sarcomere lengths beyond 3.6 μm. A third possibility is that the fundamental process is the same, but its time course depends on either muscle length or the species of the frog. Curtin and Woledge (1981) found that in *R. temporaria* the labile heat + work is produced more slowly at the long muscle length than at L_0, so only part of it is produced during the 5-s period included in this energy balance comparison. In view of the possible link between labile heat + work and unexplained energy (see Section 4.3.2.5), they speculate that unexplained energy may be produced more slowly at the long length than at L_0, and that, after a longer period of stimulation when all the unexplained energy had been produced, the quantities of unexplained energy would be the same at both muscle lengths.

4.3.2.3 *Relaxation*

The energy balance study of Curtin and Woledge (1974) showed that the energy gap does not diminish during the 3 s after the end of tetanic stimulation, that is, during the period when the tension declines back to its resting level. This finding further supports the idea that the unexplained energy is due to something other than the transition of actin and myosin between the state characteristic of rest and that of contraction. At least from the point of view of interaction, it seems that actin and myosin would have returned to their original precontraction state by the end of relaxation, whereas the reaction responsible for the unexplained energy has not been reversed by this time. Further work is needed to establish when it is reversed.

4.3.2.4 *Steady-State Contractions*

Although it is known from measurements of the enthalpy production and high-energy phosphate splitting that an energy imbalance arises *during* a tetanic contraction, and that there is an imbalance present when the amount of ATP hydrolysed during a tetanus is compared to the oxygen consumed during recovery, it has not been explicitly shown for a complete cycle of contraction–relaxation–recovery that there is an energy balance. One reason is that the contraction–relaxation–recovery cycle at 0°C is long (about 45 min) and measurements of total heat production over this length of time require extensive correction for heat loss so that their accuracy is suspect.

Paul (1983) used a design based on that of Bugnard (1934) which avoids the heat loss problem; it gives information about recovery during a "steady-state series" of contractions. Bugnard had shown that a periodically tetanized muscle attains a steady state in which the amount of heat produced during each successive contraction (initial heat) is the same, but less than that produced during the first contraction of the series. In addition, in the steady state the heat produced in each cycle (contraction + interval) is also constant. (It should be noted that during the interval the muscle returns to the state it was in before the previous contraction, but does not recover back to the state it was in before the beginning of the series.) Because the recovery heat for each cycle is evolved over the relatively short time interval between tetani (several minutes at 0°C), accurate measurements of the initial and recovery heat can be made. Paul tetanized frog sartorius muscle for 3 s every 256 s and measured the steady-state rate of oxygen consumption, the steady-state rate of heat production, the initial heat production, and the amount of high energy phosphates consumed during the contraction. From these measurements the initial enthalpy, the recovery enthalpy, and the total enthalpy per cycle could be determined. The oxygen consumption measurements were used to calculate the amount of enthalpy explained by substrate oxidation [on the assumption that 75% of the substrate was glucose and 25% fatty acid (Woledge, 1971)]. The change in creatine phosphate during the contraction was used to calculate the initial explained enthalpy production. The results of these measurements and calculations are shown in Table 4.XI.

Two important facts about the steady state are apparent from the table. First, in row A, over a complete contraction–relaxation–recovery cycle there

TABLE 4.XI

Energy balance study on periodically tetanized muscle[a, b]

	Type of contraction	Enthalpy production (mJ/g)		
		Observed	Explained	Unexplained
A	Steady-state total enthalpy	181.5 ± 4.4	180.6 ± 3.2	1.1 ± 5.9
B	Steady-state initial enthalpy	76.5 ± 2.5	38.8 ± 8.1	37.7 ± 10.2
C	Initial enthalpy (first tetanus)	97.2 ± 2.0	38.4 ± 4.8	58.8 ± 6.1

[a] From Paul (1983).

[b] The observed enthalpy is the heat and work produced by the muscles; the explained initial enthalpy is based on the extent of PCr hydrolysis during the contraction and the $\Delta H = -34$ kJ/mol PCr; and the explained total enthalpy is based on the O_2 consumption and $\Delta H_m = -460$ kJ/mol O_2.

is an energy balance; that is, the total enthalpy produced is explained by the oxidation of substrate. Second, there is a significant amount of unexplained enthalpy produced during the initial phase of a steady-state tetanus (row B). Taking these two results together, the most simple interpretation is that during the steady-state contraction a reaction produces heat but does not involve high energy phosphate hydrolysis, and during recovery this reaction is reversed at the expense of metabolic energy. Finally, the data show that a significant amount of unexplained enthalpy is produced both in the steady-state contraction (row B) and in a contraction after a long rest (row C). Although one cannot state conclusively ($0.1 < p < 0.05$) that the steady-state unexplained enthalpy is less than that in fully rested muscles, if the reversal of the process that produced unexplained enthalpy is monoexponential, the reversal of the process has a time constant of about 4 min (256 s).

4.3.2.5 *A Comparison of Unexplained Energy and the Labile Energy*

The amount of unexplained energy produced by sartorius muscles of *R. temporaria* in an isometric tetanus of 5 s (or more) is very similar to the amount of labile energy production (Fig. 4.34A) (for a definition of labile energy see Section 4.2.1). This similarity suggests that labile energy and unexplained energy may both be aspects of a single process occurring in the muscle. A number of experiments have been performed to test this idea. The results support the idea that these phenomena are related but have shown that the relation is not a simple one.

As first pointed out by Aubert and Marechal (1963a) the amount of labile heat is diminished as a result of previous contraction. Curtin and Woledge (1977) found that the unexplained energy is also diminished, and to the same extent as labile energy (Fig. 4.34B).

Although the labile energy is not influenced by muscle length as much as the stable heat, it *is* reduced at long sarcomere lengths (Aubert, 1956). Curtin and Woledge (1981) found that there is a corresponding reduction in the unexplained energy (Fig. 4.34C). Since both phenomena still occur in a stretched muscle, in which actin–myosin interaction is negligible, neither is likely to be a consequence of actin–myosin ATPase activity.

The time course of unexplained heat production during an isometric tetanus has been compared with that of the labile heat by Homsher *et al.* (1979) and by Curtin and Woledge (1979a). The results of these two studies were similar (Fig. 4.32A and B) and show that the unexplained energy production lags behind that of the labile energy. At early times in the tetanus (for example at 2 s in Fig. 4.32A), the amount of unexplained energy is significantly less than the labile energy.

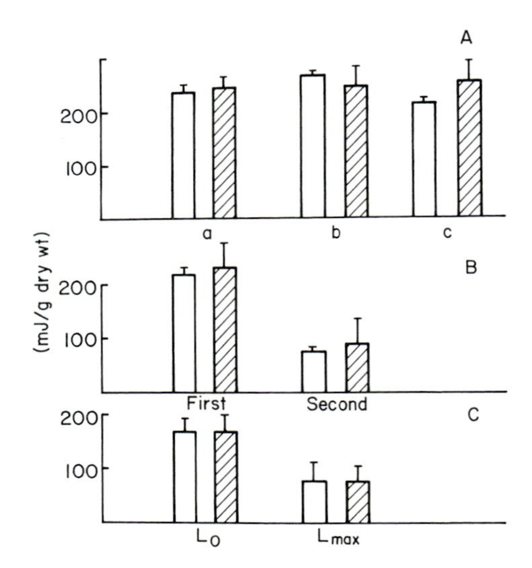

FIG. 4.34. Comparison of the mean (±SEM) unexplained energy (shaded blocks) and the labile heat (unshaded blocks) in muscle from *R. temporaria* at 0°C. (A) Three experiments with isometric tetanic contractions of sartorius muscle near L_0. (a) Curtin and Woledge (1979a), 15-s tetanus; (b) Homsher *et al.* (1979), 6-s tetanus, "first series"; (c) Homsher *et al.* (1979), 5-s tetanus, "second series." (B) A comparison of the first and second of a pair of 5-s tetani (at L_0) separated by an interval of 3 s, in sartorius muscle. (Curtin and Woledge, 1977.) (C) A comparison of 5-s tetani at two different muscle lengths, semitendinosus muscle. L_0, sarcomere length of 2.3 μm before stimulation; L_{max}, sarcomere length 3.8 μm before stimulation. (Curtin and Woledge, 1981.)

In studies of *R. pipiens*, it has been found that in both 1- and 5-s tetani the amount of unexplained energy is less than the labile energy (Homsher *et al.*, 1975). The labile energy is about the same as in *R. temporaria*, but the unexplained energy is less. It is possible that in *R. pipiens* the unexplained energy continues to increase in isometric tetani for longer than 5 s and eventually matches the labile heat, as it does in *R. temporaria.* The results of Kushmerick and Paul (1976a,b) from chemical energy balance studies of *R. pipiens* suggest that this may be so (see Section 4.3.6), but the point has not been directly investigated.

In the light of these results about the time courses, the simple hypothesis that both unexplained energy and labile energy are the result of one and the same process in the muscle is not tenable. Nevertheless, the similarities between the two phenomena are striking and it seems quite likely that there is a genuine, but not a simple, connexion between them which should be investigated further.

4.3.2.6 *Unexplained Energy and Parvalbumin*

One hypothesis for the origin of the unexplained energy produced during an isometric tetanus is that some or all of it comes from the binding of calcium ions to the protein parvalbumin and, to a lesser extent, to troponin. Parvalbumin is a small soluble protein which can bind 2 mol of calcium or magnesium. Its affinity for calcium is much higher than that for magnesium. Two of the calcium binding sites on troponin (the P sites) share these properties, whereas the other two sites (the T sites) bind calcium more weakly and have no significant affinity for magnesium. Parvalbumin is present in higher concentration in the sarcoplasm of frog muscle (0.4 μmol/g; Gosselin-Rey and Gerday, 1977) than troponin (0.07 μmol/g; Ebashi *et al.*, 1969).

Electron probe analysis, which measures total (i.e., free and bound) calcium concentration has been used by Somlyo *et al.* (1981) to study the location of calcium in resting and in tetanized frog muscle. They found that, in resting muscle, most of the calcium is in the terminal cisternae of the SR where the total concentration is about 70 m*M*. Probably, most of this calcium is bound to calsequestrin, a calcium-binding protein that is located there (Ostwald and MacLennan, 1974). When a muscle was tetanized, the total calcium concentration in the sarcoplasm was found to increase by about 900 μM. Aequorin experiments show that the free calcium concentration is only 5 μM (Blinks *et al.*, 1978); therefore most of the calcium in the sarcoplasm must be bound. Of the known binding sites, only those on parvalbumin have enough capacity to bind this large amount of calcium. Thus the electron probe studies suggest that, during a tetanus, the binding sites on parvalbumin become occupied by calcium.

A similar conclusion can be reached from consideration of the binding constants of parvalbumin for calcium and for magnesium, and of the free ion concentrations in resting and active muscle. As is shown in Table 4.XII, it would be expected from these figures that, in resting muscle, most of the parvalbumin binding sites would be occupied by magnesium. In active muscle, if the tetanus continued long enough for the equilibrium distribution of calcium and magnesium to be reached, nearly all the sites would be occupied by calcium. The rate at which this equilibrium state would be approached probably depends mainly on the rate at which magnesium dissociates from parvalbumin, because calcium cannot bind until this has occurred. The rate of parvalbumin–magnesium dissociation is about 5 s^{-1} at 20° (Robertson *et al.*, 1981) and would presumably be somewhat slower at 0°C, perhaps about 1 s^{-1}. The kinetics of calcium uptake by parvalbumin and its consequences for activation and relaxation of muscle have been discussed in detail by Robertson *et al.* (1981), Gillis *et al.* (1982), and Cannell and Allen (1984).

TABLE 4.XII

Calcium and magnesium binding to parvalbumin

The binding of Ca^{2+} to a parvalbumin site (P) can be represented by

$$P + Ca \longrightarrow PCa$$

The binding constant K_{ca} is defined by

$$K_{ca} = \frac{[PCa]}{[P][Ca]}$$

Thus, the ratio of occupied binding sites to vacant ones is given by

$$[PCa]/[P] = K_{ca}[Ca] \qquad (T\text{-}1)$$

Similarly for magnesium binding,

$$[PMg]/[P] = K_{mg}[Mg] \qquad (T\text{-}2)$$

Dividing (1) by (2) gives an equation for the ratio of sites occupied by calcium to those occupied by magnesium:

$$\frac{[PCa]}{[PMg]} = \frac{K_{ca}[Ca]}{K_{mg}[Mg]}$$

Numerical values

K_{mg}	$10^4\ M^{-1}$ (Robertson *et al.*, 1981)
K_{ca}	$10^8\ M^{-1}$ (Robertson *et al.*, 1981)

	In resting muscle	In tetanized muscle
Free [Mg]	3.3 mM (Hess *et al.*, 1982)	3.3 mM[a]
Free [Ca]	0.06 μM (Coray *et al.*, 1980)	5 μM (Cannell and Allen, 1984)
$K_{mg}[Mg]$	33	33
$K_{ca}[Ca]$	6	500
$\frac{[PCa]}{[PMg]} = \frac{K_{ca}[Ca]}{K_{mg}[Mg]}$	0.18	15
$\frac{[PCa]}{[PMg] + [PCa]}$	15%	94%

[a] Assumed to be the same as in resting muscle.

The binding of calcium to parvalbumin, in exchange for magnesium, is an exothermic process producing about 25 kJ/mol of calcium bound (Closset and Woledge, quoted in Curtin and Woledge, 1978; Moeschler *et al.*, 1980; Smith and Woledge, 1982). The release of calcium from the SR and from calsequestrin seem to be thermally neutral (T. Kodama, personal communication). Thus, the net effect of transfer of calcium from the SR to parvalbumin is to produce heat. As the concentration of binding sites is about 1 μmol/g, the maximum amount of heat produced would be about 25 mJ/g of muscle. This is, in fact, rather less than the amount of unexplained energy (30–40 mJ/g). If we assume that magnesium dissociation from parvalbumin is the rate-limiting step, this heat would be produced with an exponential time course and with a time constant of about 1 s. Thus the properties of the expected heat production are similar to those of the unexplained energy, except that the quantity of heat predicted is rather less than that observed.

There is experimental support for another aspect to this hypothesis. If the rate of calcium uptake by the SR does not vary during a tetanus, and if during the early stages of a tetanus some calcium ions are being taken up by binding to parvalbumin, then the total rate of calcium uptake should be greater in the early stages of the tetanus before the parvalbumin is saturated with calcium. Blinks *et al.* (1978) and Cannell (1983) have found, in experiments on frog muscle fibres using aequorin as a calcium indicator, that the free calcium concentration does fall more rapidly after a short tetanus than after a long one. In Cannell's experiments, at 8°C, the rate of calcium uptake fell exponentially with increasing tetanus duration with a time constant of about 0.4 s. As would be expected, when the rate of calcium uptake is less, so is the rate of fall of tension after the end of stimulation. This slowing of relaxation with increase in tetanus duration had been reported earlier (see Chapter 2, Section 2.2.5 and Fig. 2.34).

Large concentrations of parvalbumin are found in the muscles of aquatic and amphibious animals, whereas muscle of terrestrial animals contain much less parvalbumin (Kretsinger, 1980). If the hypothesis outlined above is correct, tetanus duration should have less effect on the rate of relaxation of muscles from terrestrial animals and they should produce less unexplained energy. In fact, in mouse and tortoise muscle, it is known that tetanus duration has little or no effect on rate of relaxation (Chapter 2, Section 2.2.5). Crow and Kushmerick (1982) have shown by comparison of oxygen consumption and ATP splitting measurements (Section 4.3.6 of this chapter) that in mouse muscle there is probably little or no unexplained energy produced during isometric contraction. In contrast Gower and Kretzschmar (1976) have shown that unexplained energy is produced in rat soleus muscle, and the amount is similar to that produced by frog muscle. In tortoise muscle also, there seem to be an appreciable amount of unexplained energy (Walsh and Woledge, 1970). Because of the similarities noted above between the unexplained energy and the labile energy, it is interesting to note here that rat and tortoise muscles do not produce labile heat; it has been observed only in the muscles of amphibians (frogs and toads) and of the amphibious terrapin *Pseudemys elegans* (Woledge, 1982; see also Table 4.IV, pp. 196–197.)

More detailed energy balance studies on the muscles of terrestrial animals are clearly needed, particularly to determine the time course of the unexplained energy production. It would also be interesting to make some studies on fish muscle in which the parvalbumin content is even higher than it is in frog muscles.

4.3.3 ENERGY BALANCE DURING CONTRACTIONS WITH SHORTENING

Contractions with shortening are interesting because shortening is due to filament sliding and crossbridge interaction, the basic contractile event. Shortening results in an increase in the rate of production of energy above the rate characteristic of isometric contraction. This behaviour, which is referred to as the "Fenn effect," has been described in Section 4.2.7.2, and is illustrated in Fig. 4.35 (upper curve). The rate at which mechanical work is done during shortening is shown in Fig. 4.35 by the lower curve. Their dependence on the velocity of shortening should be noted. At moderate velocities, the rate of energy production is at its maximum value and a large fraction of the energy is work; in contrast, at the maximum velocity of

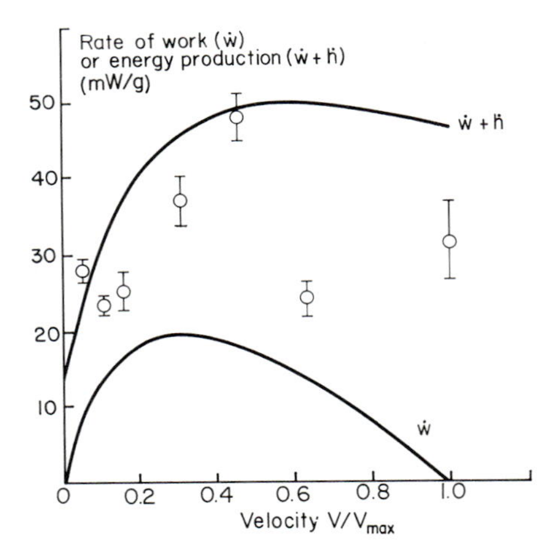

FIG. 4.35. The solid lines show the relation between the rate of energy production ($\dot{w} + \dot{h}$) and velocity (V), and between power output ($\dot{w}$) and velocity. These were calculated from Hill's formulation:

$$\dot{h} = 0.0625P_0V_{max} + (0.18P_0 + 0.16P)V$$

$$\dot{w} = PV$$

using values of P_0 and V_{max} given by Kushmerick and Davies (1969). (○) Energy calculated from the mean (±SEM) rates of ATP splitting observed by Kushmerick and Davies (1969). See text for further explanation of curves.

shortening, the rate of energy production is higher than the isometric value, but the whole of the energy is heat (no work is done).

Studies of these two extreme situations involving shortening (maximum work rate and maximum velocity of shortening) show that they are quite different from the point of view of energy balance.

The energetics of contractions involving a high rate of work have been investigated by Kushmerick and Davies (1969), Curtin *et al.* (1974), and Homsher and Yamada (1984). The experimental designs were quite different, but all the results agree in showing that the energy from ATP cleavage is sufficient to account for the mechanical work done by the muscle. There is no evidence that the energy from an unidentified process can be converted by the crossbridges into work. As described in more detail in the next section, Kushmerick and Davies (1969) found that ATP cleavage was correlated with the performance of work. The efficiency (work/free energy from ATP splitting) varied somewhat with the velocity of shortening, but it was always less than the theoretical maximum value of 1.0, which means that no other reaction is *required* to explain the work. Under the conditions used by Curtin *et al.* (1974), an unidentified process was shown to be occurring during working contractions and contributing to the production of energy. However, there was enough ATP splitting occurring to explain all the work. Homsher and Yamada (1984) chose conditions so that the unidentified process was not occurring at a significant rate before work began. The performance of work did not result in an increase in the rate of the unidentified process. In summary, it has been shown that work can be done when the unidentified process is not occurring, and there is no evidence that the unidentified process can supply energy to be converted into work.

Kushmerick and Davies (1969) measured work and ATP splitting over a range of velocities. From their results, which are shown in Fig. 4.35, it is clear that as the velocity increases beyond that optimum for the performance or work, the rate of ATP splitting diminishes. They point out that the chemical change (and thus the explained energy) behaves like the work, and not like the energy (heat + work) production, which remains quite high because of the high rate of heat production during rapid shortening (Hill, 1964). This line of argument was supported by experiments designed to show that shortening heat is not due to ATP splitting (Kushmerick *et al.*, 1969; see Section 4.2.2).

The energetic consequences of rapid shortening were further investigated by Rall *et al.* (1976) and by Homsher *et al.* (1981). In agreement with previous energy balance studies and the results of Kushmerick *et al.* (1969) concerning shortening heat, they found that the ATP splitting was not sufficient to account for the energy (mostly heat) produced during rapid shortening. These results are shown in Table 4.XIII and in Fig. 4.14A. Some other unidentified process must occur and produce some of the energy.

TABLE 4.XIII
Energy balance comparison for contractions with rapid shortening

Reference	Conditions	Enthalpy (mJ/g)		
		Observed	Explained	Unexplained
Rall *et al.* (1976)[a]	0.5 s isotonic (a)	18.0 ± 1.1	2.7 ± 4.0	15.3 ± 4.1
	0.75 s isotonic (b)	22.3 ± 1.9	8.8 ± 2.5	13.4 ± 3.2
	Isotonic and relaxation (c)	24.3 ± 2.0	16.4 ± 2.9	7.9 ± 3.6
Homsher *et al.* (1981)[b]	0.3 s isovelocity (d)	11.3 ± 0.7	4.8 ± 2.4	6.5 ± 2.5
	0.7 s isometric after shortening (e)	10.6 ± 0.7	16.8 ± 2.3	−6.2 ± 2.5

[a] In the experiment by Rall *et al.* (1976) the contractions were afterloaded isotonics (small load) and the muscles were frozen during shortening at either (a) 0.5 or (b) 0.75 s after stimulation started, or (c) after the end of relaxation.

[b] In the experiment by Homsher *et al.* (1981) the muscles were shortened at a constant rapid velocity starting after 2 s of stimulation under isometric conditions. Stimulation continued during and after the end of shortening. The first results (d) are for the 0.3 s of shortening, and the second (e) are for the next 0.7 s of contraction under isometric conditions at the short length.

In the experiment by Homsher *et al.* (1981), the tetanus started 2 s before shortening began. They showed that in the next 1-s period the rate of production of unexplained energy is not significantly different than zero if the isometric conditions are maintained. However when rapid shortening occurs during this period, the rate of production of unexplained energy is quite high.

In the studies by Rall *et al.* (1976) and Homsher *et al.* (1981), energy changes in the period *after* shortening were also considered, and for the first time a genuine balance between ATP splitting and energy (heat + work) production was achieved. Rall *et al.* (1976) found that when the entire period of contraction with rapid shortening and relaxation was considered, a larger fraction of the heat + work was explained than when shortening alone was considered [(c) vs (a) or (b) in Table 4.XIII]. This suggests that extra ATP splitting occurs during relaxation after shortening. In contrast, during relaxation after a purely isometric contraction, the amount of unexplained energy does not diminish; it actually increases (Curtin and Woledge, 1974).

Homsher *et al.* (1981) did a more complete study in which among other things the energy change during a 0.7-s period of isometric contraction immediately after shortening was measured [(e) of Table 4.XIII]. They found that the energy explained by ATP splitting was *greater* than that produced as heat + work. Thus, it seems clear that rapid shortening is accompanied by an unidentified exothermic reaction and that it is subsequently reversed. The results suggest that the reversal is driven by ATP cleavage and the coupled process is approximately thermally neutral. This striking feature of muscle energetics awaits explanation (see Section 5.2.6, p. 298).

4.3.4 EFFICIENCY AND WORK BALANCE

Only the free energy of a chemical reaction can be converted into work. The efficiency of muscular contraction is defined as the proportion of the free energy that appears as mechanical work; the rest of the free energy is degraded irreversibly to heat or converted into another form of free energy (Wilkie, 1974).

Wilkie (1960), in discussing the application of thermodynamic principles to biological systems, has described the special properties they have as a result of being at a uniform temperature. One of these is the fact that heat cannot be converted into work. He points out that it is misleading to make comparisons between biological systems and heat engines which can convert heat to work. Similarly, the ratio of work output to the sum of heat and work, $w/(h + w)$, is not useful by itself in trying to measure or understand efficiency. This is because the enthalpy change $(h + w)$ consists of two parts only one of which, a free energy change, can be converted into work. One should be forewarned that in the literature on muscle heat production, the values of $w/(h + w)$ are sometimes confusingly referred to as "mechanical efficiency" (Hill, 1939, 1964b,c, 1965). As Wilkie (1960) pointed out the relationship between the thermodynamic efficiency and the "mechanical efficiency" can be defined only if information is available about the nature of the chemical processes producing the heat + work and the free energy and enthalpy produced by these processes. At about the same time as Wilkie's article was published, methods were developed for following changes in concentrations of ATP, etc, during contraction; without the information about extents of chemical reactions it was not possible to evaluate the free energy that was available from ATP splitting in muscle. Nevertheless, even taking these things into account it is hard to understand how $w/(h + w)$ and thermodynamic efficiency had got muddled up and why A. V. Hill was so unrepentant.

The free energy made available (for conversion to work) by a chemical reaction depends on the amount of the reaction that occurs and its affinity (that is, the free energy per mole of reaction, $A = dG/d\xi$). Affinity can be calculated from the standard molar free energy (ΔG_{std}), temperature, and the activities (a = activity coefficient × concentration, C) of all the reactants and products, including all charged species:

$$dG/d\xi = \Delta G_{\text{std}} + RT \ln\left(\frac{a_{\text{product 1}}\, a_{\text{product 2}} \cdots}{a_{\text{reactant 1}}\, a_{\text{reactant 2}} \cdots}\right)$$

If the conditions for ΔG_{std} are different (in their pH, pMg, etc.) than those under which the reaction is occurring, other terms must be added to take account of these differences.

Of the identified reactions that participate in contraction, the cleavage of ATP is the one most immediately involved in energy transduction. The free energy of this reaction has been evaluated by Kushmerick (Appendix, Kushmerick and Davies, 1969) for the conditions that exist in frog muscle during brief contractions of DNFB-treated and untreated frog sartorius muscle at 0°C. Figure 4.36A shows how the affinity ($dG/d\xi$) depends on the concentrations of reactant and products, which of course change as the

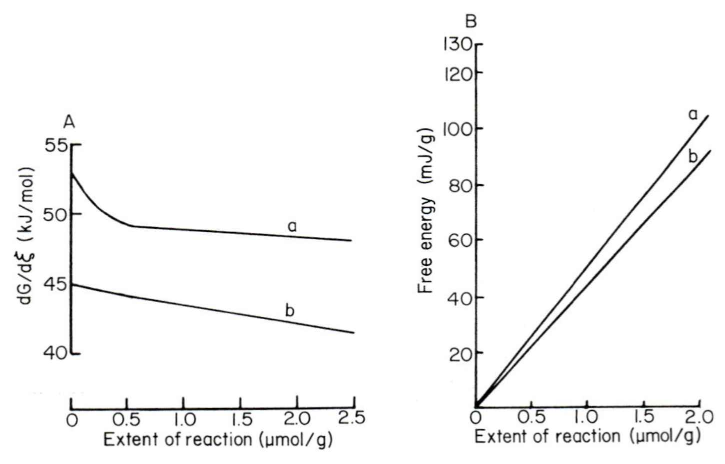

FIG. 4.36. (A) The dependence of the affinity ($dG/d\xi$, kJ/mol reaction) of ATP cleavage on the extent (ξ, μmol/g) of reaction in untreated (a) and DNFB-treated (b) frog sartorius muscle at 0°C (based on Appendix to Kushmerick and Davies, 1969). The values for the affinity are calculated from the equation

$$dG/d\xi = \Delta G_{std} - 2.3RT\{6.9 - 7.0 - \log([\mathrm{ADP}][\mathrm{P_i}]/[\mathrm{ATP}])\}$$

where ΔG_{std} is -29.3 kJ/mol (Benzinger *et al.*, 1959), 6.9 is the value assumed for the intracellular pH, and 7.0 is the pH appropriate to the ΔG_{std} value. For DNFB-treated muscle the values for ATP and P_i, 2.22 and 2.25 μmol/g, respectively, in resting muscle, were taken as the initial values (means from Kushmerick and Davies, 1969). It was assumed that the adenylate kinase reaction went to completion; that is, all the ADP that was formed was converted to ATP and AMP [the ADP level remained constant at 0.79 μmol/g (mean from Kushmerick and Davies, 1969)] and that no P_i was incorporated into hexose phosphates. In untreated muscles. the initial values for ATP and P_i were taken to be the same as in DNFB-treated muscles. It was assumed that the ADP level increased from 0.02 to 0.10 μmol/g during the occurrence of the first 0.25 μmol of reaction/g muscle and then remained constant. The creatine kinase reaction was assumed to keep the ATP level constant. (B) The amount of free energy (mJ/g) made available during the progress of ATP cleavage (μmol/g). The values of free energy are the integrals of $dG/d\xi$ (shown in A) and ξ. (a) Untreated muscle; (b) DNFB-treated muscle.

ATP cleavage proceeds. Figure 4.36B shows the total amount of free energy ($\xi \times dG/d\xi$) that is made available as a result of various amounts of ATP cleavage (the changing value of $dG/d\xi$ with extent of reaction shown in Fig. 4.36A has been taken into account). A similar calculation of free energy has been done by Dawson *et al.* (1978).

A number of studies have been done in which the amount of mechanical work performed and the extent of ATP cleavage during a single contraction was measured (see Fig. 4.37). As shown in the figure, the amount of ATP split is strongly dependent on the amount of work done. From this information and the free energy of ATP splitting (Fig. 4.36B), an "over-all" efficiency can be calculated, on the assumption that ATP splitting is the only process providing free energy. The justification for this assumption is discussed below. Figure 4.38A shows some efficiency values calculated from work, extent of ATP splitting, and free energy. The efficiency is, of course, dependent on the velocity of shortening. These calculated values are certainly smaller than the efficiency of the contractile process itself, because other processes occur during contraction that use ATP but do not produce mechanical work. The most obvious example is the pumping of calcium ions by the sarcoplasmic reticulum, which accounts for about a quarter of the ATP hydrolysis during an isometric contraction at L_0. If allowance is made for

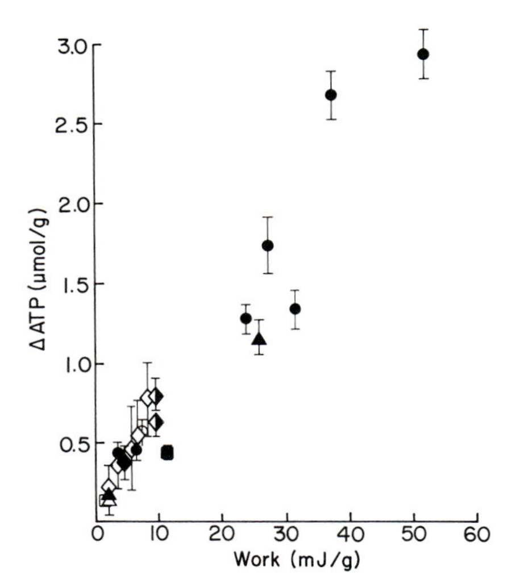

FIG. 4.37. The relation between the amount of ATP cleavage and the work done during shortening at various velocities. All the results are mean values for experiments on frog sartorius muscle at 0°C. (■) Cain and Davies (1962); (◇) Cain *et al.* (1962); (◆) Infante and Davies (1965); (half-filled ◇) Infante *et al.* (1965); (●) Kushmerick and Davies (1969); (▲) Curtin *et al.* (1974); (△) Rall *et al.* (1976); (○) Homsher *et al.* (unpublished).

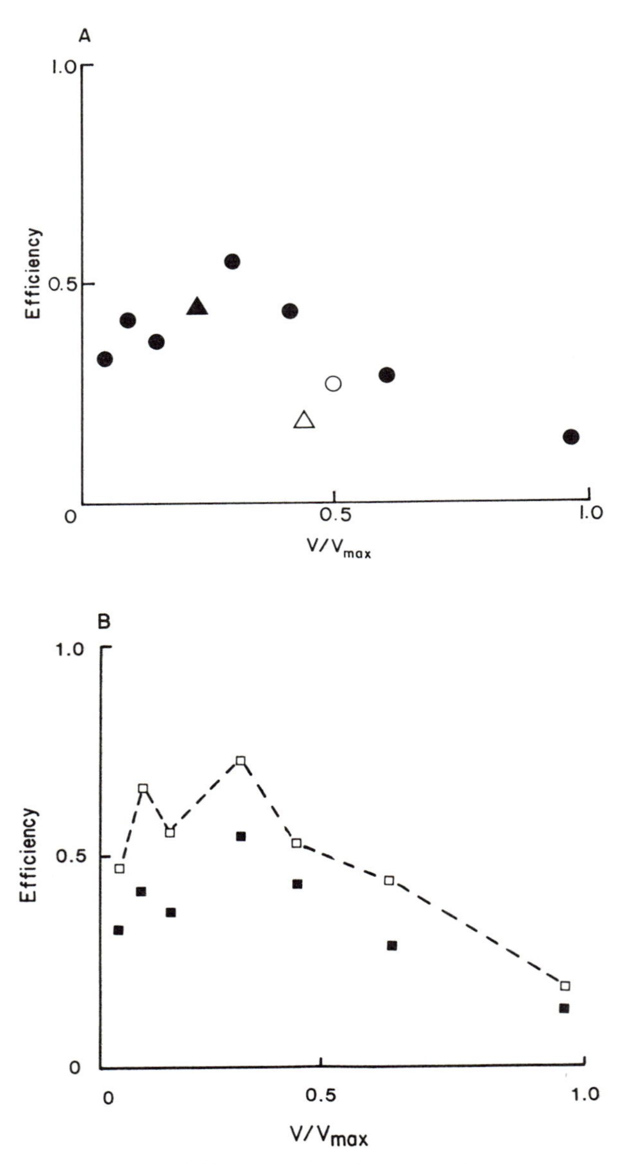

FIG. 4.38. (A) Efficiency (= work/free energy) during shortening at various velocities. The free energy is that shown in Fig. 4.36B for the extent of ATP cleavage that occurred during performance of work. Values of ATP cleavage and work are those shown in Fig. 4.37. (●) Kushmerick and Davies (1969); (▲) Curtin *et al.* (1974); (△) Rall *et al.* (1976); (○) Homsher *et al.* (unpublished). (B) Efficiency during shortening at various velocities. Filled symbols are the same as those in Fig. 4.38A and represent the efficiency calculated from the whole of the ATP cleavage during performance of work. (○) Efficiency calculated from net ATP cleavage, that is, total ATP cleavage − ATP cleavage for pumping Ca^{2+} into the sarcoplasmic reticulum.

calcium pumping, an estimate of the efficiency of the contractile process can be made (Fig. 4.38B). It should, however, be kept in mind that the exact quantity of ATP involved in the pumping of calcium is uncertain, and other processes also may be using ATP.

The maximum values of the efficiency shown in Fig. 4.38A and B are remarkably high compared to efficiencies calculated on the basis of work and free energies determined from oxygen consumption. When measurements are made of heat + work for a complete cycle of contraction, relaxation, and aerobic recovery, a reasonable assumption is that the enthalpy change comes from oxidation of glycogen, for which the free energy change is known to be 1.04 times the enthalpy change (Burk, 1929). Therefore, for such complete cycles, the thermodynamic efficiency is given by work/[1.04(heat + work)] = 0.25 for frog muscle (Wilkie, 1960). It seems that the recovery processes that occur after the work has been done are much less efficient than the energy conversion during the contractile process itself.

All the values of efficiency in Fig. 4.38A are for frog muscle. There is indirect evidence suggesting that the efficiency of tortoise (*Testudo graeca*) is higher than that of frog muscle. This evidence is that in tortoise muscle, in isotonic contractions, the ratio of work done to heat + work produced is 0.77, compared to 0.45 for frog muscle (Woledge, 1968). In isometric contraction, the ratio of heat + work produced to phosphocreatine split is about the same in both species (Walsh and Woledge, 1970). Assuming that the free energy from ATP cleavage is the same in tortoise and frog muscle, this suggests that the ratio of work done to ATP split in isotonic contractions would be found to be greater in tortoise muscle than in frog muscle. Also, the efficiency for complete cycles of contraction and aerobic recovery is greater in tortoise than it is in frog, 40 and 25%, respectively. It is desirable that a direct measurement of the efficiency of tortoise muscle during contraction should be made.

Work Balance

As described elsewhere in this chapter, it is clear that the energy (the sum of heat and work) produced during contraction under most conditions cannot be explained by the concurrent hydrolysis of ATP and the reactions that immediately resynthesize it. In other words, some other chemical reaction must be occurring, and since it has not been taken into account, there is an energy imbalance. Does this additional reaction participate in the fundamental contractile event? The energy balance comparison does not in itself answer this question because the unexplained energy may be due to non-crossbridge reactions. One approach to this question is to do a "work balance," since it is certain that work can only be produced by the contractile process. The idea is to compare the amount of work that is done by the muscle with the free energy available from the observed amount of ATP splitting. If the work exceeded the free energy from the observed chemical reactions, the unidentified reaction must participate in the contractile event, providing free energy to be converted into work. In fact as shown above (p. 261), work has never been found to exceed the available free energy; in other words, the observed efficiency, *w*/free energy, is always less than 1.0. As can be seen from Fig. 4.38A, the work requires at most 55% of the ATP splitting; the remaining 45% or more of the ATP splitting could be devoted to other processes. Kushmerick and Davies (1969) have estimated that ATP required for calcium pumping under the conditions of their experiment and calculated efficiency after taking this into account (Fig. 4.38B.) Thus, on the basis of "work balance" comparisons it is not necessary to suppose that the unidentified reaction, which certainly occurs during contraction, actually participates in the crossbridge reactions.

4.3.5 ATP CLEAVAGE DURING STRETCHING OF ACTIVE MUSCLE

The interpretation of studies of the heat production during stretching made clear predictions about how large the ATP cleavage in such tetani should be (see Section 4.2.4). A number of laboratories responded to the invitation (or provocation) to test these predictions. The chemical results were quite uniform in that no evidence was found for ATP synthesis or resynthesis during stretching (Aubert and Marechal, 1963b; Infante *et al.*, 1964; Marechal, 1964; Marechal and Beckers-Bleukx, 1965; Gillis and Marechal, 1974; Curtin and Davies, 1975).

Some results of the study by Infante *et al.* (1964) are shown in Fig. 4.39A. The design of the experiment was similar to that of Hill and Howarth (1959) in that the muscles were stretched at a constant velocity by a Levin–Wyman ergometer, but the velocity was a smaller fraction of V_{max} than that used by Hill and Howarth (1959). As shown in Fig. 4.39A, ATP splitting occurred at an approximately constant rate during 1.1 s of contraction under isometric conditions (solid line). The figure also shows the results for three sets of muscles that contracted under isometric conditions for a time and were then stretched at a constant velocity. The duration of the initial isometric part was different in the three cases. There was additional ATP split during the stretch as shown by the dotted line and the (○) symbols, which lie *above* the values of ATP split in the earlier, isometric part of each tetanus.

The second point about the experiment of Infante *et al.* (1964) is that the ATP splitting in contractions with stretch is less than that in an isometric tetanus of the same total duration. At the time the stretch ends, the ATP

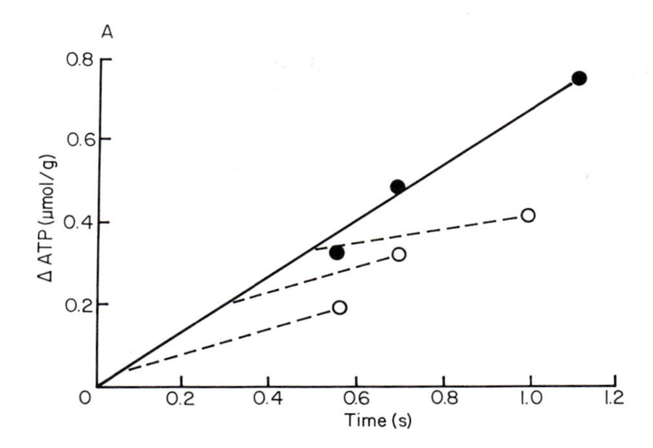

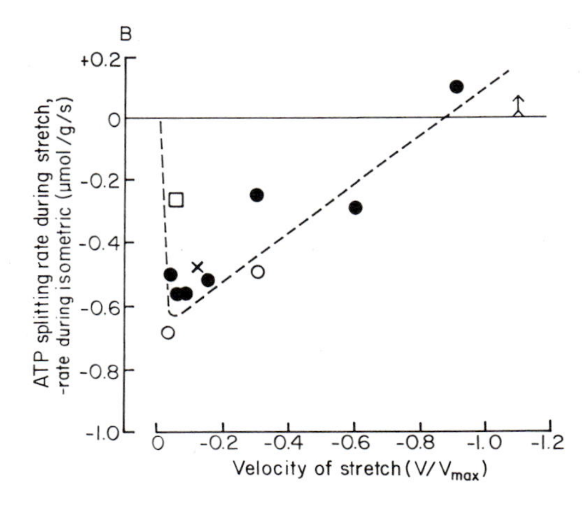

FIG. 4.39. (A) The time course of ATP splitting by frog sartorius (*R. pipiens*) at 0°C. (●) Results for isometric tetani (the solid line was drawn through them by eye). (○) Total ATP used during tetani in which the muscle was held under isometric conditions for a time and then stretched. The ATP splitting is taken to follow the solid line during the isometric period and then follow the dotted line during the stretch. The stretching was controlled by a Levin–Wyman ergometer and occurred at a constant rate, $0.24L_0$/s (about $0.1V_{max}$). (Based on Infante *et al.*, 1964, and Davies, 1965b.) (B) The difference between the rate of ATP splitting during stretching at various velocities and that during an isometric tetanus (frog sartorius at 0°C). (●) Curtin and Davies (1975), DNFB-treated muscles; (○) Curtin and Davies (1975), untreated muscles; (×) Infante *et al.* (1964); (□) Marechal (1964). Muscles performed a series of contractions; the value shown was calculated assuming that all ATP splitting occurs during stimulation. Arrow, Aubert and Marechal (1963b), indicates that the rate of ATP splitting was greater during stretching than during isometric contraction, but authors did not give the value.

split (○) is always below the solid line. Thus, the effect of stretch is to *reduce* the rate of ATP splitting, rather than to reverse it. This conclusion is consistent with the results of the heat experiments explained earlier in this chapter.

A striking feature about the rate of ATP splitting is that it is very strongly dependent on the velocity of stretch, as shown in Fig. 4.39B, which summarizes results of a number of different experiments. The results have been plotted as the difference between the rate of splitting during stretch and that during isometric contraction as measured in the same set of experiments. It seems that lowest rates of splitting occur during slow stretches. These rates are close to zero and very much less than the isometric rate. At higher velocities of stretching, the rate of ATP splitting increases so that at velocities equivalent to V_{max}, the rate may actually be equal to that during isometric contraction; however, more experiments would be needed to prove this point. The need for more experiments is further emphasized when Fig. 4.39B for the rates of ATP splitting is compared to Fig. 4.18B for the rate of energy production (see Section 4.2.3). These two variables would be expected to have the same dependence on velocity if no other processes were involved. The fact that the ATP splitting increases at higher velocities of stretch, whereas the energy production does not, implies that energy is being stored in some other form, probably by coupling to an endothermic process.

A possible flaw in this reasoning is that although there may actually be ATP synthesis during stretching, the simultaneous splitting that occurs may be so large that the net effect, which is all that can be detected in the chemical experiments described above, is splitting. Experiments on glycerinated fibres and with ^{32}P-labelled phosphate have been done to test this idea. The results indicate that a small amount of ATP synthesis does occur, but the synthesis is no greater during stretching than during shortening or during isometric contraction (Gillis and Marechal, 1974).

The Relation between Force and ATP Cleavage after Stretching

Active muscle produces more force during stretching than when isometric at the same length (see Chapter 2, Section 2.2.2). In contrast, the net energy produced and the chemical changes are smaller during stretching, as explained earlier. These results suggest that the crossbridge cycle itself is affected during stretching. Also, it has been clearly established that stretching changes the mechanical properties of muscle so that it continues to produce higher forces after the stretching has been completed (see Chapter 2, Section 2.2.3). The heat production and ATP hydrolysis during the period after stretch have been studied by Curtin and Woledge (1979b) (Fig. 4.19). ATP hydrolysis was measured only during the period from 1.5 to 6.5 s of stimulation, that is, after the stretch was complete. As expected from earlier studies, more force was produced following stretch than in a purely isometric tetanus. Somewhat more ATP was split after stretch than in the isometric control. Thus, it seems that events are different from those *during* stretch, when extra force is produced with less ATP cleavage, than under isometric conditions. If the extra force after stretch were due to more of the same cross-bridge cycles as occur under isometric conditions, then ATP cleavage and (on the simplest assumptions) energy production should increase in proportion to the tension–time integral ($\int Pdt$). Measurement of these parameters showed that after stretch the ratio, $\int Pdt{:}\Delta\text{ATP}$, was greater than that for a purely isometric tetanus. However, the difference was not large compared to its standard error ($P = 0.112$). The corresponding comparison for energy production, $\int Pdt/(h + w)$, was greater after stretch than in a purely isometric tetanus. In this case the difference was quite convincing, $P < 0.031$. Taken together, these results suggest that stretching causes long-lasting changes in the way the muscle produces force. After stretching the muscle seems to have intermediate properties; that is, it can produce less force/ATP hydrolysed than during stretch, but more than during a purely isometric tetanus.

4.3.6 CHEMICAL ENERGY BALANCE

Another approach to energy balance for contraction is that based on biochemical information (see Kushmerick, 1984, for a more complete description). It is based on the idea that the biochemical recovery processes, for example glycogenolysis and oxidation, which occur *after* a brief contraction, effectively reverse the ATP cleavage and the creatine kinase reaction which occur as a result of contraction. In other words, recovery processes balance with the initial processes. For muscle constrained to metabolize carbohydrate, this idea can be summarized as

$$k\xi_{\Delta O_2} + l\xi_{\Delta \text{lact}} = \xi_{\sim \text{P}} \qquad (1)$$

where ξ is the extent (moles) of the reaction indicated by the subscript; $\xi_{\sim \text{P}}$ refers to usage of high-energy phosphate compounds; $\xi_{\Delta O_2}$, to oxidation; and $\xi_{\Delta \text{lact}}$, to the glycolytic pathway. k and l are the coupling coefficients indicating stoichiometry. This equation is analogous to that used in the myothermal energy balance studies

$$\text{heat} + \text{work} = \xi_{\sim \text{P}} \Delta H_{\sim \text{P}} \qquad (2)$$

Scopes (1973, 1974) studied the glycolytic pathway using isolated enzymes under the conditions believed to exist in muscle. A value of $l = 1.5$ was observed, which is the expected stoichiometric value. This means that for each glycosyl unit metabolized to 2 lactate molecules, 3 molecules of ATP are formed. Kushmerick (1984) has summarized the uncertainties about the exact value of k, the coefficient for oxidation, and the reasons for supposing that the value 6 applies in frog muscle. Taking this value and other processes into account, the complete oxidation of 1 mol of glycosyl unit results in the formation of 36 mol of ATP. Thus, under adequately aerobic conditions ATP formation from glycolytic reactions would be very small compared to that from oxidation, approximately 3/36 per mole of glycosyl unit; the contribution to recovery from glycolytic reactions can often be taken to be negligible. Thus Eq. (1) reduces to

$$k\xi_{\Delta O_2} = \xi_{\sim P} \tag{3}$$

The biochemical energy balance approach was used by Kushmerick and Paul (1976a,b) in studies of sartorius of *R. pipiens* at 0°C. Their main finding was that comparison of $\sim$P usage ($\xi_{\sim P}$) during stimulation, with O_2 usage during recovery ($\xi_{\Delta O_2}$) gave ratios of 4 rather than the expected value of $k = 6$ discussed above. Isometric contractions of durations up to 20 s consistently gave a ratio of 4. Assuming that recovery does return the muscle to its initial precontraction state, this result means that $\sim$P usage during stimulation underestimates the real energetic cost of contraction, probably because there is extra usage of $\sim$P after stimulation is over.

The extra $\sim$P usage would have to amount to about half of that during stimulation. However, in experiments intended to detect postcontractile usage of $\sim$P, it has been elusive for reasons that are not clear. In a number of studies, no net change in $\sim$P compounds was detected after contraction of muscle in which recovery processes were prevented (Marechal and Mommaerts, 1963; Dawson *et al.*, 1977; Kushmerick and Paul, 1976b). However, in other studies, a variable amount of reaction has been detected after contraction (Lundsgaard, 1934; Spronck, 1965, 1970; Wilkie, 1975; De Furia and Kushmerick, 1977). A study of initial and recovery processes (usage of $\sim$P and O_2) for the entire period of contraction and recovery seems necessary to solve this question.

Do the results of these biochemical energy balance experiments throw any light on the questions raised by the myothermal energy imbalance? The myothermal studies show that more $h + w$ is produced during contraction and relaxation than can be explained by the simultaneous initial processes, ATP cleavage, and the creatine kinase reaction. An unidentified reaction must occur and produce part of the $h + w$ during the initial period, and it has been suggested that it is reversed during subsequent recovery at the expense of additional $\sim$P usage. Is this idea consistent with the biochemical results? As already described, the O_2 consumption could provide more $\sim$P energy than is needed to explain the initial processes detected as ATP splitting. When the O_2 usage that resynthesizes the ~P used in contraction is deducted from the total O_2 consumption, is there enough excess oxidation to account for the unexplained $h + w$ during contraction that is detected in the myothermal studies? To answer this question, some assumptions must be made. If the stoichiometry of the reversal of the unidentified reaction by ~P is 1:1, its extent is 1.07 μmol/g in a 5-s tetanus (total ~P from oxidation minus ~P used during contraction). If the ΔH_m of the unidentified reaction is assumed to be −24 kJ/mol, then the observed oxidation (O_2 usage) would produce just enough ~P to account for all the $h + w$ produced during a 5-s contraction of *R. pipiens* and the reversal of the unidentified reaction afterwards (see Table 4.XIV). This example shows that it is not necessary to make unusual or unreasonable assumptions about the unidentified process in proposing a scheme that gives an over-all energy balance and is consistent with the results of both the myothermal and biochemical balance studies (see also Section 4.2.4 concerning recovery heat and its interpretation in relation to energy balance).

There is another parallel between the results of the biochemical and myothermal energy balance studies. Kushmerick and Paul (1976b) found that there is a component of the oxygen usage during recovery which is "labile"; that is, the amount of oxygen usage after a series of tetani is smaller

TABLE 4.XIV
Comparison of biochemical and myothermal results for a 5-s tetanus of *R. pipiens*, 0°C[a]

	~*P* (μmol/g)			
ξ_{O_2} (μmol/g)	Calculated total from ΔO_2	Observed usage during tetanus	Calculated excess	Unexplained $h + w$ (mJ/g)
0.555 ± 0.62 (8)	3.33 ± 0.37	2.26 ± 0.75 (30)	1.07 ± 0.84	26

[a] The extent of O_2 usage (ξ_{O_2}) is the total O_2 used for recovery (Kushmerick and Paul, 1976a,b). The amount of ~P produced by oxidation (column 3) is calculated as $6 \times \xi_{O_2}$. Column 4 shows the amount of ~P used during tetanus (Kushmerick and Paul, 1976a,b). Column 5 is the difference between columns 3 and 4 and is the amount of ~P produced by oxidation, but not used in the tetanus. The last column is the amount of energy produced during the tetanus that cannot be explained by simultaneous usage of ~P (Homsher and Kean, 1978). An unidentified reaction with a ΔH_m of about 27 kJ/mol could produce the unexplained energy, and be reversed subsequently coupled to usage of ~P with a stoichiometry of 1:1.

when the intervals between tetani is brief than the usage when the intervals are long (200 s). It seems that the cause of this part of the O_2 usage is depleted during contraction (and therefore called "labile") and a long time is required for it to be reestablished. In myothermal energy balance experiments on two contractions separated by a brief interval (3 s), it was found that there was less unexplained $h + w$ in the second of the contractions (Curtin and Woledge, 1977); thus, the reaction responsible for the unexplained $h + w$ also seems to be "labile."

There is, however, a feature of the biochemical and myothermal results that should be clarified: the relation between the time course of the production of unexplained $h + w$ in *R. pipiens* and the amount of excess O_2 consumption during recovery. The biochemical balance studies show that the amount of O_2 usage during recovery exceeds that required to resynthesize $\sim$P used during stimulation; the important point is that the amount of the excess O_2 is proportional to $\sim$P during tetani lasting up to 20 s. If the excess of O_2 usage is financing the unexplained $h + w$, this result means that the unidentified reaction that is the source of the $h + w$ should continue for at least 20 s. In contrast, myothermal studies of *R. temporaria* show that the unidentified reaction proceeds for between 5 and 10 s and then stops. So it seems from the biochemical results for *R. pipiens* and the myothermal results for *R. temporaria* that the time course of the unidentified reaction we have been discussing is different in these two species. This is not an unreasonable idea; the first 5 s of an isometric contraction of *R. pipiens* produces much less unexplained $h + w$ than *R. temporaria* does (see Fig. 4.33). It could be that the unexplained $h + w$ is produced for a much longer time (at least 20 s) by *R. pipiens* than *R. temporaria*. If this were the case the results for the biochemical and myothermal studies would agree very well. Additional experiments—either myothermal energy balance studies using *R. pipiens* for long contraction or measurements of recovery oxygen consumption using *R. temporaria*—are needed.

A biochemical energy balance study has also been done by Crow and Kushmerick (1982) on extensor digitorum longus and soleus muscle of the mouse at 20°C. The results were quite different than those we have been discussing for frog muscle. Crow and Kushmerick found that in mouse muscle there *is* a balance between initial processes and recovery processes for tetanic isometric contractions lasting up to 15 s. A balance was also found for soleus muscles in which initial processes and oxygen consumption had reached a steady state during a tetanus of long duration. These results suggest that the initial processes would balance energy produced as heat and work during contraction under these conditions too, but this has not been tested experimentally. In the one myothermic energy balance study that has been done on mammalian muscle, Gower and Kretzschmar (1976) found that in rat soleus only 70% of the initial heat + work could be accounted for by chemical changes during a 10-s tetanus; clearly more experiments are needed on mammalian muscles. If mouse muscles do not produce unexplained energy during contraction as suggested by the results of Crow and Kushmerick, it would be useful to establish which characteristics that distinguish these muscles from frog muscle are associated with this difference in energetic properties.

5 Crossbridge Theories of Muscle Contraction

5.1 A. F. Huxley's 1957 Theory

A. F. Huxley (1957) was the first to explore in detail the predictions about muscle behaviour that can be made from simple assumptions about the kinetic properties of individual crossbridges. The ideas he put forward have been very influential and have been elaborated and commented on by many authors. Many of the more recent theories are easily recognized as variants of the 1957 theory. Some thermodynamic and kinetic constraints which are inherent in these theories have been restated by T. Hill (1974, 1977). Later authors have often presented their ideas in the form of the diagrams popularized by these papers and used also in this chapter. We shall now give a simple account of the original form of Huxley's theory (referred to as the 1957 theory) and list its successes and failures in describing the behaviour of muscle.

There are three premises in the 1957 theory:

1. A crossbridge can exist in one of a number of states. The word *crossbridge* is used to denote a myosin head (an S-1), whether or not it is attached to actin. The 1957 theory has only two states (A, attached, and D, detached: Fig. 5.1A); in later theories additional states are included. It is supposed that a crossbridge splits an ATP when it passes in sequence through the states (clockwise in Fig. 5.1A). Consequently, the energy from ATP splitting drives cyclic attachment and detachment of the crossbridges.

2. The crossbridge exerts force on the thin filament only when attached. The amount of force exerted (F) changes, if the thick and thin filaments are displaced with respect to each other. This displacement is represented by the variable x (Fig. 5.1B). F is assumed to be proportional to x ($x = 0$ when $F = 0$) and can be either positive or negative (Fig. 5.2B). As the muscle shortens x decreases. The distance between sites on the actin filament to which the crossbridge can attach (l in Fig. 5.1B) is assumed to be large enough that only one such site at a time is "within range" of the crossbridge.

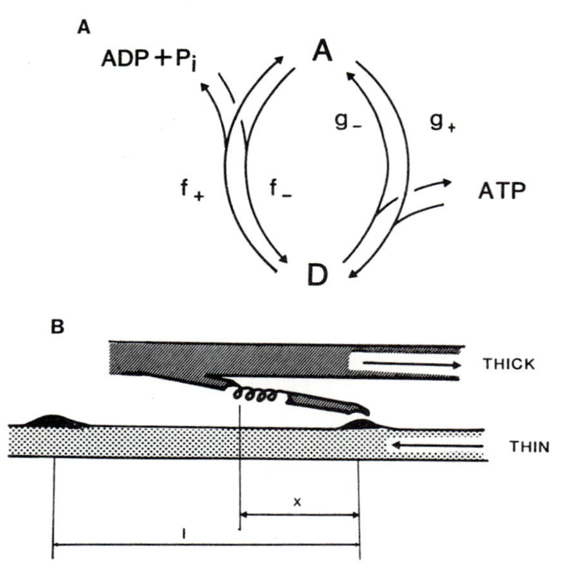

FIG. 5.1. (A) Huxley's 1957 theory postulates the existence of two crossbridge states: attached (A) and detached (D). The cyclic attachment and detachment is driven by ATP splitting. f_+ and g_+ are the rate constants for the forward (clockwise) processes; f_- and g_- are those for the reverse processes. (B) The 1957 theory supposes that the thin filament has sites spaced at a distance l from one another to which a crossbridge can attach. The crossbridge contains a spring so that the attachment site can move with respect to the thin filament. The variable x describes this movement, and is zero when the force in the spring is zero. The arrows show the direction of the relative movement of the filaments during shortening.

3. The rate constants for the transitions between the states are functions of x. The functions are chosen such that the crossbridges attach more rapidly at positive values of x than at negative values, while the converse is true for detachment. This asymmetry causes shortening and the development of force (Fig. 5.2A).

There are four rate constants for a two state theory: f_+, f_-, g_+, g_- (Fig. 5.1A). The concentrations of ATP, ADP, and P_i are assumed to be constant so that all the rate constants can be made pseudo-first order. The values of f_+ and g_+ assumed are shown in Fig. 5.2A. The values of f_- and g_- cannot be chosen independently but are fixed by the following considerations:

(a) The rate constants are related to the change in free energy for ATP splitting (e in Huxley's notation)

$$\begin{aligned} e = \Delta G_{\mathrm{ATP}} &= kT \ln(f_+/f_-)(g_+/g_-) \\ &= kT \ln(f_+/f_-) + kT \ln(g_+/g_-) \\ &= \Delta A_f + \Delta A_g \end{aligned}$$

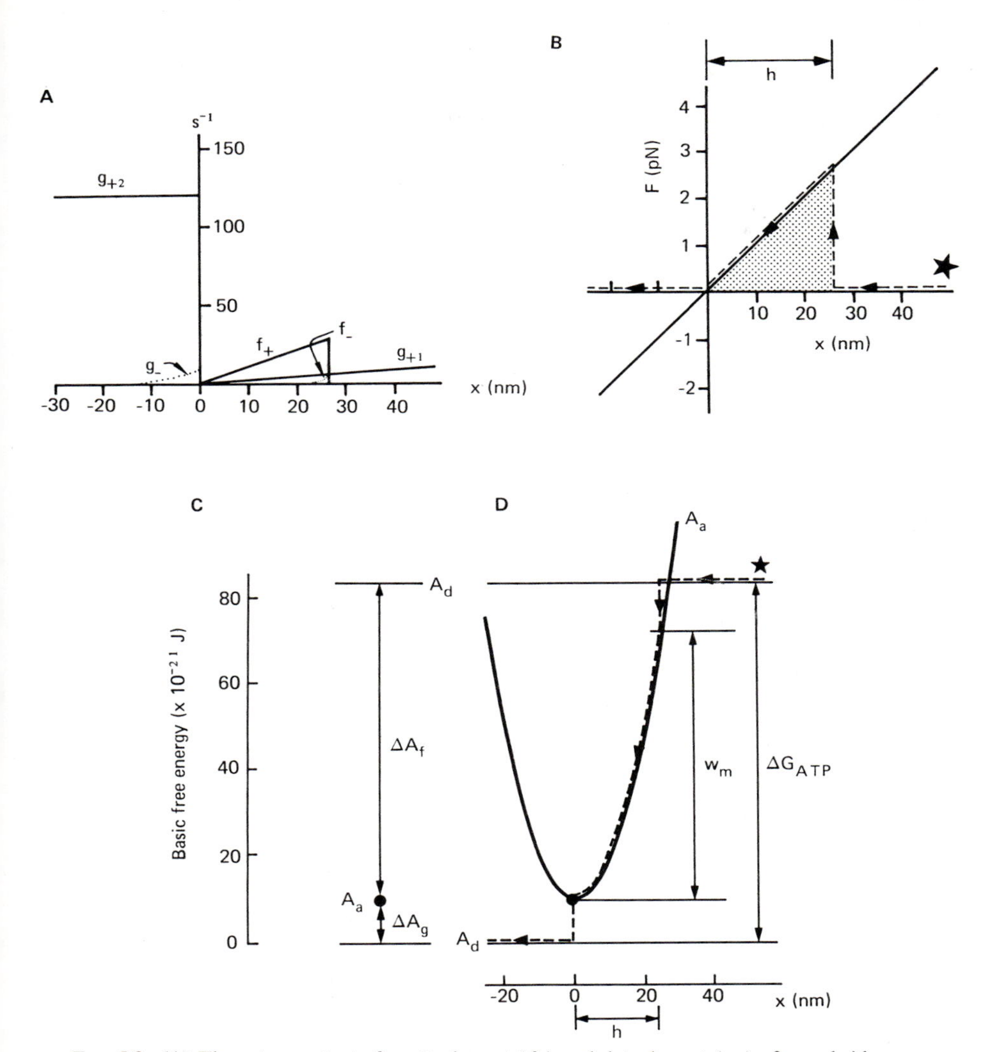

FIG. 5.2. (A) The rate constants for attachment (f_+) and detachment (g_+) of crossbridges are assumed to be functions of x as shown here. The values are chosen to give an appropriate description of the behaviour of frog muscle at 0°C (see text). The dotted lines are the values of the rate constants for the reverse reactions calculated as described in the text. (B) The force in the crossbridge is a linear function of x and can be either positive or negative. The absolute value of F is chosen so that the shaded area (w_m) is $0.75\Delta G_{ATP}$ (see text). (C) The basic free energy change of a system of one crossbridge (at $x = 0$) and a large supply of ATP. A_d is the free energy with the bridge detached; A_a with it attached. ΔA_f and ΔA_g are the free energy changes for the attachment and detachment processes. (D) The x dependence of the quantities shown in (C). h is the range of x over which attachment is possible. The dashed line shows the path a crossbridge follows during shortening when it does the maximum possible amount of work.

where k is Boltzmann's constant and T is absolute temperature. The terms $kT \ln(f_+/f_-)$ and $kT \ln(g_+/g_-)$ are referred to as the "basic" free energy change for the attachment processes and are represented by the symbols ΔA_f and ΔA_g. These quantities, unlike ΔG_{ATP}, are functions of x. The basic energy differs from the standard free energy in that the latter is the value for 1 M concentration of reactants and products, the former for other specified, constant concentrations, which are usually chosen to be close to those thought to exist *in vivo*.

(b) The relation between F and x specified in premise 2 above determines how the free energy content of the attached state (A_a) varies with x. Because the free energy content of the detached state (A_d) is independent of x, this variation in A_a causes an x dependence in ΔA_f, and ΔA_g, and, therefore, in the ratios f_+/f_- and g_+/g_-.

It is next necessary to assign values to these constants, and a method of choosing these values is illustrated in Fig. 5.2C and D. Figure 5.2C is for the simplest case in which there is no force in the crossbridge ($x = 0$). The diagram refers to a "system" consisting of one crossbridge and a large supply of ATP. The two horizontal lines (A_d) represent the free energy content of the system in a detached state, before (upper) and after (lower) splitting one ATP, and are thus separated by ΔG_{ATP}. The symbol (●) shows the free energy (before the ATP splitting) when the crossbridge is attached. ΔA_g and ΔA_f are also shown. In Fig. 5.2D the dependence of these quantities on x is shown. As x is increased, the free energy content of the attached state (A_a) rises along the parabola because of the work (w) done on the crossbridge,

$$w = \int_0^x F\,dx = \int_0^x kx\,dx = kx^2/2$$

Therefore ΔA_f decreases and ΔA_g rises.

Consider an actin site moving past the crossbridge from * (Fig. 5.2B,C, and D). If the crossbridge attaches as soon as the site is within the region where f_+ is not zero (at $x = h$), and detaches at $x = 0$ (as shown by the broken line), the amount of work done in this crossbridge cycle is w_m. This work is the maximum that can be done in a single cycle. Huxley used the value of $w_m = 0.75\Delta G_{ATP}$, as illustrated. For this value of w_m, the values of the rate constants for g_- and f_- are shown in Fig. 2A and B. As the value of f_- is small compared to f_+, and the value of g_- is small compared to g_+, Huxley did not include these back reactions in his quantitative treatment. If w_m were larger, the values of ΔA_f at $x = h$ and of ΔA_g at $x = 0$ would be smaller. The crossbridge would be working in regions of x where ΔA_f or

ΔA_g would be quite small and in these regions f_- and g_- would therefore be larger and would have to be included in any quantitative analysis.

The 1957 theory is completely specified by the six constants (Fig. 5.2A,B, and D and Table 5.I): h, l, f_+, g_{+1}, g_{+2}, and w_m. It remains to describe how the predictions of the theory compare with the behaviour of muscle. Some insight into the way the theory works can be gained from Fig. 5.3, in which Fig. 5.3A shows the proportion of crossbridges attached (n) for each x value during steady-state shortening at various velocities. Figure 5.3B shows the tension contributed by bridges at each x value. The total force (P) exerted by all the bridges is proportional to the area enclosed by this curve.

In isometric contraction, $v = 0$, bridges can only attach within the region $0 < x < h$ because $f_+ = 0$ elsewhere. During shortening there are fewer attached crossbridges in this region, particularly at values of x approaching h. The force exerted is less because of this, and also because there are crossbridges in the region $x < 0$, which are exerting negative force. At v_{max} the total force exerted is zero, because this negative force is equal to the positive force from the region $0 < x < h$.

As bridges can only attach within the region $0 < x < h$ the turnover of bridges is proportional to $\int_0^h (1 - n) f_+ \, dx$. At higher velocity, n is less within this region and therefore the rate of turnover of bridges is greater.

Huxley obtained the equations (given here in a simple form in Table 5.I) for the force (P) and rate of free energy output (E) of this system as a function of shortening velocity (v). These equations can be used to compare the theory with the observed behaviour of frog sartorius muscle at 0°C. The comparison is best described in two stages: first (points A-1 through A-4 below) using the dimensionless, normalized muscle properties P/P_{max}, v/v_{max}, $E_0/P_{max}v_{max}$, E/E_0, and the efficiency Pv/E (where E is the rate of free energy output, as work and heat, and E_0 is the value of E for isometric contraction). Second (points A-5 and A-6 below), the absolute values for v, P, and E can be compared with those observed.

A-1. The predicted force velocity curve is similar in shape to that observed for frog muscle if $g_{+2}/(f_{+1} + g_{+1})$ is given a value of 4 (Fig. 5.4). Larger values of this ratio give more curved force–velocity curves.

A-2. The predicted ratio of energy output in isometric contraction (E_0) to the product $P_0 v_{max}$ is 0.063 if f_+/g_{+1} is given a value of 4.3. The observed value of the ratio of the stable rate of heat production in an isometric tetanus to $P_0 v_{max}$ is also 0.063. This heat production is all explained by ATP splitting and the rephosphorylation of ADP by PCr (see Chapter 4). For these processes the free energy dissipated is greater than the heat production by 45% (Kushmerick and Davies, 1969; Curtin and Woledge, 1978), but on the other hand about 25% of the ATP is probably split by ATPases other

TABLE 5.I
Equations describing the behaviour of the 1957 theory

Symbols	Definitions	Units
f_+	Rate constant for attachment (Fig. 5.3A). f_{+1} is the maximum value	s^{-1}
g_+	Rate constant for detachment (Fig. 5.3A). g_{+1} and g_{+2} are the values for $x > 0$ and $x < 0$, respectively	s^{-1}
x	Distance from myosin to actin site (Figs. 5.2 and 5.3B)	nm
l	Distance apart of actin sites (Fig. 5.2)	nm
h	Maximum value of x at which attachment can occur (Fig. 5.2)	nm
s	Sarcomere length	nm
e	Free energy for splitting an ATP molecule	10^{-21} J
w_m	Maximum work that can be done by one crossbridge cycle	10^{-21} J
v	Velocity of shortening (for one half-sarcomere)	nm/s
P	Force produced (per crossbridge site)	pN
E	Rate of free energy output (per crossbridge site)	10^{-21} J/s
v'	Velocity relative to maximum value (v_{max})	—
P'	Force relative to maximum value (P_0)	—
E'	Rate of free energy output relative to the product $P_0 v_{max}$	nm/s
ϕ	$(f_{+1} + g_{+1})h$	—
d	$(f_{+1} + g_{+1})/g_{+2}$	—
c	ϕ/v_{max}	—

The following equations can be easily derived from those given by Huxley (1957) using the definitions above

$$E = e\frac{h}{2l}\frac{f_{+1}}{f_{+1} + g_{+1}}\left[g_{+1} + f_{+1}\frac{v}{\phi}(1 - e^{-\phi/v})\right]$$

$$P = \frac{w_{max}}{2l}\frac{f_{+1}}{f_{+1} + g_{+1}}\left[1 - \frac{v}{\phi}(1 - e^{-\phi/v})\left(1 + \frac{vd^2}{2\phi}\right)\right]$$

from which it can be shown that $c \approx d$.

Using this and the appropriate definitions above we obtain

$$P' = 1 - \frac{v'}{c}(1 - e^{-c/v'})\left(1 - \frac{d^2v'}{2c}\right) \approx 1 - \frac{v'}{d}(1 - e^{-d/v'})\left(1 - \frac{dv'}{2}\right)$$

$$E' = \frac{e}{w}\,\frac{d\left(\dfrac{g_{+1}}{f_{+1}} + \dfrac{v'}{d}\right)(1 - e^{-d/v'})}{(g_{+1}/f_{+1} + 1)}$$

The maximum velocity is given by

$$v_{max} = \frac{g_{+2}h}{2}$$

The rate of ATP splitting, per myosin site, in isometric contraction (k_{cat}, in s^{-1}) is given by

$$k_{cat} = \frac{h}{2l}\frac{f_{+1}g_{+1}}{(f_{+1} + g_{+1})}$$

The isometric force (per myosin site) is given by

$$P_0 = \frac{w_m}{2l}\frac{f_{+1}}{f_{+1} + g_{+1}}$$

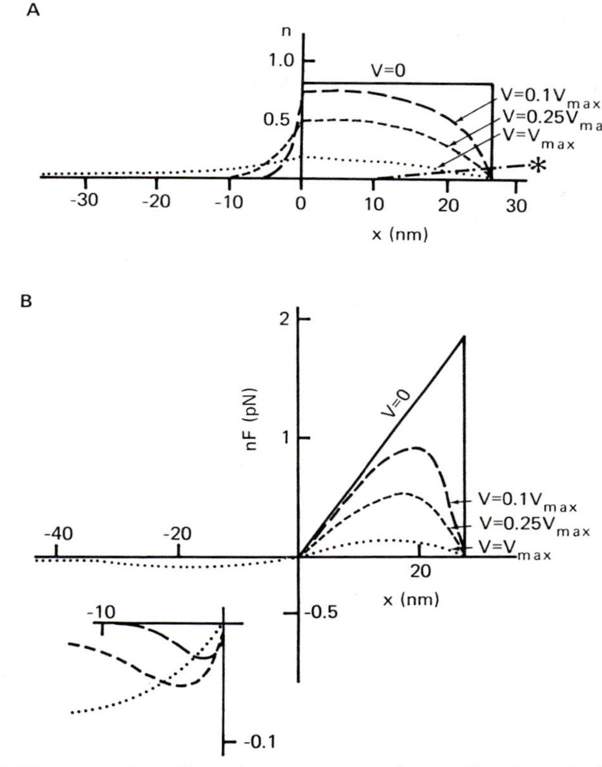

FIG. 5.3. (A) The proportion of crossbridges attached (n) as a function of x for four different velocities of shortening (v). The line marked with the asterisk shows for $v = v_{max}$ the proportion of crossbridges that might be still attached to the previous site. (B) The total force exerted by the bridges at each value of x. This is the product of n and F (see Fig. 5.2B). The inset shows part of the curve on larger scale.

than the crossbridges. Thus the observed value of $E_0/P_0 v_{max}$ for the crossbridges is in fact about $0.063 \times 1.45 \times 0.75 = 0.068$.

A-3. As w_m/e is assumed in the theory to have a value of 0.75 (see Fig. 2C and D), (a) the predicted rate of free energy dissipation when shortening at v_{max} is five times greater than in isometric contraction; (b) E/E_0 is approximately a linear function of $(P_0 - P)$; (c) the rate of free energy dissipation $(E - Pv)$ is predicted to be an approximately linear function of v (Fig. 5.5A); (d) the predicted efficiency has a maximum value of 0.44. These four properties are similar to those considered in 1957 to be characteristic of frog muscle. More recent studies of muscle energetics have altered this picture, as is described below (p. 287).

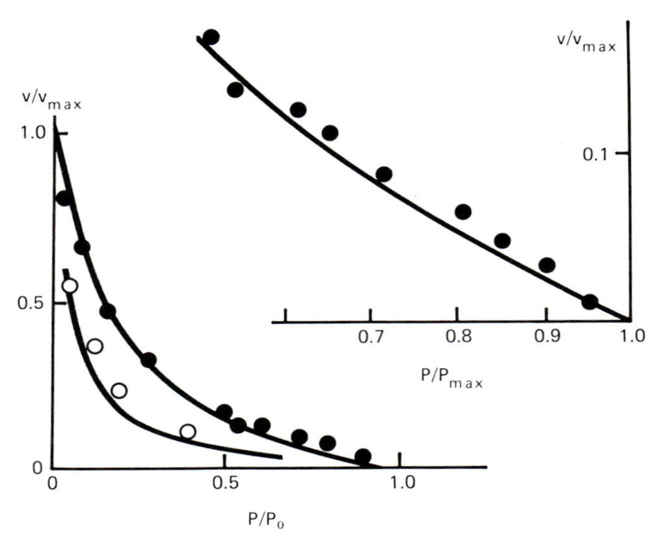

FIG. 5.4. Predicted and observed force–velocity curves. (●) Data points for frog muscle from Edman and Hwang (1977). The line drawn through these points is from the appropriate equation in Table 5.I with $g_{+2}/(f_{+1} + g_{+1}) = 3.9$. (○) Data for tortoise muscle from Woledge (1968). The line through these points is from the same equation, with $g_{+2}/(f_{+1} + g_{+1}) = 10$.

A-4. The values of these normalized parameters in tortoise muscle can also be described quite well if $g_{+2}/(f_{+} + g_{+1})$ is given a value of 10 and the values of f_{+}/g_{+1} and w_m/e are 14 and 0.95, respectively. This is illustrated in Fig. 5.5B. The model thus successfully predicts that, with w_m/e constant, the very curved force–velocity curve of tortoise muscle and its low normalized maintenance heat rate should be accompanied by a higher efficiency than that of frog muscle.

In considering absolute values of the rate of shortening, of the rate of ATP splitting, and of the force exerted, some useful equations are given in Table 5.I. Values of the distances l and h can be obtained as follows: the pitch of the actin helix 0.035 μm is taken as l, the distance between actin sites to which a crossbridge can attach, thus assuming that there is only one possible attachment site per turn of the helix, and that the crossbridge can only attach to one thin filament. It is not clear from structural evidence whether or not these simple assumptions are reasonable. h is taken as 0.027 μm from the observation that a moderately rapid release of half this distance is sufficient to drop the tension to zero (see Fig. 2.27). For the moment we ignore the effects of very rapid releases but shall return to this point later.

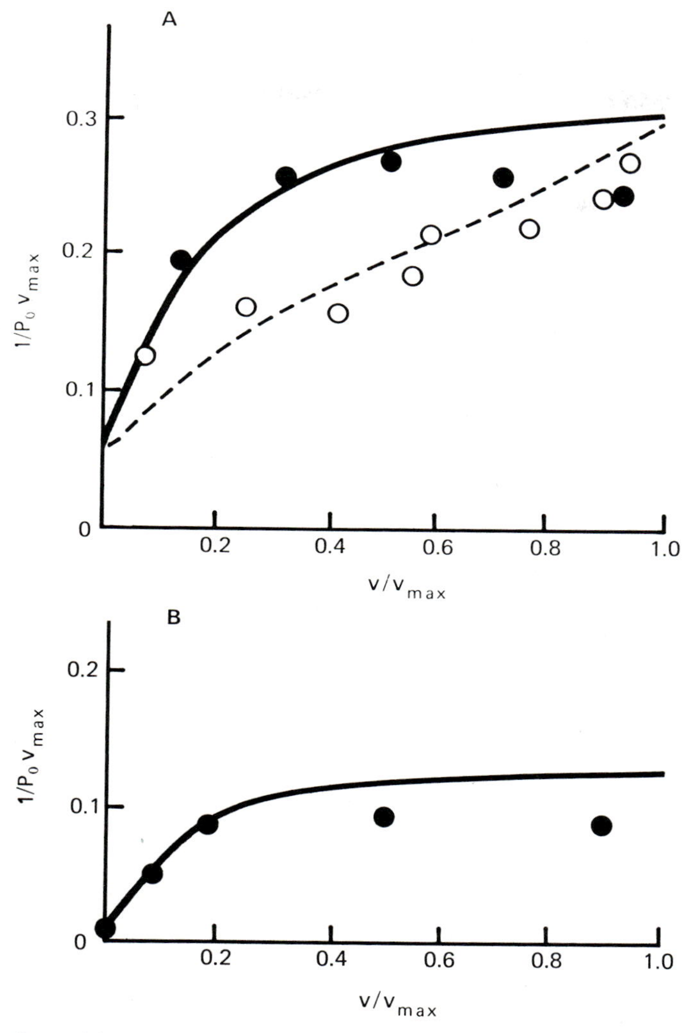

FIG. 5.5A. Rate of free energy output E (●, and full lines) and of free energy dissipation $E - PV$ (○, and broken lines) as function of the normalized shortening velocity (v/v_{max}). The free energy rates are normalized by $P_0 v_{max}$. The lines are calculated from the equations in Table 5.1. The points are the results of A. V. Hill (1964b) for frog sartorius muscle at 0°C replotted as described in Chapter 4, Section 4.2.2 with the additional assumption that the ratio of free energy change to enthalpy change is 1.45. (B) Predicted and observed free energy ouput rates for tortoise muscle, plotted on the same scale as (A). The line drawn is the prediction (see text) and the points are the observations of Woledge (1968) on the assumption that the ratio of free energy output to enthalpy output is 1.45.

A-5. The isometric force exerted per unit cross-sectional area of muscle can now be predicted from

$$P_0 = 0.305(s/l)me$$

where s is the sarcomere length and m is the number of crossbridges per unit volume of muscle. Taking m as 0.28 nmol/mm^3 (Ebashi *et al.*, 1969), the value of P_0 predicted is 0.24 N/mm^2. The observed values range from 0.18 to 0.30.

A-6. The equations for v_{max} and for k_{cat} (the rate of hydrolysis of ATP per crossbridge in isometric contraction) will be considered together because both contain g_{+1}, which is a parameter to which we have not yet assigned a value. We shall proceed by obtaining a value of g_{+1} from the observed v_{max} and then predict a value of k_{cat}. v_{max} is 1.6 μm/s per half-sarcomere (Edman, 1979) giving $g_{+1} = 6.1$ s^{-1}. The predicted value of k_{cat} is 1.8 s^{-1}. This can be compared to the value of 2.5 s^{-1}, obtained by dividing the observed ATPase rate (in mol g^{-1} s^{-1}) by the estimated myosin content (mol/g). The agreement is good.

A-7. The value of g_{+1} (the rate constant for bridge breaking) might also be compared to the rate of relaxation. Relaxation, however, is a complex process in which the rate of removal of calcium, leading to a decrease in the rate of bridge formation, is a major rate-limiting process. There are also redistributions of length to be considered, in which parts of the relaxing fibre shorten and stretch other parts (Huxley and Simmons, 1973) and which greatly influence the rate of tension decline. Nevertheless, it is possible that toward the end of relaxation there may be a time when the length redistributions are over and new bridges are forming at a negligible rate. If so, the remaining tension should decay with a rate constant of the same order as g_{+1} (6 s^{-1} for frog muscle). A more precise prediction is difficult also because g_+ is a function of x. In fact, the last part of relaxation has been found to be exponential with a rate constant of 6 s^{-1} (Ritchie and Wilkie, 1958). The similarity also exists in tortoise muscle for which g_{+1} is 0.12 s^{-1} and the time constant of the last part of relaxation is 0.15 s^{-1}.

A-8. As we have described in Chapter 2, a discontinuity in the force–velocity curve is observed at $v = 0$. This feature is correctly predicted by the theory. Force increases with velocity for slow lengthening about four times faster than it decreases for slow shortenings. However, the validity of the theory in this region is probably rather restricted, as we discuss below (point B-3).

A-9. The compliance of muscle was not explicitly discussed in the 1957 theory but recent research has shown that nearly all the compliance of muscle is located in the crossbridges (Chapter 2). The theory does make predictions about how the number of crossbridges varies with the velocity of shortening (Fig. 5.3A) and therefore predicts that the compliance of muscle should rise as velocity increases. Experiments show that this is indeed the case (Julian and Sollins, 1975). However, the measurements necessary to test this prediction quantitatively have not yet been reported (see discussion by A. F. Huxley, 1981). Evidence from X-ray diffraction studies has also been interpreted as compatible with a decrease in the number of crossbridges attached as velocity rises (H. E. Huxley, 1979). Another set of observations suggesting that fewer crossbridges are attached during rapid shortening is of the "pullout" phenomenon (A. F. Huxley, 1971). When the force on a muscle is rapidly increased to a sufficient level, a very rapid shortening ensues, as though all the attached bridges have been broken off. The load required to bring this about is less during rapid shortening than it is in isometric contraction.

The 1957 theory cannot, in its original form, account for the following observations.

B-1. A. V. Hill (1964b) found that the rate of heat + work production was maximum at $0.5v_{max}$ (Fig. 4.15). Kushmerick and Davies (1969) found the rate of ATP splitting was also maximum at about this velocity (Fig. 4.35). Thus the rate of free energy liberation E is not a monotonically increasing function of v. This feature cannot be duplicated by the 1957 theory, because the rate of bridge turnover is proportional to the number of unattached bridges in the region $0 < x < 1$. This number increases at higher velocity because the time available for bridge attachment is less. The predicted rate of bridge turnover is therefore necessarily a monotonically increasing function of v.

B-2. At high velocity of shortening, the rate of ATP splitting is low even though the heat rate is relatively high. Unless we are to suppose that the high heat rate during shortening is not related to crossbridges at all, which seems perverse, these observations can only be explained by postulating another crossbridge state that is formed during rapid shortening and therefore cannot be explained within the confines of a two-state model. However, it is quite possible that the occupancy of the extra states that must be postulated will be negligible except during rapid shortening, so that the two-state model would suffice for other conditions.

B-3. The rapid transient changes in velocity in response to a change in load, or changes of force in response to a change in length, are not predicted correctly by the theory. This can be seen quite simply from the fact that the highest rate constant, g_{+2}, has a value of 130 s^{-1}, whereas the rate constant for the rapid recovery of force after a large, very quick, release is 10 times greater than this (Ford *et al.*, 1977). The 1957 theory predicts that after a sudden change in tension the velocity should approach a new steady state

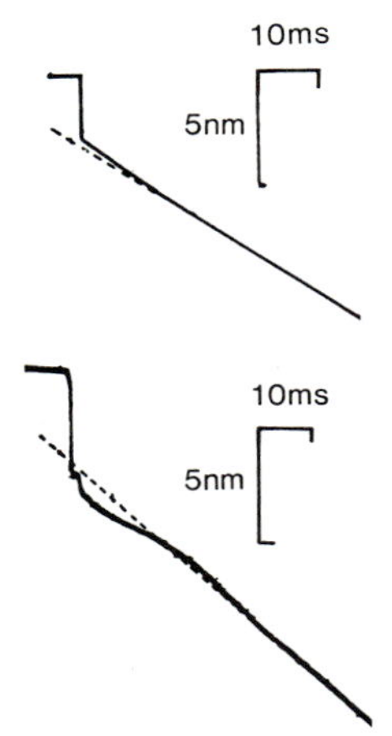

FIG. 5.6. Length changes per half-sarcomere after a reduction of force to 63% of its initial value. The upper line is the prediction of the 1957 theory; the lower line is the observed behaviour. The dotted lines are extrapolations of the later, linear phase of shortening. (From Civan and Podolsky, 1966.)

monotonically (Fig. 5.6), whereas it is observed that there are oscillations in velocity before the steady value is reached.

B-4. The energetic consequences of lengthening are not correctly predicted by the theory. It predicts that during lengthening the rate of crossbridge turnover, and therefore of ATP splitting, should increase, because the number of unattached bridges in the region $0 < x < h$ will be greater than for isometric contraction. It is observed, however, that the rate of ATP splitting is much less during lengthening than during isometric contraction (Curtin and Davies, 1975). During lengthening, bridges formed in the region $0 < x < h$ are carried into the region $x > h$. The tension in these bridges then becomes very high, a factor that would be expected to promote their breaking off by the reversal of the reaction by which they formed, that is, without splitting ATP. This process is specifically excluded in the 1957 theory (as stated here) since as f_+ is zero in this region, f_- is also zero.

B-5. The 1957 theory does not predict the after-effects of a period of lengthening (Chapter 2, Section 2.2.3): an enhancement of the ability to exert tension and to do work in a subsequent shortening.

The points of agreement between this theory and experimental observations are remarkable for a theory which is so simple and contains rather few adjustable parameters. Although, as we have seen, there are quite a number of observations that the theory cannot explain adequately, it is natural to conclude that the points of agreement are not fortuitous and that the theory should not be lightly discarded. For this reason, many attempts have been made to extend Huxley's theory to cover these later observations without losing its many attractive features. As these attempts usually involve increasing the number of crossbridge states, they inevitably lead to a theory more complicated than the original. Some of these attempts are described in the next section of this chapter.

5.2 Extensions to the 1957 Theory

As we have seen above, the 1957 theory cannot explain what happens in muscle during rapid shortening, during rapid transients, or during lengthening, but none of these objections necessarily throws doubt on the explanations offered for what happens in isometric contraction or during shortening at slow and moderate speeds. An attractive type of theory would therefore be one that extended the 1957 theory in these areas, while reducing to the original theory where that is already successful. Two ways in which such an extension might be achieved are illustrated by the schemes in Fig. 5.7.

In Fig. 5.7B an extra state is added that is in equilibrium with another, that is, $k_{+1}, k_{-1} \gg g_+, f_+$. The transitions between A_1 and A_2 need only be considered on very short time scales, for instance when rapid transient behaviour is discussed. At other times these two states can be replaced by a single state representing the equilibrium mixture of A_1 and A_2 (see T. L. Hill, 1977, p. 197, for further discussion of this sort of reduction).

In Fig. 5.7C an extra state is added that is a transient intermediate, that is, $j_+ \gg g_+, f_+$. The occupancy of state D_1 would be negligible for many purposes but could be significant when g_+ was increased, for instance, by rapid shortening.

5.2.1 HUXLEY'S PROPOSAL OF 1973

Huxley suggested in 1973 that the difficulty (described in point B-1 above) in providing a maximum in the relation between E and v could be overcome if it was supposed that crossbridge attachment takes place in two stages (Fig. 5.7D). The first stage does not lead to the development of force and is readily reversible without ATP being split. The second stage does lead to force development and its reversal is negligible. After this second step ATP splitting is therefore required to dissociate the crossbridge. Huxley shows that this model can predict a relation between energy output and velocity similar to that observed by A. V. Hill in 1964. The forward rates of both attachment steps are of a similar order, and therefore this theory does *not* reduce to the 1957 theory at low velocities of shortening. No full analysis

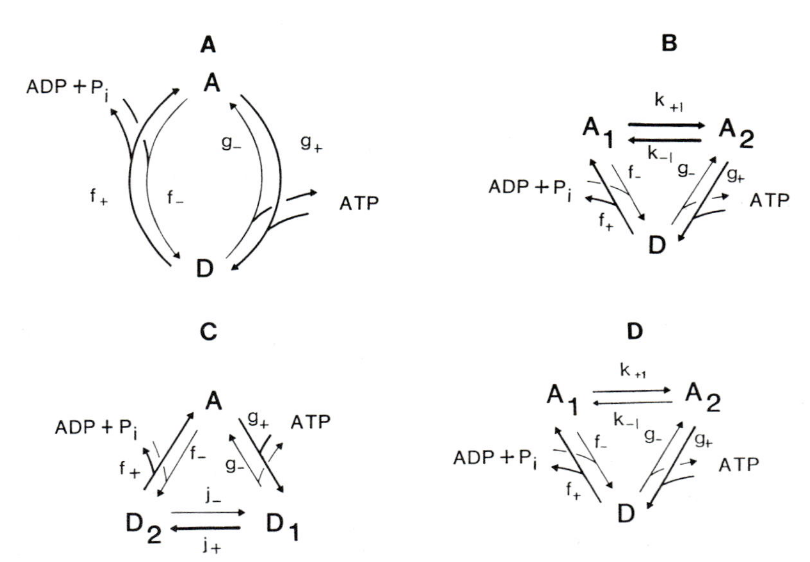

FIG. 5.7. Possible extensions to the 1957 theory by addition of extra states that are negligible under some conditions. In these diagrams, reactions that occur at a negligible rate are indicated by thin arrows, and the very thick arrows show very fast reactions. (A) The original scheme from Fig. 5.1A. (B) Two states in rapid equilibrium: A_1 and A_2. (C) A transient intermediate D_1. (D) The proposal of Huxley (1973). A_1 and A_2 are two attached states, but only A_2 produces force. f_+, k_+, and f_- are all similar in magnitude but f_-, k_-, and g_- are negligible.

of the properties of this model have been published, but because it has more degrees of freedom than the 1957 model, it would be expected that, with a suitable choice of constants, it could explain everything that the 1957 theory explains, in addition to giving a more realistic relation between E and v. A point in favour of further consideration of this idea is that X-ray diffraction studies have suggested that crossbridges may attach to the thin filament before tension is developed (H. E. Huxley, 1979).

5.2.2 EFFECTS OF CLOSE SPACING OF ATTACHMENT SITES

An alternative way of providing a maximum in the relation between E and v is suggested by the spacing of the attachment sites used in fitting the 1957 theory to the observed behaviour of frog muscle. The value of l (35 nm) used was not much greater than of h (27 nm). This is in fact a violation of one of the premises of the 1957 theory, that the attachment sites are far enough apart that a crossbridge is within range of only one such site. With these values of l and h it would be expected that, during rapid shortening, there would be some bridges unable to attach to the site they are passing because they are still attached to the previous site (see Fig. 5.3A). The effect would be to reduce the ATP turnover at high rates of shortening. The magnitude of the effect would be very dependent on the relative size of l and h. It is not known whether a good quantitative account of the properties of muscle can be given by this type of theory, which is an example of a three-state theory with the third state only significant during rapid shortening.

5.2.3 PODOLSKY–NOLAN THEORIES

Podolsky and Nolan (1973) and T. Hill *et al.* (1975) have examined the possibilities of improving the performance of a two-state model by changing the x dependence of f_+, g_+, and F. Their particular aim was to model the rapid transient behaviour. An example of one of their models is shown in Fig. 5.8. These models differ from the 1957 model in that the value of f_+ is

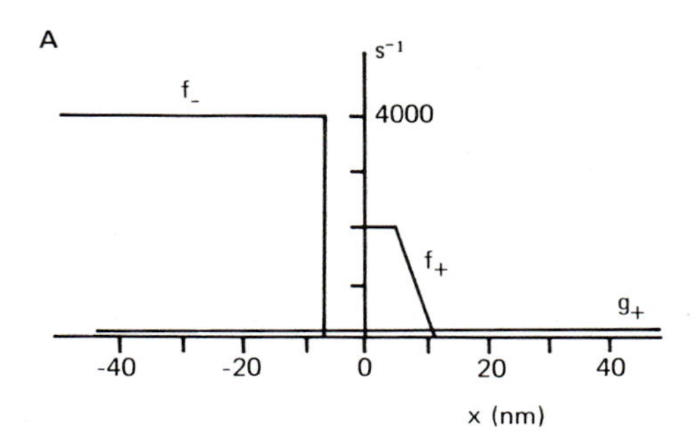

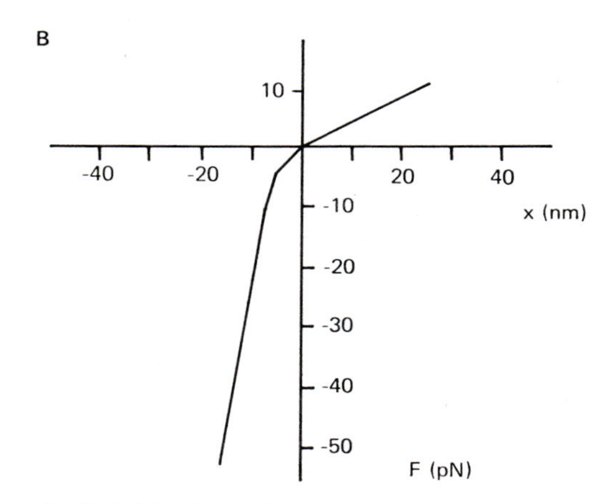

FIG. 5.8. A representative Podolsky–Nolan theory (1973). These graphs are similar to those in Fig. 5.2 for the Huxley (1957) theory, but note that both the ordinate scales are much reduced.

very much larger and the value of h is smaller. Crossbridge attachment is possible only within a restricted range of bridge positions, but within this range is almost certain. The mechanism by which the bridges break in the region $x < 0$ is also different: the large g_{+2} value at $x < 0$ in the 1957 model is replaced by a large f_- value at $x < -6$ nm. These features, together with the asymmetry of the force–extension curve (Fig. 5.8), allow these models to give good predictions of the response to load steps or length steps, brought about by rapid changes in the number of attached bridges. They do not, however, give as good descriptions of the force–velocity curve, or of the relation between E and v, as does the 1957 model. In particular they predict a very low efficiency. Some improvement in these properties of the Podolsky–Nolan theories might be possible, as suggested by T. Hill *et al.* (1975), but they suffer also from another serious disadvantage: they predict that the number of attached crossbridges should rise as the velocity increases, a consequence of the high value of f_+ and low value of h. The available evidence (see point A-9 above) suggests that the number of crossbridges decreases as velocity rises. These theories also predict that the stiffness should be greater shortly after a release than in isometric contraction. Huxley and Simmons (1973) tested this point, and found that there was no increase. These difficulties are direct consequences of the way the rapid transients are explained and thus cannot be overcome by modifying the two-state model.

5.2.4 THE HUXLEY–SIMMONS THEORY

A quite different explanation of the rapid recovery of tension after a quick length step was proposed by Huxley and Simmons (1971b, 1973). They suggested that it results from a change in the distribution of attached crossbridges between a small number of states without any attachment or detachment (as in Fig. 5.7B). The essence of their idea is illustrated in Fig. 5.9. The crossbridge consists of two parts joined by a hinge. The tail part contains an undamped spring, as in the original theory. The head part of the crossbridge can attach to the actin at a number of different angles and because of the presence of the spring it can move between these different positions without detaching, and without relative movement of the actin and myosin. The force exerted by the bridge (essentially equal to the force in the spring) now depends on the two variables x and z. Relative movement of the filaments changes x, but not z; rotation of the head changes z, but not x. The force in the spring (F) is given by $p(x + z)$ (Fig. 5.10A), and the energy stored in the spring is $e_s = p(x + z)^2$ (Fig. 5.10B). The potential energy of the bridge includes also the energy of binding, e_b. Binding is assumed to be strong for a few specific values of z but weak elsewhere. Therefore, there are narrow potential energy "wells" when e_b is plotted against z (Fig. 5.10D). The energy

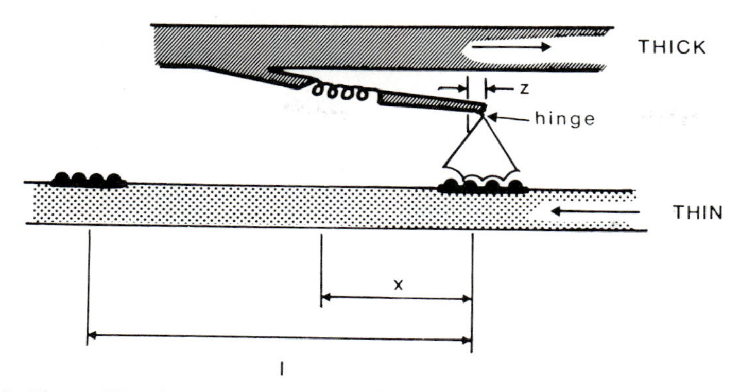

FIG. 5.9. The modification to the 1957 theory suggested by Huxley and Simmons (1971b). Compare with Fig. 5.1B which shows the original version. The head of the crossbridge can bind to the actin in a number of stable configurations at different angles, and therefore with different values of z. The length of the spring in the crossbridge is now dependent on two variables, x and z.

well at $z = 6$ is deeper than that at $z = 0$; that is, binding has been assumed to be tighter in this higher tension state. The total energy for the crossbridge, $e_s + e_b$, is shown as a function of z (for two x values) in Fig. 5.10C and as a function of x (for two z values) in Fig. 5.10E. For a population of crossbridges (at a particular x value) that are in equilibrium between the two z states, the distribution between the two states will be a function of x. Figure 5.10F shows the proportion in each state as a function of x; and the force exerted by the crossbridges at each value of x is shown by the dotted line in Fig. 5.10A.

The shape of the T_1 and T_2 curves is explained as follows. If a muscle is released very rapidly, so that there is no time for rotation of crossbridge heads to occur, the observed stress–strain curve is that of the spring itself. This gives the T_1 curve, which is found experimentally to be nearly linear. If the release is made more slowly, so that rotation can occur, keeping the population of crossbridges in an equilibrium between their stable states, then the stress–strain curve for bridges at each initial x value (the value of x at the start of the release) would follow the dashed curve in Fig. 5.10A. The muscle contains crossbridges over a range of initial x values and so the predicted T_2 force is the sum of the forces over this range. The dashed line in Fig. 5.10A is the predicted T_2 curve and it clearly has a shape generally resembling an observed T_2 curve. In fact, a good quantitative explanation of the shape of the T_2 curve along these lines requires at least three stable z positions.

As well as giving this simple explanation for the shape of the T_2 curve, the Huxley–Simmons theory leads to a natural explanation of why tension

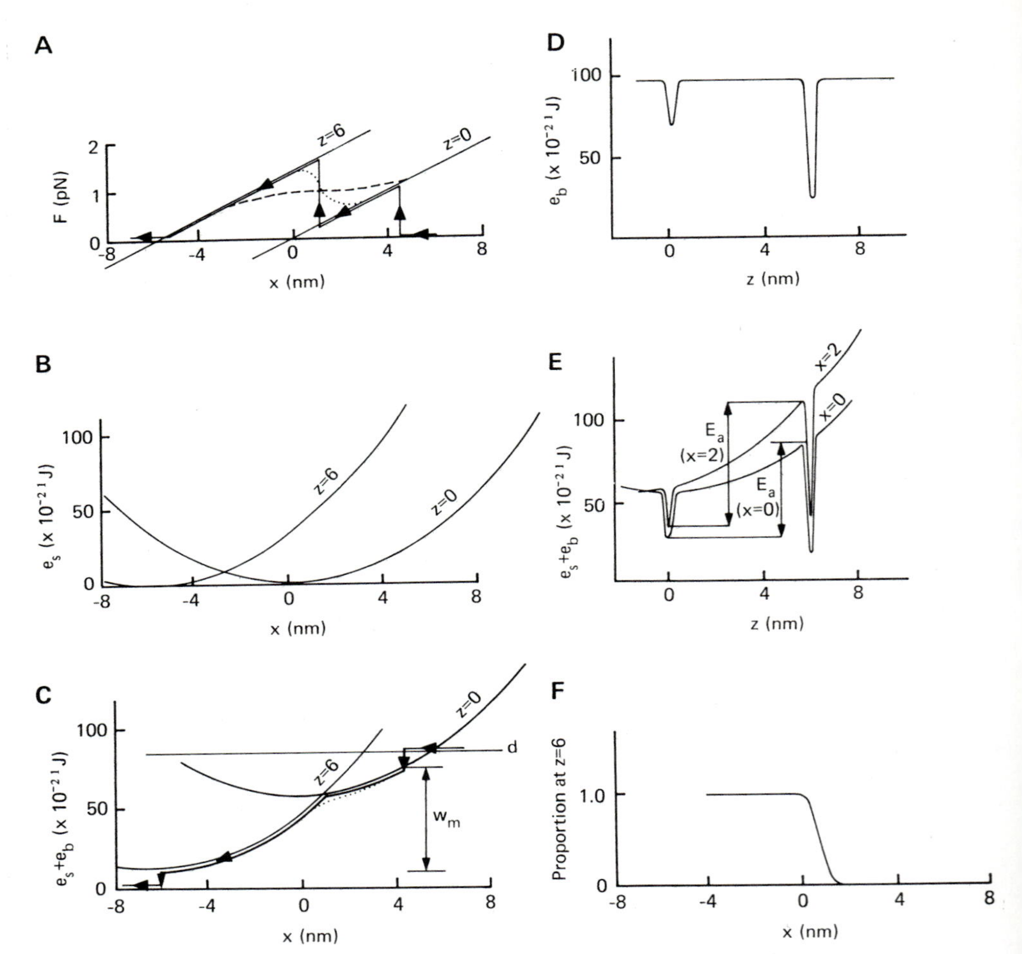

FIG. 5.10. Some properties of a simplified version of Huxley and Simmons theory (1971b) with two stable attached states, labelled here $z = 0$ and $z = 6$. (In fact three or more such states are needed to give good predictions of the rapid transient behaviour.) (A) Tension produced by the two crossbridge states. The heavy line shows the path followed by a crossbridge performing the maximum work as it passes the actin site. The dotted line is the tension produced by the equilibrium mixture of the two states. The dashed line is the predicted T_2 curve. (B) Energy in the spring (e_s) for the two crossbridge states. (C) Basic free energy of the crossbridge, the sum of e_s and e_b (see text for an explanation of e_b) as a function of x. The dotted line is for the equilibrium mixture of the two states. The heavy line shows the path followed by a crossbridge performing the maximum work as it passes the actin site. The basic free energy of the detached state is shown by the horizontal line (d). (D) e_b as a function of z. (E) Basic free energy of a crossbridge as a function of z. The two curves are for two different values of x. The vertical arrows show the activation energy (E_a) for the transition from $z = 0$ to $z = 6$; note it is greater for the larger value of x. (F) Proportion of attached crossbridges in the $z = 6$ state.

recovery from T_1 to T_2 is faster after large releases than after small ones. This can be appreciated by considering the activation energy for the transition from state 1 to state 2 indicated by the arrows labelled E_a in Fig. 5.10E. E_a is larger when the force in the crossbridge is larger ($x = 2$), and therefore the recovery process is slower after a small release. This explanation can also be made to fit quantitatively with the observation of these time constants if three attached states are considered (Huxley and Simmons, 1971b).

In summary, the strengths of this theory are that it provides a natural way of explaining the two nonlinearities observed in the length step experiments: the shape of T_2 curve and the dependence of the rate constants of phase 2 on extent of release. It also correctly predicts that the instantaneous stiffness after recovery of tension to T_2 should be unchanged. These predictions have of course been obtained by an ad hoc hypothesis, that the crossbridge head can attach to actin at a small number of different angles. There is, as yet, no independent evidence that this is true.

As Podolsky–Nolan theories have been criticized for their unrealistic steady-state properties, particularly their energetic properties, it is natural to ask whether Huxley–Simmons theories do any better in this respect. Unfortunately, the implications for steady-state behaviour of the interpretation offered by Huxley and Simmons of transient behaviour have not been explored in detail, although it is likely that a satisfactory hybrid theory could be constructed retaining many of the advantages of the original theory, in particular the high efficiency with which work is done.

One further point about this type of theory is that it seems to predict that the shape of the T_2 curve would be different when a release is made with the muscle shortening, rather than isometric. This is because it is necessary, to explain the plateau on the T_2 curve, to suppose that during isometric contraction there are few crossbridges in the region $0 < x < 4$ nm, from which no further tension development would be possible by rotation of the crossbridge head. However, during shortening, crossbridges would be carried into this region and the plateau of the T_2 curve should therefore be less flat than during isometric contraction.

5.2.5 THE HILL–EISENBERG THEORIES

The theories considered under this heading are those described by Eisenberg *et al.* (1980) and by Eisenberg and Greene (1980). Although in many ways similar to Huxley–Simmons theories, these are four-state theories: they postulate that the crossbridge can exist in two detached states, D_1 and D_2, and two attached states, A_1 and A_2 (Fig. 5.11A). As in the Huxley–Simmons theory, the transitions between A_1 and A_2 are fast and are important only in determining the rapid transient behaviour of the model. At values of x

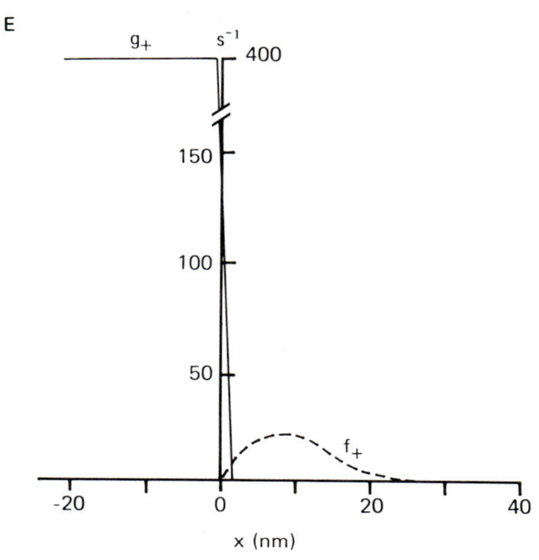

FIG. 5.11 (A) The four crossbridge states proposed in the Eisenberg–Hill theories. The thickness of the arrows indicates qualitatively the assumed speed of the reactions. To a first approximation the steady-state behaviour is similar to that of a two-state model, in which the rate of attachment is governed by f_+, and the rate of detachment by g_+. The x dependence of these rate constants is shown in FIG. 5.11E. (B), (C), and (D). The properties of the Eisenberg–Hill theory presented in the same way as Fig. 5.10 which is for the Huxley–Simmons theory. (**B**) Tension produced by the two attached states. The dashed line is the predicted T_2 curve. Compare with Fig. 5.10A. (C) Basic free energy of the two attached and the two detached states, d_1 and d_2. Compare with Fig. 5.10C. (D) Proportion of attached bridges in the state A_2. Compare with Fig. 5.10F. It is hard to draw diagrams for this theory comparable to Fig. 5.10B, D, and E. This is because the distinction between e_s and e_b in the Huxley–Simmons theory has been dropped to allow the assumption of an arbitrary x dependence of the rates of transition between A_1 and A_2. (E) A simplified version of the rate constants for the attachment and detachment processes in the Eisenberg–Hill theory. The rate constant f_+ is for the transition from D_1 to A_1 (A). Compare with Figs. 5.2A, showing the assumptions of the Huxley (1957) theory, and 5.8, showing the Podolsky–Nolan theory.

at which bridges can attach, the state D_2 is a transient intermediate, because the rate of its conversion to A_1 is much greater than its rate of formation from D_1. The steady-state behaviour of the model can therefore be understood, to a first approximation, by considering only two states, D_1 and the equilibrium of A_1 and A_2.

The other assumptions used in these theories are summarized in Fig. 5.11B,C,D, and E. The similarity to the Huxley–Simmons theory is obvious. The major difference is in the activation energy required to pass from one attached state to the next. The energy wells for these states are much broader than in the Huxley–Simmons theory, and have an arbitrary

x dependence. The advantage of this is that, with only two attached states, the theory can give a good account of the way in which the time course of a tension transient depends on the size of a length step. However, this account is obtained only by ad hoc assumptions about the x dependence of the rates of transition between A_1 and A_2.

A comparison of Fig. 5.11E with Fig. 5.2A emphasizes that Eisenberg–Hill theories explain the steady-state properties of muscle in essentially the same way as the 1957 theory. It is not surprising therefore that similar predictions are made about the force–velocity curve, about the relation between ATP turnover and velocity, and about the efficiency. These predictions are largely satisfactory but suffer from the shortcomings of points B-1, B-2, B-4, and B-5 described above.

These theories thus seem to demonstrate the successful hybridization of the ideas of Huxley (1957) with many of those of Huxley and Simmons (1971b). The theories are notable for another reason, that an attempt has been made in them to relate the physiological states they postulate to the biochemical states used to explain the kinetics of actomyosin *in vitro*. These ideas are briefly described in Section 5.3 of this chapter.

5.2.6 THEORETICAL REQUIREMENTS FOR RAPID SHORTENING

No theory has yet been published that successfully predicts the behaviour of muscle when shortening rapidly (point B-2 above), that is, rapid heat production, but a slow turnover of ATP, followed after the shortening by delayed ATP splitting without heat production. This is surprising because this striking behaviour of muscle has been known, in outline at least, for 20 years. To illustrate some of the requirements of such a theory we will consider how two-state crossbridge theories could be extended to explain these facts. Low ATP turnover can be due either to the number of crossbridges interacting being very small, with each cycle of attachment and detachment splitting one ATP, or to a larger number of crossbridge cycles, in many of which ATP is not split. This type of cycle occurs when a crossbridge is broken by a process that does not complete the ATPase cycle, for instance by reversal of the crossbridge-forming reaction. Evidence about the number of crossbridges attached during rapid shortening (see point A-9 above) suggests that this number is less than in isometric contraction, but not very small. This means that crossbridge turnover during rapid shortening is higher than in isometric contraction. Therefore, most of the crossbridge cycles must occur without ATP splitting, as in the Podolsky–Nolan theories. However, the crossbridges do not, as in these theories, break by a reversal of the process by which they attach, because such crossbridge cycles would produce no heat (other than the dissipation of work done on them by other crossbridges). The observed heat production can only be explained if the crossbridges are, after detachment, in a state different from the initial one. We must therefore hypothesize another detached state (D_1 in Fig. 5.12) from which recovery to A is relatively slow. The delayed ATP splitting could then be explained as the accompaniment of this recovery process.

We could thus explain what happens during shortening by the hypothesis that there are two different crossbridge cycles: *y* and *z* in the scheme shown in Fig. 5.12. One wonders whether it would not be simpler to suppose that there is only a single cycle (*z*) and that the accumulation of D_1 is observed only in rapid shortening because the cycle is then running faster (as proposed for example by Kodama and Yamada, 1979). If this were the case, the accumulation of D_1 would be even more marked during shortening at moderate speed, when the ATPase rate is higher than during rapid shortening. But this is not observed: there is no large amount of unexplained heat production during shortening at moderate speed, nor has any delayed ATP splitting been detected (see Chapter 4, Section 4.3.3). It must, therefore, be supposed that there are two different crossbridge cycles, that cycle *z* occurs predominately during rapid shortening, whereas in isometric and moderate speed shortening *y* is the predominant cycle. This has the advantage of reducing to the familiar two-state 1957 theory for isometric contractions and for moderate velocities of shortening, when cycle *z* will be negligible.

This type of theory offers another possible explanation of why the observed rate of energy output does not monotonically increase with velocity. At high velocities there would be a progressive accumulation of D_1 at the expense of A and D_2 and, therefore, a reduction in bridge turnover. Of course if the measurements could be made for very small amounts of shortening this effect would be minimized, but such measurements have not been made, and would be very difficult. It does seem, however, that the rate of

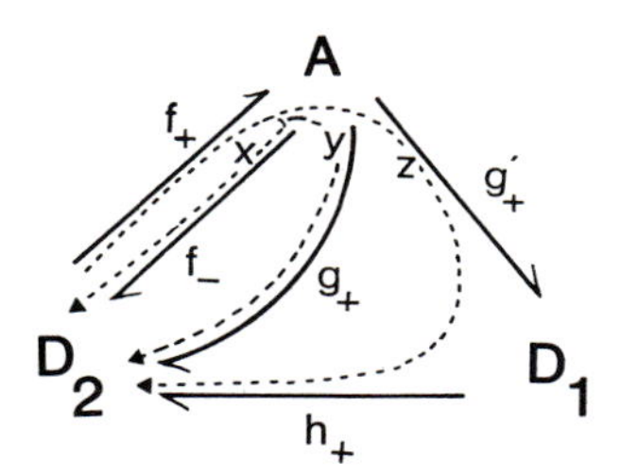

FIG. 5.12. Reaction schemes with two detached states (D_1, D_2) and one attached state (A). The dotted arrows *x*, *y*, and *z* show three different routes that could be taken by a crossbridge cycle.

shortening heat production declines during a period of rapid shortening (see Chapter 4, Section 4.2.2) and this could be an effect due to accumulation of D_1.

5.2.7 THEORETICAL REQUIREMENTS FOR LENGTHENING

None of the theories so far discussed predicts the behaviour during lengthening better than the 1957 theory does. The problem is to account for the very high tension produced with a very low rate of ATP splitting. The obvious suggestion (put forward in fact by Huxley in the 1957 paper) is that bridges are broken during stretch by the reversal of the process that forms them. The predominant crossbridge cycle is that labelled x in Fig. 5.12. This would explain the low rate of ATP splitting, and predicts that the work done on the crossbridge by the stretching force would be dissipated as heat, and the change in internal energy (heat produced and work done) would be zero. This may be what happens in most crossbridge cycles during slow stretches, although the energetics of more rapid stretches seem more complex (see Chapter 4, Section 4.2.3). Because there have been no studies of energy balance during stretches, it is not worthwhile to speculate further about their energetics and its interpretation.

The modifications required to the 1957 theory to make cycle x the predominant process during stretching are simple. If f_+ is given a nonzero value in the region $x > h$, it follows that f_- increases steeply in this region, and because it is much greater than g_+ the x cycle predominates over the y cycle in this region, but the predicted behaviour of the muscle while shortening is almost unchanged. The quantitative aspects of this sort of idea have not yet been explored, so it is not known whether or not the large influence of slow lengthening on the rate of ATP splitting can be explained in this way.

Further theoretical consideration of the after-effects of a period of lengthening is probably not worthwhile until more progress has been made in understanding what happens during lengthening.

5.2.8 SIMPLICITY IN CROSSBRIDGE THEORIES

A possibly valid criticism of the type of theory we have been discussing is that it has too much freedom. Perhaps it is possible to explain virtually anything by a suitable choice and x distribution of a small number of rate constants which are not exposed to the hazard of direct observation. If so, these theories do not really give any insight into the behaviour of muscle. Two ways to counter this type of criticism are discussed here:

1. To expose the theories to new types of experimental tests. As the theories make assumptions about states of crossbridges and the rates of transitions

between them, it is natural to want to compare these assumptions with the known kinetic behaviour of actomyosin in solution. Can the theories be constrained by what is known of these kinetic properties? This is a difficult area, to which the final section of this chapter is devoted.

2. To simplify the theories so that they have fewer degrees of freedom. A notable feature of muscle is its high efficiency, particularly in tortoise muscle. This implies that the "route" taken by a crossbridge, on a diagram such as Fig. 5.2D, as it passes a single site during shortening is relatively consistent. This in turn suggests that we might be able, at least for some purposes, to regard this route as fixed. The amount of work done by each crossbridge in such a pass is then constant. Cooke and Bialek (1979) have explored some of the implications of this idea and suggest that it may lead to good predictions for the force–velocity curve and the relationship between energy output and load. As a detailed account of this theoretical work has not yet been published, it cannot be decided whether their theory really is much simpler than the original 1957 theory.

Are other simplifications possible? What would happen if one replaced the usual assumptions about the spatial distribution of f and g with the assumptions that the attachment reaction is kept at equilibrium near $x = h$ and the detachment reaction is kept at equilibrium near $x = 0$ (see Fig. 5.2A and B). Such a theory is in fact not realistic because it leads to the following predictions:

1. The efficiency is constant at all velocities, and equal to w_m/e.
2. The rate of turnover of crossbridges and of ATP splitting is zero in isometric contraction.
3. The rate of ATP turnover during shortening is proportional to velocity.
4. The power output is also proportional to velocity.
5. The force exerted is a constant, independent of velocity, because crossbridges cannot exert a negative force, and the number of crossbridges forming is independent of velocity.

As these predictions are so unrealistic, we can conclude that it *is* necessary in crossbridge theories to allow the crossbridge some choice of whether to attach as it passes an attachment site, and/or whether to detach as it passes $x = 0$.

5.3 Relation of Crossbridge Theories to Kinetics of Actomyosin ATPase

The crossbridge theories discussed in this chapter are concerned largely with postulating crossbridge states that are mechanically distinct from one another, that is, those that produce different forces at a particular x value.

The study of actomyosin ATPase, in contrast, is concerned with identifying states of the system that are kinetically distinct, in particular those that may be rate limiting under particular conditions. What relation should we expect between the states postulated in these two kinds of study? Is it possible to combine the two approaches to give a theory that can explain both the properties of muscle and those of actomyosin in solution? Is the attempt to do this a good way of evaluating a crossbridge theory? Definite answers to these question cannot be given in the present state of our understanding of muscle, but nevertheless some aspects are worth discussion. The following comments are based largely on the ideas of Huxley (1980), Eisenberg and Hill (1978), Eisenberg *et al.* (1980), Eisenberg and Greene (1980), and Sleep and Smith (1981), where more detailed treatment of these questions can be found.

Whereas there are only two to four crossbridge states in the theories discussed above, the number of intermediates used to explain the properties of actomyosin in solution is rather greater. This suggests that each mechanically distinct state is identified with several biochemically distinct states. A. F. Huxley (1980) suggests the useful idea that biochemical events should be regarded as occurring alternately with mechanical events around the crossbridge cycle. A particular stage of the biochemical cycle *permits* the next stage of the mechanical cycle, and vice versa. In this view it would be reasonable to associate each state in the mechanical cycle with two (or more) states in the chemical cycle. This idea perhaps offers some intuitive understanding of why ATPase activity is "coupled" to the performance of work in muscle. If the performance of work is prevented, by a large load on the muscle, the next stage in the ATPase cycle is unable to occur. A fairly tight coupling of this kind is in fact essential to explain the relatively high efficiency of muscle in performing mechanical work.

As we have seen in Chapter 3, most of the kinetic information has been obtained on experiments on actomyosin in solution. It is, in principle, possible to obtain similar information for actomyosin in a structured system that is subject to the mechanical constraint of coupling to an external load, but, so far, few such experiments have been reported. At present we therefore have to compare the mechanical cycle in the structured system with the biochemical cycle as it occurs in solution. We would expect the structured system to compare to that in solution as follows.

1. The rates of transitions between detached states would be almost the same as in solution.

2. The basic free energy changes between attached states would be similar to those in solution, if we consider only the minimum values with respect to x. This implies that the chemical processes that lead to a state in which

mechanical work is done would have a large free energy change in solution.

3. The rates of transitions between attached states may be quite different in the structured system, than they are in solution. In particular, the rates of the changes in the attached state that do mechanical work, or are consequent on work being done, would be much slower than in solution. This may be true even when the muscle is not doing work against an external load, because in some theories the crossbridges are doing work, when shortening at v_{max}, against other crossbridges, and this constraint is of course not present in solution.

4. Rates and equilibrium constants of attachment processes would be quite different from their values in solution. In the structured system these can be considered first-order processes, because there is only one actin site at a time within range of a crossbridge. In solution attachment processes are, of course, second order. The equilibrium constant for an attachment process in a structured system (K_{str}) can be related to that in solution (K_{sol}) by

$$K_{str} = A_{eff} K_{sol}$$

where A_{eff} is called the effective actin concentration. It would be expected that A_{eff} would be the same for all attachment processes.

It seems from these considerations that there is no general answer to the question of how the maximum steady-state ATPase observed in solution should be compared to the rate of ATP splitting by muscle. There does not seem to be a physiological state in which, in general, ATP splitting should occur at the same rate as in solution. In isometric contraction the constraint on ATP splitting is at its greatest, which suggests that the differences from the behaviour in solution would be large. During shortening at v_{max} the behaviour of the muscle is complex (see point B-2 in Section 5.1) and is not explained by any of the theories we have considered, so this state can hardly be a good basis for comparison. Perhaps the best that can be done is to make the comparison with the maximum observed rate of ATP splitting *in vivo* which occurs when muscle is shortening at an intermediate rate. On the basis of specific theories of contraction, however, more justifiable comparison may be possible.

In Huxley's 1957 theory, as originally written and as presented here, the products of reaction dissociate from myosin at the time of attachment. The state that does the work is actomyosin without any bound nucleotide. This implies that product dissociation should be associated with a large change in free energy in solution. This has turned out to be the case, more specifically, for dissociation of P_i (White and Taylor, 1976). However, if we wished to write the theory in a way that corresponded more closely to ideas of actomyosin kinetics we would leave ADP dissociation to a latter stage, immediately preceding association of ATP and detachment of the crossbridge.

On this theory it is the association process that limits the maximum rate of ATP splitting, but the detachment process that limits the rate during isometric contraction. In solution the rapid detachment process, which *in vivo* is only available at $x < 0$, would presumably be continuously in operation, so that attachment would be the rate-limiting process and the ATPase would be similar to the maximum *in vivo* rate. This is in fact what is found to be the case (Ferenczi *et al.*, 1978a). The 1957 theory suggests that in the absence of ATP crossbridges would be attached, thus giving an account of the rigor state. It suggests that when ATP is added to muscle in rigor, detachment (relaxation) would occur at a rate similar to that of ATP turnover in isometric contraction. It is now known that "relaxation" from rigor is much faster than this (Chapter 2, Section 2.4.2), showing that the rate-limiting step is elsewhere in the cycle.

One way in which the 1957 theory could be related to a biochemical cycle is shown in Fig. 5.13A, a scheme which would be in accordance with the

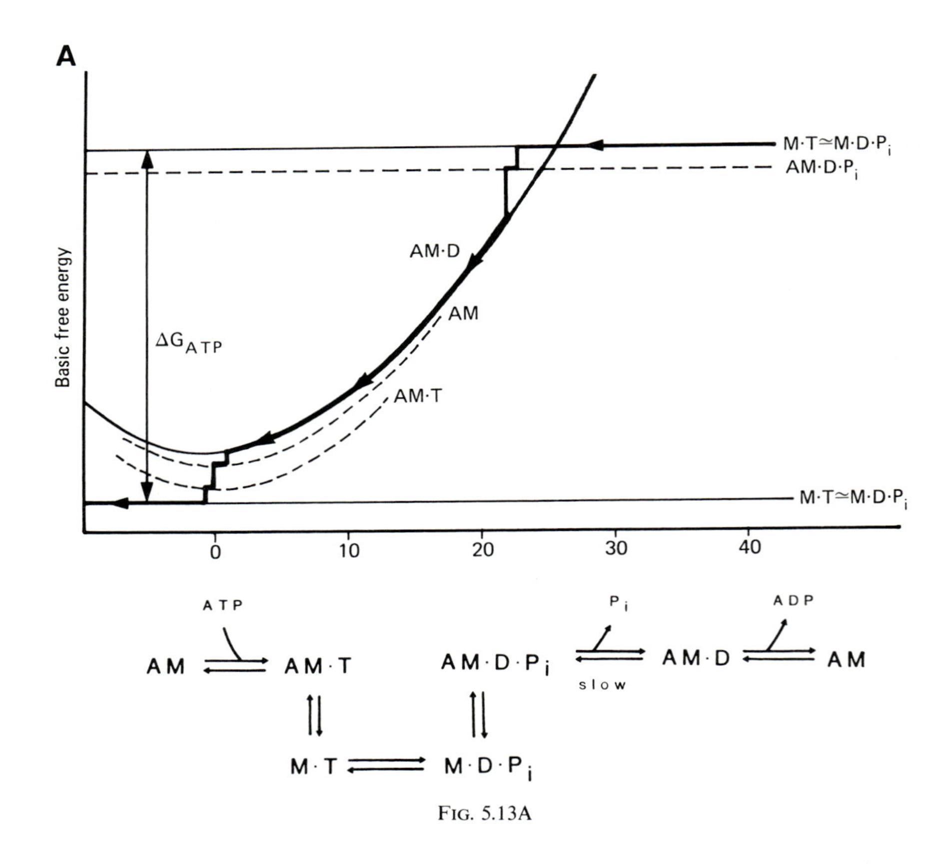

FIG. 5.13A

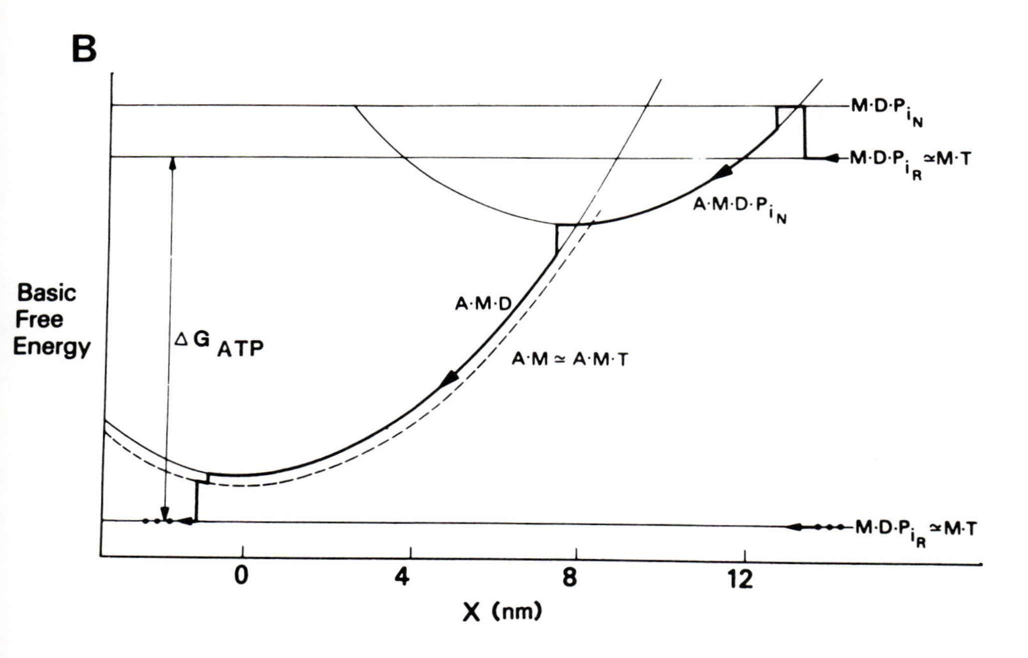

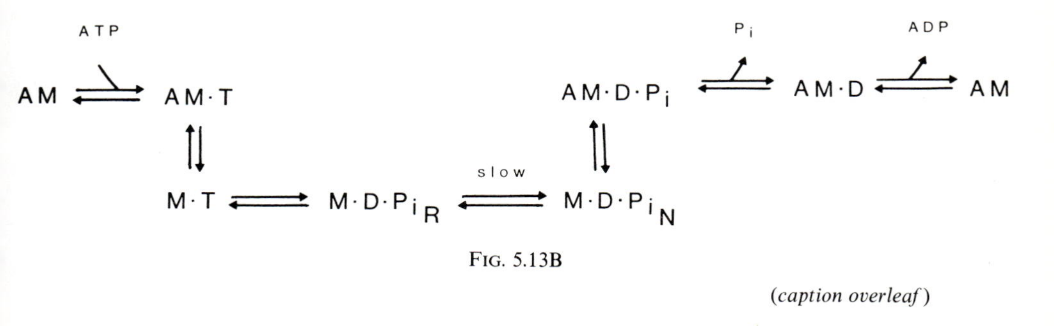

FIG. 5.13B

(*caption overleaf*)

principles enumerated above, and which leaves the predictions of the theory about the behaviour of intact muscle unaltered. It is suggested in this scheme that the state AM·ADP·P_i, although an attached state, is unable to exert force and therefore is grouped with detached states in considering the relation of the biochemical scheme to the two-state mechanical cycle. The transition to the state in which work is done requires the dissociation of P_i. The rate of this reaction corresponds to the rate constant f_+, and limits the maximum rate of ATP splitting *in vivo* and also in solution. The rate constant

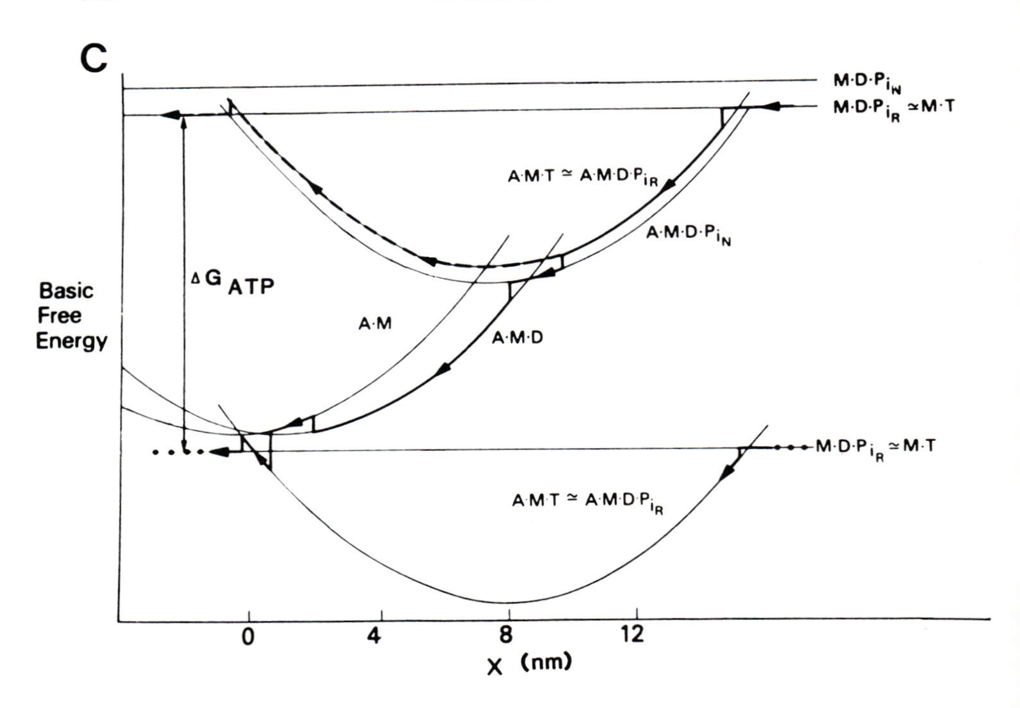

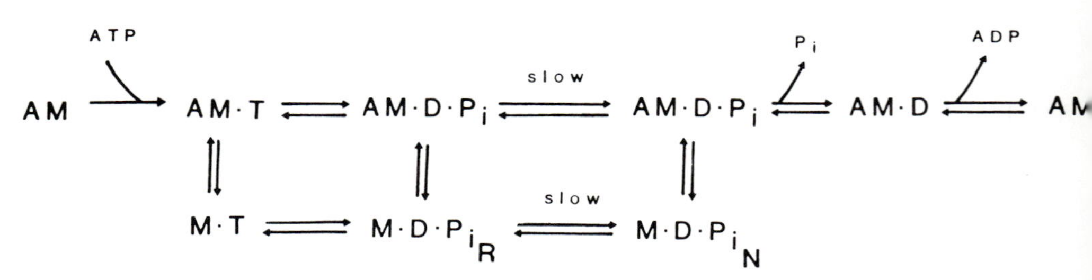

FIG. 5.13. Three illustrations of how theories of contraction can be related to reaction schemes for actomyosin ATPase. (A) The Huxley (1957) theory; (B) the theory of Eisenberg *et al.* (1980); (C) the theory of Eisenberg and Greene (1980). The graphs above each scheme show the basic free energy for the various crossbridge states as a function of x. The dotted lines are for states that are transient. Solid lines are for states that preceed a process that is, under some circumstances, rate limiting. The heavy lines show the path followed in a crossbridge cycle in which the maximum amount of work is done. Because of the high efficiency of muscle it has to be supposed that, during shortening at moderate speeds, most crossbridge cycles follow a path fairly similar to this one. They would be constrained to do so by the x dependence of the rate of the appropriate reaction, which is not shown in these graphs. The actomyosin reaction sequence for each theory is also shown. The reaction that is considered to be rate limiting in solution is labelled "slow." In labelling the states of the crossbridges the abbreviation D is used for ADP, and T for ATP. The symbol $\simeq$ denotes the equilibrium mixture of two species.

g_+ is supposed to correspond to the dissociation of ADP. It is the rates of these two reactions therefore that are considered to be x dependent. In solution there is, as required, a large free energy change associated with P_i dissociation (White and Taylor, 1976). To explain the high efficiency of contraction, the free energy drop when the crossbridge detaches at $x = 0$ must be kept small; this can be achieved by supposing that the basic free energy for the attachment of actin to myosin is similar to that for attachment of ATP. This would mean that the effective actin concentration would have to be about 1000 times greater than that of ATP, but this is not obviously impossible.

As a final point, we discuss how theories with two or more, mechanically distinct, attached states can be related to the biochemical scheme. Eisenberg *et al.* (1980) have suggested how this can be done (Fig. 5.13B). In this scheme, which has been worked out in considerable quantitative detail, the biochemical basis for the model is a reaction scheme in which crossbridge detachment is obligatory before ATP splitting and in which the rate-limiting step occurs in the detached crossbridge after ATP splitting. The two tension-producing states are identified with AM·ADP·P_i and AM·ADP, and most of the work is associated with the latter. The rate of the attachment process itself is considered fast but the effective rate of attachment is limited by the same transition in the detached crossbridge that is rate limiting in solution. Thus, the maximum rates of ATPase are similar in solution and *in vivo* as in the above consideration of the 1957 theory.

Eisenberg and Greene (1980) have suggested a different scheme, with the reaction sequence proposed by Stein *et al.* (1979) as a biochemical basis. In this sequence, there is a nondissociating pathway for the hydrolysis of ATP and the rate-limiting step can occur with the myosin either attached to actin, or detached (Fig. 5.13C). The mechanical model thus has to provide a mechanism for the cyclic detachment of the crossbridge as the muscle shortens, although ATP splitting can occur in solution without detachment. The scheme also has to explain how the large free energy change on ATP binding to actomyosin in solution is utilized to perform work. The solution suggested to these problems is illustrated in Fig. 5.13C. The major difference between this scheme and the first is in the role given to the state AM·T, which is now supposed to have mechanical properties similar to AM·ADP·P_i, with which it is equilibrium, rather than like AM·ADP. This enables the mechanical cycle to follow a path, shown by the heavy line in Fig. 5.13C, that does not involve any excessive inefficiencies.

Thus, we can conclude that there is no difficulty in providing plausible biochemical parallels to the mechanical theories that have been proposed. The composite theories can be said to make predictions about the behaviour of both structured systems and actomyosin in solution. We are left, however,

References

Abbott, B. C. (1951). The heat production associated with the maintenance of a prolonged contraction and the extra heat produced during large shortening. *J. Physiol (London)* **112**, 438–445.

Abbott, B. C. and Aubert, X. M. (1951). Changes of energy in muscle during very slow stretches. *Proc. R. Soc. London Ser. B* **139**, 104–117.

Abbott, B. C., and Aubert, X. (1952). The force exerted by active striated muscle during and after change of length. *J. Physiol. (London)* **117**, 77–86.

Abbott, B. C., Aubert, X. M., and Hill, A. V. (1951). The adsorption of work by a muscle stretched during a single twitch or a short tetanus. *Proc. R. Soc. London Ser. B* **139**, 86–104.

Allen, D. G. (1978). Shortening of tetanized skeletal muscle causes a fall of intra-cellular calcium concentration. *J. Physiol. (London)* **275**, 63P.

Aubert, X. (1956). "Le Couplage Energetique de la Contraction Musculaire." Arscia, Brussels.

Aubert, X. (1968). *In* "Symposium on Muscle" (E. Ernst and F. B. Straub, Eds.), pp. 187–189. Akademiai Kiado, Budapest.

Aubert, X., and Gilbert, S. (1980). Variation in the isometric maintenance heat rate with muscle length near that of maximum tension in frog striated muscle. *J. Physiol. (London)* **303**, 1–8.

Aubert, X., and Lebacq, J. (1971). The heat of shortening during the plateau of tetanic contraction and at the end of relaxation. *J. Physiol. (London)* **216**, 181–200.

Aubert, X., and Maréchal, G. (1963a). La fraction labile de la thermogenese associée au maintien de la contraction isometrique. *Arch. Int. Physiol. Biochim.* **71**, 282–283.

Aubert, X., and Maréchal, G. (1963b). Le bilan energétique des contractions musculaires avec travail positif ou negatif. *J. Physiol. (Paris)* **55**, 186–187.

Bagshaw, C. R., and Trentham, D. R. (1973). The reversibility of adenosine triphosphate cleavage by myosin. *Biochem. J.* **133**, 323–328.

Bagshaw, C. R., and Trentham, D. R. (1974). The characterisation of myosin-product complexes and of product release steps during the magnesium ion-dependent adenosine triphosphatase reaction. *Biochem. J.* **141**, 331–349.

Bagshaw, C. R., Eccleston, J. F., Trentham, D. R., and Yates, D. W. (1973). Transient kinetic studies of the Mg^{++} -dependent ATPase of myosin and its proteolytic subfragments. *Cold Spring Harbor Symp. Quant. Biol.* **37**, 127–135.

Bagshaw, C. R., Eccleston, J. F., Eckstein, F., Goody, R. S., Gutfreund, H., and Trentham, D. R. (1974). The magnesium ion dependent adenosine triphosphatase of myosin. *Biochem. J.* **141**, 351–364.

Bates, R. G., and Acree, S. F. (1945). pH of aqueous mixtures of potassium dihydrogen phosphate and disodium hydrogen phosphate at 0°C to 60°C. *J. Res. Natl. Bur. Stand.* **34**, 371–394.

BENDALL, J. R. (1969). "Muscles, Molecules and Movement: An Essay in the Contraction of Muscles." Heinemann, London.

BENDALL, J. R., and TAYLOR, A. A. (1970). The Meyerhof quotient and the synthesis of glycogen from lactate in frog and rabbit muscle. *Biochem. J.* **118**, 887–893.

BENZINGER, T., KITZINGER, C., HEMS, R., and BURTON, K. (1959). Free-energy changes of the glutaminase reaction and the hydrolysis of the terminal pyrophosphate bond of adenosine triphosphate. *Biochem. J.* **71**, 400–407.

BERNHARD, S. A. (1956). Ionization constants and heats of tris(hydroxymethyl)aminomethane and phosphate buffers. *J. Biol. Chem.* **218**, 961–969.

BLANGE, T., and STIENEN, G. T. M. (1979). Isometric tension transients in skeletal muscle before and after inhibition of ATP synthesis. *In* "Cross-bridge Mechanism in Muscle Contraction" (H. Sugi and G. H. Pollack, Eds.), pp. 211–222. Univ. of Tokyo Press, Tokyo.

BLINKS, J. R., PRENDERGAST, F. G., and ALLEN, D. G. (1976). Photoproteins as biological calcium indicators. *Pharmacol. Rev.* **28**, 1–93.

BLINKS, J. R., RÜDEL, R., and TAYLOR, S. R. (1978). Calcium transients in isolated amphibian skeletal muscle fibres: Detection with aequorin. *J. Physiol. (London)* **277**, 291–323.

BLINKS, J. R., WIER, W. G., HESS, P., and PRENDERGAST, F. G. (1982). Measurement of Ca^{2+} concentrations in living cells. *Prog. Biophys.* **40**, 1–114.

BOLTON, T. B., and VAUGHAN-JONES, R. D. (1977). Continuous direct measurement of intracellular chloride and pH in frog skeletal muscle. *J. Physiol. (London)* **270**, 801–833.

BOZLER, E. (1930). The heat production of smooth muscle. *J. Physiol. (London)* **69**, 442–462.

BRADY, A. J. (1966). Onset of contractility in cardiac muscle. *J. Physiol. (London)* **184**, 560–580.

BREMEL, R. D., and WEBER, A. (1972). Cooperation within actin filament in vertebrate skeletal muscle. *Nature (London) New Biol.* **238**, 91–101.

BRESSLER, B. H. (1981). Isometric contractile properties and instantaneous stiffness of amphibian skeletal muscle in the temperature range of 0 to 20°C. *Can. J. Physiol. Pharmacol.* **59**, 548–554.

BRESSLER, B. H., and CLINCH, N. F. (1974). The compliance of contracting skeletal muscle. *J. Physiol. (London)* **237**, 477–493.

BRIDEN, K. L., and ALPERT, N. R. (1972). The effect of shortening on the time course of active state decay. *J. Gen. Physiol.* **60**, 202–220.

BRIDGE, J. A. B. (1976). A thermochemical investigation of chicken latissimus dorsi muscle. Ph.D. thesis, University of California, Los Angeles, California.

BROCKLEHURST, L. M. (1973). The response of active muscle to stretch. Ph.D. thesis, University of London.

BRONK, D. (1930). The energy expended in maintaining a muscular contraction. *J. Physiol. (London)* **69**, 306–315.

BROWN, T., CHANCE, E. M., DAWSON, M. J., GADIAN, D. G., RADDA, G. K., and WILKIE, D. R. (1980). The activity of creatine kinase in frog skeletal muscle studied by saturation transfer nuclear magnetic resonance. *J. Physiol. (London)* **305**, 84P.

BUGNARD, L. (1934). The relation between total and initial heat in single muscle twitches. *J. Physiol. (London)* **82**, 509–519.

BURK, D. (1929). The free energy of glycogen–lactic acid breakdown in muscle. *Proc. R. Soc. London Ser. B* **104**, 153–170.

CAIN, D. F. (1960). The immediate energy source for muscular contraction. Ph.D. thesis, University of Pennsylvania.

CAIN, D. F., and DAVIES, R. E. (1962). Breakdown of adenosine triphosphate during a single contraction of working muscle. *Biochem. Biophys. Res. Commun.* **8**, 361–366.

CAIN, D. F., and DAVIES, R. E. (1964). Rapid arrest of metabolism with melting freon. *In* "Rapid Mixing and Sampling Techniques in Biochemistry" (B. Chance, R. H. Eisenhardt, Q. H. Gibson, and K. K. Lonberg-Holm, Eds.), pp. 229–237. Academic Press, New York.

CAIN, D. F., INFANTE, A. A., and DAVIES, R. E. (1962). Chemistry of muscle contraction. *Nature (London)* **196**, 214–217.

CANFIELD, S. P. (1971). The mechanical properties and heat production of chicken latissimus dorsi muscles during tetanic contractions. *J. Physiol. (London)* **219**, 281–302.

CANFIELD, P. and MARÉCHAL, G. (1973). Equilibrium of nucleotides in frog sartorius muscle during an isometric tetanus at 20°C. *J. Physiol. (London)* **232**, 453–466.

CANFIELD, P., LEBACQ, J., and MARÉCHAL, G. (1973). Energy balance in frog sartorius muscle during an isometric tetanus at 20°C. *J. Physiol. (London)* **232**, 467–483.

CANNELL, M. B. (1982). Intracellular calcium during relaxation in frog single muscle fibres. *J. Physiol. (London)* **326**, 70P–71P.

CANNELL, M. B. (1983). The intracellular calcium concentration during relaxation of frog skeletal muscle. Ph.D. thesis, University of London.

CANNELL, M. B., and ALLEN, D. G. (1984). A model of calcium movements during activation in the sarcomere of frog skeletal muscle. *Biophys. J.* **45**, 913–925.

CAPELLOS, C., and BIELSKI, B. H. J. (1972). "Kinetic Systems: Mathematical Description of Kinetics in Solution." Wiley (Interscience), New York.

CARLSON, F. D. (1963). The mechanochemistry of muscular contraction, a critical revaluation of *in vivo* studies. *Prog. Biophys. Mol. Biol.* **13**, 262–314.

CARLSON, F. D., and SIGER, A. (1959). The creatine phosphoryltransfer reaction in iodoacetate-poisoned muscle. *J. Gen. Physiol.* **43**, 301–313.

CARLSON, F. D., and SIGER, A. (1960). The mechanochemistry of muscular contraction. I. The isometric twitch. *J. Gen. Physiol.* **44**, 33–60.

CARLSON, F. D., and WILKIE, D. R. (1974). "Muscle Physiology." Prentice-Hall, New York.

CARLSON, F. D., HARDY, D., and WILKIE, D. R. (1963). Total energy production and phosphocreatine hydrolysis in the isotonic twitch. *J. Gen. Physiol.* **46**, 851–882.

CARLSON, F. D., HARDY, D. J., and WILKIE, D. R. (1967). The relation between heat produced and phosphorylcreatine split during isometric contraction of frog's muscle. *J. Physiol. (London)* **189**, 209–235.

CARPENTER, T. M. (1939). "Tables for Respiratory Exchange." Carnegie Institution of Washington Publications.

CASELLA, C. (1950). Tensile force in total striated muscle, isolated fibre and sarcolemma. *Acta Physiol. Scand.* **21**, 380–401.

CATTELL, MCK. (1934). The delayed heat production of isolated muscle in single and tetanic contractions. *J. Cell. Comp. Physiol.* **5**, 115–122.

CAVAGNA, G. A., and CITTERIO, G. (1974). Effect of stretching on the elastic characteristics and the contractile component of frog striated muscle. *J. Physiol. (London)* **239**, 1–14.

CECCHI, G., COLOMO, F., and LOMBARDI, V. (1978). Force–velocity relation in normal and nitrate-treated frog single muscle fibres during rise of tension in an isometric tetanus. *J. Physiol. (London)* **285**, 257–273.

CECCHI, G., COLOMO, F., and LOMBARDI, V. (1981). Force–velocity relation in deuterium oxide-treated frog single muscle fibres during the rise of tension in an isometric tetanus. *J. Physiol. (London)* **317**, 207–221.

CHANCE, B., ELEFF, S., and LEIGH, J. S., Jr. (1980). Non-invasive, non-destructive approaches to cell bioenergetics. *Proc. Natl. Acad. Sci. U.S.A.* **77**, 7430–7434.

CHAPLAIN, R. A. (1971). The importance of energy storage for the late phase of a muscle twitch. *Acta Biol. Med. Ger.* **26**, 1–8.

CHINET, A., CLAUSEN, T., and GIRARDIER, L. (1977). Microcalorimetric determination of energy expenditure due to active sodium–potassium transport in the soleus muscle and brown adipose tissue of the rat. *J. Physiol. (London)* **265**, 43–61.

CHOCK, S. P., and EISENBERG, E. (1974). Heavy meromyosin Mg-ATPase: Presteady-state and steady-state H^+ release. *Proc. Natl. Acad. Sci. U.S.A.* **71**, 4915–4919.

CHOCK, S. P., CHOCK, P. B., and EISENBERG, E. (1976). Pre-steady state kinetic evidence for a cyclic interaction of myosin subfragment-one with actin during the hydrolysis of adenosine 5′-triphosphate. *Biochemistry* **15**, 3244–3253.

CIVAN, M. M., and PODOLSKY, R. J. (1966). Contraction kinetics of striated muscle fibres following a quick change in load. *J. Physiol. (London)* **184**, 511–534.

CLEWORTH, D., and EDMAN, K. A. P. (1969). Laser diffraction studies on single skeletal muscle fibres. *Science* **163**, 296–298.

CLEWORTH, D. R., and EDMAN, K. A. P. (1972). Changes in sarcomere length during isometric tension development in frog skeletal muscle. *J. Physiol. (London)* **227**, 1–17.

CLINCH, N. F. (1968). On the increase in rate of heat production caused by stretch in frog's skeletal muscle. *J. Physiol. (London)* **196**, 397–414.

CLOSE, R. (1964). Dynamic properties of fast and slow skeletal muscles of the rat during development. *J. Physiol. (London)* **173**, 74–95.

CLOSE, R. (1965). Force:velocity properties of mouse muscle. *Nature (London)* **206**, 718–719.

CLOSE, R. I. (1972a). Dynamic properties of mammalian skeletal muscles. *Physiol. Rev.* **52**, 129–197.

CLOSE, R. I. (1972b). The relations between sarcomere length and characteristics of isometric twitch contractions of frog sartorius muscle. *J. Physiol. (London)* **220**, 745–762.

CLOSE, R. I., and LUFF, A. R. (1974). Dynamic properties of inferior rectus muscle of the rat. *J. Physiol. (London)* **236**, 259–270.

CONNOLLY, R., GOUGH, W., and WINEGRAD, S. (1971). Characteristics of the isometric twitch of skeletal muscle immediately after a tetanus. *J. Gen. Physiol.* **57**, 697–709.

COOKE, R., and BIALEK, W. (1979). Contraction of glycerinated muscle fibers as a function of the ATP concentration. *Biophys. J.* **28**, 241–258.

CORAY, A., FRY, C. H., HESS, G., MCGUIGAN, J. A. S., and WEINGART, R. (1980). Resting calcium in sheep cardiac tissue and in frog skeletal muscle measured with ion-selective microelectrodes. *J. Physiol. (London)* **305**, 60P.

COSTANTIN, L. L. (1975a). Electrical properties of the transverse tubular system. *Fed. Proc., Fed. Am. Soc. Exp. Biol.* **34**, 1390–1394.

COSTANTIN, L. L. (1975b). Contractile activation in skeletal muscle. *Prog. Biophys. Mol. Biol.* **29**, 197–224.

CRAIG, R., and OFFER, G. (1976). Axial arrangement of crossbridges in thick filaments of vertebrate striated muscle. *J. Mol. Biol.* **102**, 325–332.

CROW, M. T., and KUSHMERICK, M. J. (1982). Chemical energetics of slow- and fast-twitch muscles of the mouse. *J. Gen. Physiol.* **79**, 147–166.

CURTIN, N. A. (1976). The prolongation of relaxation after an isometric tetanus. *J. Physiol. (London)* **258**, 80P–81P.

CURTIN, N. A., and DAVIES, R. E. (1973). Chemical and mechanical changes during stretching of activated frog skeletal muscle. *Cold Spring Harbor Symp. Quant. Biol.* **37**, 619–626.

CURTIN, N. A., and DAVIES, R. E. (1975). Very high tension with very little ATP breakdown by active skeletal muscle. *J. Mechanochem. Cell Motil.* **3**, 147–154.

CURTIN, N. A., and WOLEDGE, R. C. (1974). Energetics of relaxation in frog muscle. *J. Physiol. (London)* **238**, 437–446.

CURTIN, N. A., and WOLEDGE, R. C. (1975). Energy balance in DNFB-treated and untreated frog muscle. *J. Physiol. (London)* **246**, 737–752.

CURTIN, N. A., and WOLEDGE, R. C. (1977). A comparison of the energy balance in two successive isometric tetani of frog muscle. *J. Physiol. (London)* **270**, 455–471.

CURTIN, N. A., and WOLEDGE, R. C. (1978). Energy changes and muscular contraction. *Physiol. Rev.* **58**, 690–761.

CURTIN, N. A., and WOLEDGE, R. C. (1979a). Chemical change and energy production during contraction in frog muscle: How are their time courses related? *J. Physiol.* (*London*) **288**, 353–366.

CURTIN, N. A., and WOLEDGE, R. C. (1979b). Chemical change, production of tension and energy following stretch of active muscle of frog. *J. Physiol.* (*London*) **297**, 539–550.

CURTIN, N. A., and WOLEDGE, R. C. (1981). Effect of muscle length on energy balance in frog skeletal muscle. *J. Physiol.* (*London*) **316**, 453–468.

CURTIN, N. A., GILBERT, C., KRETZSCHMAR, K. M., and WILKIE, D. R. (1974). The effect of the performance of work on total energy output and metabolism during muscular contraction. *J. Physiol.* (*London*) **238**, 455–472.

CURTIN, N. A., HOWARTH, J. V., and WOLEDGE, R. C. (1983). Heat production by single fibres of frog muscle. *J. Muscle Res. Cell Motil.* **4**, 207–222.

DAVIES, R. E. (1965a). On the mechanism of muscular contraction. *In* "Essays in Biochemistry" (P. N. Campbell and G. D. Greville, Eds.), Vol. 1, pp. 29–55. Academic Press, New York.

DAVIES, R. E. (1965b). The role of ATP in muscle contraction. *In* "Muscle" (W. M. Paul, E. E. Daniel, C. M. Kay, and G. Monckton, Eds.), pp. 49–65. Pergamon, Oxford.

DAWSON, M. J., GADIAN, D. G., and WILKIE, D. R. (1977). Contraction and recovery of living muscles studied by ^{31}P nuclear magnetic resonance. *J. Physiol.* (*London*) **267**, 703–735.

DAWSON, M. J., GADIAN, D. G., and WILKIE, D. R. (1978). Muscular fatigue investigated by phosphorous nuclear magnetic resonance. *Nature* (*London*) **274**, 861–866.

DAWSON, M. J., GADIAN, D. G., and WILKIE, D. R. (1980). Mechanical relaxation rate and metabolism studied in fatiguing muscle by phosphorous nuclear magnetic resonance. *J. Physiol.* (*London*) **299**, 465–484.

DE FURIA, R. R., and KUSHMERICK, M. J. (1977). ATP utilization associated with recovery metabolism in anaerobic frog muscle. *Am. J. Physiol.* **232**, C30–C36.

DÉLÈZE, J. B. (1961). The mechanical properties of the semitendinosus muscle at lengths greater than its length in the body. *J. Physiol.* (*London*) **158**, 154–164.

DE MEIS, L., and VIANNA, A. L. (1979). Energy interconversion by the Ca^{2+}-dependent ATPase of the sarcoplasmic reticulum. *Annu. Rev. Biochem.* **48**, 275–292.

DEWEY, M. M., LEVINE, R. J. C., COLFLESH, D., WALCOTT, B., BRANN, L., BALDWIN, A., and BRINK, P. (1979). Structural changes in thick filaments during sarcomere shortening in *Limulus* striated muscle. *In* "Cross-Bridge Mechanism in Muscle Contraction" (H. Sugi and G. H. Pollack, Eds.), pp. 3–22. Univ. of Tokyo Press, Tokyo.

DICKINSON, V. A., and WOLEDGE, R. C. (1973). The thermal effects of shortening in tetanic contractions of frog muscle. *J. Physiol.* (*London*) **233**, 659–671.

DUBUISSON, M. (1939). Studies on the chemical processes which occur in muscle before, during and after contraction. *J. Physiol.* (*London*) **94**, 461–482.

DUBUISSON, M. (1954). "Muscular Contraction." Thomas, Springfield, Illinois.

DYDYNSKA, M., and WILKIE, D. R. (1966). The chemical and energetic properties of muscles poisoned with fluorodinitrobenzene. *J. Physiol.* (*London*) **184**, 751–769.

EBASHI, S. (1975). Regulatory mechanism of muscle contraction with special reference to the Ca–troponin–tropomyosin system. *Essays Biochem.* **10**, 1–36.

EBASHI, S., and ENDO, M. (1968). Calcium ion and muscle contractions. *Prog. Biophys. Mol. Biol.* **18**, 125–183.

EBASHI, S., ENDO, M., and OHTSUKI, I. (1969). Control of muscle contraction. *Q. Rev. Biophys.* **2**, 351–384.

EDMAN, K. A. P. (1966). The relation between sarcomere length and active tension in isolated semitendinosus fibres of the frog. *J. Physiol.* (*London*) **183**, 407–417.

Edman, K. A. P. (1975). Mechanical deactivation induced by active shortening in isolated muscle fibres of the frog. *J. Physiol. (London)* **246**, 255–275.

Edman, K. A. P. (1979). The velocity of unloaded shortening and its relation to sarcomere length and isometric force in vertebrate muscle fibres. *J. Physiol. (London)* **291**, 143–159.

Edman, K. A. P. (1980). Depression of mechanical performance by active shortening during twitch and tetanus of vertebrate muscle fibres. *Acta Physiol. Scand.* **109**, 15–26.

Edman, K. A. P., and Flitney, F. W. (1982). Laser diffraction studies of sarcomere dynamics during "isometric" relaxation in isolated muscle fibres of the frog. *J. Physiol. (London)* **329**, 1–20.

Edman, K. A. P., and Hwang, J. C. (1977). The force–velocity relationship in vertebrate muscle fibres at varied tonicity of the extracellular medium. *J. Physiol. (London)* **269**, 255–272.

Edman, K. A. P., and Kiessling, A. (1971). The time course of the active state in relation to sarcomere length and movement studied in single skeletal muscle fibres of the frog. *Acta Physiol. Scand.* **81**, 182–196.

Edman, K. A. P., and Mattiazzi, A. R. (1981). Effects of fatigue and altered pH on isometric force and velocity of shortening at zero load in frog muscle fibres. *Muscle Res. Cell Motil.* **2**, 321–334.

Edman, K. A. P., and Reggiani, C. (1982). Differences in maximum velocity of shortening along frog muscle fibres. *J. Physiol. (London)* **329**, 47P–48P.

Edman, K. A. P., and Reggiani, C. (1983). The descending limb of the length–tension relation studied in short, length-clamped segments of frog muscle fibres. *J. Physiol. (London)* **341**, 41P–42P.

Edman, K. A. P., Mulieri, L. A., and Scubon-Mulieri, B. (1976). Non-hyperbolic force–velocity relationship in single muscle fibres. *Acta Physiol. Scand.* **98**, 143–156.

Edman, K. A. P., Elzinga, G., and Noble, M. I. M. (1978). Enhancement of mechanical performance by stretch during tetanic contractions of vertebrate skeletal muscle fibres. *J. Physiol. (London)* **281**, 139–155.

Edman, K. A. P., Elzinga, G., and Noble, M. I. M. (1979). The effect of stretch of contracting skeletal muscle fibres. *In* "Cross-Bridge Mechanism in Muscle Contraction" (H. Sugi and G. H. Pollack, Eds.), pp. 297–309. Univ. of Tokyo Press, Tokyo.

Edman, K. A. P., Elzinga, G., and Noble, M. I. M. (1982). Residual force enhancement after stretch of contracting frog single muscle fibres. *J. Gen. Physiol.* **80**, 769–784.

Edwards, R. H. T., Hill, D. K., and Jones, D. A. (1975a). Metabolic changes associated with the slowing of relaxation in fatigued mouse muscle. *J. Physiol. (London)* **251**, 287–301.

Edwards, R. H. T., Hill, D. K., and Jones, D. A. (1975b). Heat production and chemical changes during isometric contractions of the human quadriceps muscle. *J. Physiol. (London)* **251**, 303–315.

Eisenberg, E., and Greene, L. E. (1980). The relation of muscle biochemistry to muscle physiology. *Annu. Rev. Physiol.* **42**, 293–309.

Eisenberg, E., and Hill, T. L. (1978). A cross-bridge model of muscle contraction. *Prog. Biophys. Mol. Biol.* **33**, 55–82.

Eisenberg, E., and Kielley, W. W. (1973). Evidence for a refractory state of heavy meromyosin and subfragment-1 unable to bind to actin in the presence of ATP. *Cold Spring Harbor Symp. Quant. Biol.* **37**, 145–152.

Eisenberg, E., and Moos, C. (1968). The adenosine triphosphatase activity of acto-heavy meromyosin. A kinetic analysis of actin activation. *Biochemistry* **7**, 1486–1489.

Eisenberg, E., and Moos, C. (1970). Actin activation of heavy meromyosin adenosine triphosphatase. Dependence on adenosine triphosphate and actin concentrations. *J. Biol. Chem.* **245**, 2451–2456.

Eisenberg, E., Dobkin, L., and Kielley, W. W. (1972). Heavy meromyosin: Evidence for a refractory state unable to bind to actin in the presence of ATP. *Proc. Natl. Acad. Sci. U.S.A.* **69**, 667–671.

EISENBERG, E., HILL, T. L., and CHEN, Y. (1980). Cross-bridge model of muscle contraction. *Biophys, J.* **29**, 195–227.

ELLIOTT, G. F., ROME, E. M., and SPENCER, M. (1970). A type of contraction hypothesis applicable to all muscles. *Nature* (*London*) **226**, 417–420.

ELZINGA, G., HUISMAN, R. M., and WIECHMANN, A. H. C. A. (1981). Oxygen consumption of frog skeletal muscle fibres. *J. Physiol.* (*London*) **317**, 9P.

ELZINGA, G., PECKHAM, M., and WOLEDGE, R. C. (1982). Variation in isometric heat rate with sarcomere length in extensor digitorum longus IV of the frog (*Rana temporaria*). *J. Muscle Res. Cell. Motil.* **3**, 456–457.

ELZINGA, G., LANGEWOUTERS, G. J., WESTERHOF, N., and WIECHMANN, A. H. C. A. (1984). Oxygen uptake of frog skeletal muscle fibres following tetanic contractions at 18°C. *J. Physiol.* (*London*) **346**, 365–378.

ENDO, M. (1977). Calcium release from the sarcoplasmic reticulum. *Physiol. Rev.* **57**, 71–108.

EULER, U. S. V. (1935). Some factors influencing the heat production of muscle after stretching. *J. Physiol.* (*London*) **84**, 1–14.

FALK, G., and FATT, P. (1964). Linear electrical properties of striated muscle fibres observed with intracellular electrodes. *Proc. R. Soc. London Ser. B* **160**, 69–123.

FENG, T. P. (1932). The effect of length on the resting metabolism of muscle. *J. Physiol.* (*London*) **74**, 441–454.

FENN, W. O. (1923). A quantitative comparison between the energy liberated and the work performed by the isolated sartorius of the frog. *J. Physiol.* (*London*) **58**, 175–203.

FENN, W. O. (1924). The relation between work performed and the energy liberated in muscular contraction. *J. Physiol.* (*London*) **58**, 373–395.

FENN, W. O., and LATCHFORD, W. B. (1933). The effect of muscle length on the energy for maintenance of tension. *J. Physiol.* (*London*) **80**, 213–219.

FERENCZI, M. A. (1978). Kinetics of contraction in frog muscle. Ph.D. thesis, University of London.

FERENCZI, M. A., GOLDMAN, Y. E., and SIMMONS, R. M. (1979). The relation between maximum shortening velocity and the magnesium adenosine triphosphate concentration in frog skinned muscle fibre. *J. Physiol.* (*London*) **292**, 71P–72P.

FERENCZI, M. A., HOMSHER, E., SIMMONS, R. M., and TRENTHAM, D. R. (1978a). Reaction mechanism of the magnesium ion-dependent adenosine triphosphatase of frog muscle myosin and subfragment 1. *Biochem. J.* **171**, 165–175.

FERENCZI, M. A., HOMSHER, E., TRENTHAM, D. R., and WEEDS, A. G. (1978b). Preparation and characterization of frog muscle myosin in subfragment 1 and actin. *Biochem. J.* **171**, 155–163.

FERENCZI, M. A., SIMMONS, R. M., and SLEEP, J. A. (1982). General considerations of cross-bridge models in relation to the dependence of MgATP concentration of mechanical parameters of skinned fibers from frog muscle. *In* "Basic Biology of Muscles: A Comparative Approach" (B. M. Twarog, R. J. C. Levine, and M. M. Dewey, Eds.), Society of General Physiologists Series, Vol. 37, pp. 91–107. Raven, New York.

FERENCZI, M. A., HOMSHER, E., and TRENTHAM, D. R. (1983). A new method of measuring the kinetics of adenosine triphosphate (ATP) cleavage in single glycerinated rabbit muscle fibres. *J. Physiol.* (*London*) **341**, 37P.

FERENCZI, M. A., GOLDMAN, Y. E., and SIMMONS, R. M. (1984). The dependence of force and shortening velocity on substrate concentration and skinned fibres from frog muscle. *J. Physiol.* (*London*) **350**, 519–543.

FERSHT, A. (1977). "Enzyme Structure and Mechanism." Freeman, Reading, Massachusetts.

FISCHER, E. (1931). The oxygen consumption of isolated muscles for isotonic and isometric twitches. *Am. J. Physiol.* **96**, 78–88.

FLITNEY, F. W., and HIRST, D. G. (1978). Cross-bridge detachment and sarcomere "give" during stretch of active frog's muscle. *J. Physiol.* (*London*) **276**, 449–465.

FLOYD, K., and SMITH, I. C. H. (1971). The mechanical and thermal properties of frog slow muscle fibres. *J. Physiol.* (*London*) **213**, 617–631.

FORD, L. E., HUXLEY, A. F., and SIMMONS, R. M. (1974). Mechanism of early tension recovery after a quick release in tetanized muscle fibres. *J. Physiol.* (*London*) **240**, 42P–43P.

FORD, L. E., HUXLEY, A. F., and SIMMONS, R. M. (1977). Tension responses to sudden length change in stimulated frog muscle fibres near slack length. *J. Physiol.* (*London*) **269**, 441–515.

FORD, L. E., HUXLEY, A. F., and SIMMONS, R. M. (1981). The relation between stiffness and filament overlap in stimulated frog muscle fibres. *J. Physiol.* (*London*) **311**, 219–249.

FROST, A. A., and PEARSON, R. G. (1960). "Kinetics and Mechanism," 2nd Ed. Wiley, New York.

FURUSAWA, K., and HARTREE, W. (1926). The anaerobic delayed heat production in muscle. *J. Physiol.* (*London*) **62**, 203–210.

GASSER, H. S., and HILL, A. V. (1924). The dynamics of muscular contraction. *Proc. R. Soc. London Ser. B* **96**, 398–437.

GAUTHIER, G. F., LOWEY, S., and HOBBS, A. W. (1978). Fast & slow myosin in developing muscle fibres. *Nature* (*London*) **274**, 25–29.

GELLERT, M., and STURTEVANT, J. M. (1960). The enthalpy change in the hydrolysis of creatine phosphate. *J. Am. Chem. Soc.* **82**, 1497–1499.

GIBBS, C. L., and GIBSON, W. R. (1972). Energy production of rat soleus muscle. *Am. J. Physiol.* **223**, 864–871.

GIBBS, C. L., RICCHIUTI, N. V., and MOMMAERTS, W. F. H. M. (1966). Activation heat in frog sartorius muscle. *J. Gen. Physiol.* **49**, 517–535.

GILBERT, C., KRETZSCHMAR, K. M., WILKIE, D. R., and WOLEDGE, R. C. (1971). Chemical change and energy output during muscular contraction. *J. Physiol.* (*London*) **218**, 163–193.

GILBERT, S. H. (1982). Time course of tension and heat production in response to small shortening ramps in frog skeletal muscle. *J. Physiol.* (*London*) **326**, 71P–72P.

GILBERT, S. H., and MATSUMOTO, Y. (1976). A reexamination of the thermoelastic effect in active striated muscle. *J. Gen. Physiol.* **68**, 81–94.

GILLIS, J. M. (1972). "Le role du calcium dans le controle intracellulaire de la contraction musculaire." Vander, Louvain.

GILLIS, J. M., and MARÉCHAL, G. (1974). The incorporation of radioactive phosphate into ATP in glycerinated fibres stretched or released during contraction. *J. Mechanochem. Cell Motil.* **3**, 55–68.

GILLIS, J. M., THOMASON, D., LEFEVRE, J., and KRETSINGER, R. H. (1982). Parvalbumins and muscle relaxation: A computer simulation study. *J. Muscle Res. Cell Motil.* **3**, 377–398.

GLYNN, I. M., and KARLISH, S. J. D. (1975). The sodium pump. *Annu. Rev. Physiol.* **37**, 13–55.

GODFRAIND-DE BECKER, A. (1972). Heat production and fluorescence changes of toad sartorius muscle during aerobic recovery after a short tetanus. *J. Physiol* (*London*) **223**, 719–734.

GODFRAIND-DE BECKER, A. (1973). "La Restauration Post-tetanique du Muscle Strie Thermogenese et Fluorescence." Vander, Louvain.

GODT, R. E., and LINDLEY, B. D. (1982). Influence of temperature upon contractile activation and isometric force production in mechanically skinned muscle fibres of the frog. *J. Gen. Physiol.* **80**, 279–297.

GODT, R. E., and MAUGHAN, D. W. (1977). Swelling of skinned muscle fibers of the frog. Experimental observations. *Biophys. J.* **19**, 103–116.

GOLDMAN, Y. E., and SIMMONS, R. M. (1978). Stiffness measurements on frog skinned muscle fibers at varying interfilamentary separation. *Biophys. J.* **21**, 86a.

GOLDMAN, Y. E., and SIMMONS, R. M. (1979). A diffraction system for measuring muscle sacromere length. *J. Physiol* (*London*) **292**, 5P.

GOLDMAN, Y. E., HIBBERD, M. G., MCCRAY, J. A., and TRENTHAM, D. R. (1982). Relaxation of muscle fibres by photolysis of caged ATP. *Nature* (*London*) **300**, 701–705.

GONZALEZ-SERRATOS, H. (1971). Inward spread of activation in vertebrate muscle fibres. *J. Physiol.* (*London*) **212**, 777–799.

GORDON, A. M., HUXLEY, A. F., and JULIAN, F. J. (1966a). Tension development in highly stretched vertebrate muscle fibres. *J. Physiol.* (*London*) **184**, 143–169.

GORDON, A. M., HUXLEY, A. F., and JULIAN, F. J. (1966b). The variation in isometric tension with sarcomere length in vertebrate muscle fibres. *J. Physiol.* (*London*) **184**, 170–192.

GORDON, A. M., GODT, R. E., DONALDSON, S. K. B., and HARRIS, C. E. (1973). Tension in skinned frog muscle fibers in solutions of varying ionic strength and neutral salt composition. *J. Gen. Physiol.* **62**, 550–574.

GOSSELIN-REY, C., and GERDAY, C. (1977). Parvalbumins from frog skeletal muscle. Isolation and characterisation. Structural modifications associated with calcium binding. *Biochim. Biophys. Acta* **492**, 53–63.

GOWER, D., and KRETZSCHMAR, K. M. (1976). Heat production and chemical change during isometric contraction of rat soleus muscle. *J. Physiol.* (*London*) **258**, 659–671.

GRATZER, W. B., and LOWEY, S. (1969). Effect of substrate on the conformation of myosin. *J. Biol. Chem.* **244**, 22–25.

GRIFFITHS, P. J., GÜTH, K., KUHN, H. J., and RUEGG, J. C. (1980). Crossbridge slippage in skinned frog muscle fibres. *Biophys. Struct. Mech.* **7**, 107–124.

GULATI, J., and BABU, A. (1982). Tonicity effects on intact single muscle fibres: Relation between force and cell volume. *Science* **215**, 1109–1112.

GUTFREUND, H. (1972). "Enzymes: Physical Principles." Wiley (Interscience), New York.

GUTFREUND, H. (1975). Kinetic analysis of the properties and reactions of enzymes. *Prog. Biophys. Mol. Biol.* **29**, 161–195.

GÜTH, K., and KUHN, H. J. (1978). Stiffness and tension during and after sudden length changes of glycerinated rabbit psoas muscle fibres. *Biophys. Struct. Mech.* **4**, 223–236.

HANSON, J., and LOWY, J. (1963). The structure of F-actin and of actin filament isolated from muscle. *J. Mol. Biol.* **6**, 46–60.

HARTREE, W. (1928). The identical source of work and heat in muscular contraction. *J. Physiol.* (*London*) **65**, 385–388.

HARTREE, W. (1932). The analysis of the delayed heat production of muscle. *J. Physiol.* (*London*) **75**, 273–287.

HARTREE, W., and HILL, A. V. (1921). The nature of the isometric twitch. *J. Physiol.* (*London*) **55**, 389–411.

HARTREE, W., and HILL, A. V. (1922). The recovery heat production of muscle. *J. Physiol.* (*London*) **56**, 367–381.

HASELGROVE, J. C. (1975). X-ray evidence for conformational changes in the myosin filaments of vertebrate striated muscle. *J. Mol. Biol.* **92**, 113–143.

HASELGROVE, J. C., and REEDY, M. K. (1978). Modeling rigor cross-bridge patterns in muscle. I. Initial studies of the rigor lattice of insect flight muscle. *Biophys. J.* **24**, 713–728.

HEINL, P., KUHN, H. J., and RUEGG, J. C. (1974). Tension responses to quick length change of glycerinated skeletal muscle fibres from the frog and tortoise. *J. Physiol.* (*London*) **237**, 243–258.

HESS, A., and PILAR, G. (1963). Slow fibres in the extraocular muscles of the cat. *J. Physiol.* (*London*) **169**, 780–798.

HESS, P., METZGER, P., and WEINGART, R. (1982). Free magnesium in sheep, ferret and frog striated muscle at rest measured with ion-selective microelectrodes. *J. Physiol.* (*London*) **333**, 173–188.

HIGHSMITH, S. (1977). The effects of temperature and salts on myosin subfragment-1 and F-actin association. *Arch. Biochem. Biophys.* **180**, 404–408.

Highsmith, S. (1978). Heavy meromyosin binds actin with negative cooperativity. *Biochemistry* **17**, 22–26.

Highsmith, S., Mendelson, R. A., and Morales, M. F. (1976). Affinity of myosin S-1 for F-actin, measured by time resolved fluorescence anisotropy. *Proc. Natl. Acad. Sci. U.S.A.* **73**, 133–137.

Hill, A. V. (1925). Length of muscle, and the heat and tension developed in an isometric contraction. *J. Physiol.* (*London*) **60**, 237–263.

Hill, A. V. (1930). The heat production in isometric and isotonic twitches. *Proc. R. Soc. London Ser. B* **107**, 115–131.

Hill, A. V. (1931). Myothermic experiment on a frog gastrocnemius. *Proc. R. Soc. London Ser B* **109**, 267–303.

Hill, A. V. (1937). Methods of analyzing the heat production of muscle. *Proc. R. Soc. London Ser. B* **124**, 114–136.

Hill, A. V. (1938). The heat of shortening and the dynamic constants of muscle. *Proc. R. Soc. London Ser. B* **126**, 136–195.

Hill, A. V. (1939). The mechanical efficiency of frogs' muscle. *Proc. R. Soc. London Ser. B* **127**, 434–451.

Hill, A. V. (1949a). The heat of activation and the heat of shortening in a muscle twitch. *Proc. R. Soc. London Ser. B* **136**, 195–211.

Hill, A. V. (1949b). The energetics of relaxation in a muscle twitch. *Proc. R. Soc. London Ser. B* **136**, 211–219.

Hill, A. V. (1949c). Myothermic methods. *Proc. R. Soc. London Ser. B* **136**, 228–241.

Hill, A. V. (1950a). Does heat production precede mechanical response in muscular contraction? *Proc. R. Soc. London Ser. B* **137**, 268–273.

Hill, A. V. (1950b). The dimensions of animals and their muscular dynamics. *Proc. R. Inst. G.B.* **34**, 450–473.

Hill, A. V. (1950c). The series elastic component of muscle. *Proc. R. Soc. London Ser. B* **137**, 237–280.

Hill, A. V. (1953). The "instantaneous" elasticity of active muscle. *Proc. R. Soc. London Ser. B* **141**, 161–178.

Hill, A. V. (1956). The design of muscles. *Br. Med. Bull.* **12**, 165–166.

Hill, A. V. (1958). The relation between force developed and energy liberated in an isometric twitch. *Proc. R. Soc. London Ser. B* **149**, 58–62.

Hill, A. V. (1961a). The negative delayed heat production in stimulated muscle. *J. Physiol.* (*London*) **158**, 178–196.

Hill, A. V. (1961b). The heat produced by a muscle after the last shock of a tetanus. *J. Physiol.* (*London*) **159**, 518–545.

Hill, A. V. (1964a). The variation of total heat production in a twitch with velocity of shortening. *Proc. R. Soc. London Ser. B* **159**, 596–605.

Hill, A. V. (1964b). The effect of load on the heat of shortening of muscle. *Proc. R. Soc. London Ser. B* **159**, 297–318.

Hill, A. V. (1964c). The efficiency of mechanical power development during muscular shortening and its relation to load. *Proc. R. Soc. London Ser. B* **159**, 319–324.

Hill, A. V. (1965). "Trails and Trials in Physiology." Arnold, London.

Hill, A. V. (1970). "First and Last Experiments in Muscle Mechanics." Cambridge Univ. Press, London and New York.

Hill, A. V., and Hartree, W. (1920). The four phases of heat-production of muscle. *J. Physiol.* (*London*) **54**, 84–128.

Hill, A. V., and Howarth, J. V. (1957a). Alternating relaxation heat in muscle twitches. *J. Physiol.* (*London*) **139**, 466–473.

Hill, A. V., and Howarth, J. V. (1957b). The effect of potassium on the resting metabolism of the frog's sartorius. *Proc. R. Soc. London Ser. B* **147**, 21–43.

Hill, A. V., and Howarth, J. V. (1959). The reversal of chemical reactions in contracting muscle during an applied stretch. *Proc. R. Soc. London Ser. B* **151**, 169–193.

Hill, A. V., and Woledge, R. C. (1962). An examination of absolute values in myothermic measurement, *J. Physiol. (London)* **162**, 311–333.

Hill, D. K. (1940a). The time course of the oxygen consumption of stimulated frog's muscle. *J. Physiol. (London)* **98**, 207–227.

Hill, D. K. (1940b). The time course of evolution of oxidative recovery heat of frog's muscle. *J. Physiol. (London)* **98**, 454–459.

Hill, D. K. (1940c). Hydrogen-ion concentration changes in frog's muscle following activity. *J. Physiol. (London)* **98**, 467–479.

Hill, D. K. (1968). Tension due to interaction between the sliding filaments in resting striated muscle. The effect of stimulation. *J. Physiol. (London)* **199**, 637–684.

Hill, L. (1977). A-band length, striation spacing and tension change on stretch of active muscle. *J. Physiol. (London)* **266**, 677–685.

Hill, T. L. (1974). Theoretical formalisation for the sliding filament model of contraction of striated muscle. Part I. *Prog. Biophys. Mol. Biol.* **28**, 267–340.

Hill, T. L. (1977). "Free Energy Transduction in Biology." Academic Press, New York.

Hill, T. L., Eisenberg, E., Chen, Y., and Podolsky, R. J. (1975). Some self-consistent two-state sliding filament models of muscle contraction. *Biophys. J.* **15**, 335–372.

Homsher, E., and Kean, C. J. (1978). Skeletal muscle energetics and metabolism. *Annu. Rev. Physiol* **40**, 93–131.

Homsher, E., and Rall, J. A. (1973). Energetics of shortening muscles in twitches and tetanic contractions. I. A reinvestigation of Hills' concept of shortening heat. *J. Gen. Physiol.* **62**, 663–676.

Homsher, E., Mommaerts, W. F. H. M., Ricchiuti, N. V., and Wallner, A. (1972). Activation heat, activation metabolism and tension-related heat in frog semitendinosus muscles. *J. Physiol. (London)* **220**, 601–625.

Homsher, E., Mommaerts, W. F. H. M., and Ricchiuti, N. V. (1973). Energetics of shortening muscles in twitches and tetanic contractions. II. Force-determined shortening heat. *J. Gen. Physiol.* **62**, 677–692.

Homsher, E., Briggs, F. N., and Wise, R. M. (1974). The effects of hypertonicity on resting and contracting frog skeletal muscles. *Am. J. Physiol.* **226**, 855–863.

Homsher, E., Rall, J. A., Wallner, A., and Ricchiuti, N. V. (1975). Energy liberation and chemical change in frog skeletal muscle during single isometric tetanic contractions. *J. Gen. Physiol.* **65**, 1–21.

Homsher, E., Kean, C. J., Wallner, A., and Garibian-Sarian, V. (1979). The time course of energy balance in an isometric tetanus. *J. Gen. Physiol.* **73**, 553–567.

Homsher, E., Irving, M., and Wallner, A. (1981). High-energy phosphate metabolism and energy libration associated with rapid shortening in frog skeletal muscle. *J. Physiol. (London)* **321**, 423–436.

Homsher, E., Irving, M., and Lebacq, J. (1983). The variation in shortening heat with sacromere length in frog muscle. *J. Physiol. (London)* **345**, 107–121.

Homsher, E., Yamada, T., Wallner, A., and Tsui, J. (1984). Energy balance studies in frog skeletal muscles shortening at one-half maximal velocity. *J. Gen. Physiol.* **84**, 347–360.

Hoult, D. I., Busby, S. J. W., Gadian, D. G., Radda, G. K., Richards, R. E., and Seely, P. J. (1974). Observation of tissue metabolites using ^{31}P nuclear magnetic resonance. *Nature (London)* **252**, 285–287.

HOWARTH, J. V. (1958). The behaviour of frog muscle in hypertonic solutions. *J. Physiol. (London)* **144**, 167–175.

HOWARTH, J. V., KEYNES, R. D., RITCHIE, J. M., and VON MURALT, A. (1975). The heat production associated with the passage of a single impulse in pike olfactory nerve fibres. *J. Physiol. (London)* **249**, 349–368.

HUTTER, O. F., and TRAUTWEIN, W. (1956). Neuromuscular facilitation by stretch of motor nerve-endings. *J. Physiol. (London)* **133**, 610–625.

HUXLEY, A. F. (1957). Muscle structure and theories of contraction. *Prog. Biophys. Biophys. Chem* **7**, 255–318.

HUXLEY, A. F. (1964). Muscle. *Annu. Rev. Physiol.* **26**, 131–152.

HUXLEY, A. F. (1971). The activation of striated muscle and its mechanical response. (The Croonian Lecture, 1967.) *Proc. R. Soc. London Ser. B* **178**, 1–27.

HUXLEY, A. F. (1973). A note suggesting that the cross-bridge attachment during muscle contraction may take place in two stages. *Proc. R. Soc. London Ser. B* **183**, 83–86.

HUXLEY, A. F. (1974). Muscular contraction. *J. Physiol. (London)* **243**, 1–43.

HUXLEY, A. F. (1980). "Reflections on Muscle." (The Sherrington Lecture XIV.) Liverpool Univ. Press, Liverpool.

HUXLEY, A. F. (1981). The mechanical properties of cross-bridges and their relation to muscle contraction. *Adv. Physiol. Sci.* **5**, 1–12.

HUXLEY, A. F., and LOMBARDI, V. (1980). A sensitive force transducer with resonant frequency 50 kHz. *J. Physiol. (London)* **305**, 15P–16P.

HUXLEY, A. F., and NIEDERGERKE, R. (1954). Interference microscopy of living muscle fibres. *Nature (London)* **173**, 971–973.

HUXLEY, A. F., and PEACHEY, L. D. (1961). The maximum length for contraction in vertebrate striated muscle. *J. Physiol. (London)* **156**, 150–165.

HUXLEY, A. F., and SIMMONS, R. M. (1970a). A quick phase in the series-elastic component of striated muscle, demonstrated in isolated fibres from the frog. *J. Physiol. (London)* **208**, 52P–53P.

HUXLEY, A. F., and SIMMONS, R. M. (1970b). Rapid "give" and the tension "shoulder" in the relaxation of frog muscle fibres. *J. Physiol. (London)* **210**, 32P–33P.

HUXLEY, A. F., and SIMMONS, R. M. (1971a). Mechanical properties of the cross-bridges of frog striated muscle. *J. Physiol. (London)* **218**, 59P–60P.

HUXLEY, A. F., and SIMMONS, R. M. (1971b). Proposed mechanism of force generation in striated muscle. *Nature (London)* **233**, 533–538.

HUXLEY, A. F., and SIMMONS, R. M. (1973). Mechanical transients and the origin of muscle force. *Cold Spring Harbor Symp. Quant. Biol.* **37**, 669–680.

HUXLEY, H. E. (1960). Muscle cells. *In* "The Cell, Biochemistry, Physiology, Morphology" (J. Brachet and A. E. Mirsky, Eds.), Vol. IV, pp. 365–481. Academic Press, New York.

HUXLEY, H. E. (1971). The structural basis of muscular contraction. (The Croonian Lecture 1970.) *Proc. R. Soc. London Ser. B* **178**, 131–149.

HUXLEY, H. E. (1979). Time resolved X-ray diffraction studies on muscle. *In* "Cross-Bridge Mechanism in Muscle Contraction" (H. Sugi and G. H. Pollack, Eds.), pp. 391–405. Univ. of Tokyo Press, Tokyo.

HUXLEY, H. E., and HANSON, J. (1954). Changes in the cross-striations of muscle during contraction and stretch and their structural interpretation. *Nature (London)* **173**, 973–976.

HUXLEY, H. E., FARUQI, A. R., KRESS, M., BORDAS, J., and KOCH, M. H. J. (1982). Time-resolved X-ray diffraction studies of the myosin layer-line reflections during muscle contraction. *J. Mol. Biol.* **158**, 637–684.

IKEMOTO, N. (1982). Structure and function of the calcium pump protein of sarcoplasmic reticulum. *Annu. Rev. Physiol.* **44**, 297–317.

INFANTE, A. A., and DAVIES, R. E. (1965). The effect of 2,4-dinitrofluorobenzene on the activity of striated muscle. *J. Biol. Chem.* **240**, 3996–4001.

INFANTE, A. A., KLAUPIKS, D., and DAVIES, R. E. (1964). Adenosine triphosphate: Changes in muscles doing negative work. *Science* **144**, 1577–1578.

INFANTE, A. A., KLAUPIKS, D., and DAVIES, R. E. (1965). Phosphorylcreatine consumption during single-working contractions of isolated muscle. *Biochim. Biophys. Acta* **94**, 504–515.

INFANTE, A. A., KLAUPIKS, D., and DAVIES, R. E. (1967). Length, tension, and metabolism during isometric contractions of frog sartorius muscles. *Biochim. Biophys. Acta* **88**, 215–217.

IRVING, M., and WOLEDGE, R. C. (1981a). The dependence on extent of shortening of the extra energy liberated by rapidly shortening frog skeletal muscle. *J. Physiol.* (*London*) **321**, 411–422.

IRVING, M., and WOLEDGE, R. C. (1981b). The energy liberation of frog skeletal muscle in tetanic contractions containing two periods of shortening. *J. Physiol.* (*London*) **321**, 401–410.

IRVING, M., WOLEDGE, R. C., and YAMADA, K. (1979). The heat produced by frog muscle in a series of contractions with shortening. *J. Physiol.* (*London*) **293**, 103–118.

IWAZUME, T. (1979). A new field theory of muscle contraction. *In* "Cross-Bridge Mechanism in Muscle Contraction" (H. Sugi and G. H. Pollack, Eds.), pp. 611–632. Univ. of Tokyo Press, Tokyo.

JEWELL, B. R., and WILKIE, D. R. (1958). An analysis of the mechanical components in frog's striated muscle. *J. Physiol.* (*London*) **143**, 515–540.

JEWELL, B. R., and WILKIE, D. R. (1960). The mechanical properties of relaxing muscle. *J. Physiol.* (*London*) **152**, 30–47.

JOHNSON, K. A., and TAYLOR, E. W. (1978). Intermediate states of subfragment 1 and acto subfragment 1 ATPase: Reevaluation of the mechanism. *Biochemistry* **17**, 3432–3442.

JORGENSEN, A. O., KALNINS, V., and MACLENNAN, D. H. (1979). Localisation of sarcoplasmic reticulum proteins in rat skeletal muscle by immunofluorescence. *J. Cell Biol.* **80**, 372–384.

JULIAN, F. J., and MORGAN, D. L. (1979a). Intersarcomere dynamics during fixed-end tetanic contractions of frog muscle fibres. *J. Physiol.* (*London*) **293**, 365–378.

JULIAN, F. J., and MORGAN, D. L. (1979b). The effect on tension of non-uniform distribution of length changes applied to frog muscle fibres. *J. Physiol.* (*London*) **293**, 379–392.

JULIAN, F. J., and MOSS, R. L. (1981). Effects of calcium and ionic strength on shortening velocity and tension development in frog skinned muscle fibres. *J. Physiol.* (*London*) **311**, 179–199.

JULIAN, F. J., and SOLLINS, M. R. (1973). Regulation of force and speed of shortening in muscle contraction. *Cold Spring Harbor Symp. Quant. Biol.* **37**, 635–646.

JULIAN, F. J., and SOLLINS, M. R. (1975). Variation of muscle stiffness with force at increasing speeds of shortening. *J. Gen. Physiol.* **66**, 287–302.

JULIAN, F. J., SOLLINS, M. R., and MOSS, R. L. (1978). Sarcomere length non-uniformity in relation to tetanic responses of stretched skeletal muscle fibres. *Proc. R. Soc. London Ser. B* **200**, 109–116.

JULIAN, F. J., MOSS, R. L., and WALLER, G. S. (1981). Mechanical properties and myosin light chain composition of skinned muscle fibres from adult and new-born rabbits. *J. Physiol.* (*London*) **311**, 201–218.

KATZ, B. (1939). The relation between force and speed in muscular contraction. *J. Physiol.* (*London*) **96**, 45–64.

KATZ, B. (1966). "Nerve, Muscle and Synapse." McGraw-Hill, New York.

KAWAI, M., and BRANDT, P. W. (1976). Two rigor states in skinned crayfish single muscle fibres. *J. Gen. Physiol.* **68**, 267–280.

KAYE, G. W. C., and LABY, T. H. (1973). "Tables of Physical and Chemical Constants," 14th Ed. Longmans, London.

Kean, C., Homsher, E., Sarian-Garibian, V., and Zemplenyi, J. (1976). Phosphocreatine (PC) splitting by stretched frog muscle. *Physiologist* **19**, 250.

Kean, C., Homsher, E., and Sarian-Garibian, V. (1977). The energy balance of crossbridge cycling in frog skeletal muscle. *Biophys. J.* **17**, 202a.

Keynes, R. D., and Aidley, D. J. (1981). "Nerve and Muscle." Cambridge Univ. Press, London and New York.

Kodama, T., and Woledge, R. C. (1979). Enthalpy changes for intermediate steps of the ATP hydrolysis catalyzed by myosin subfragment-1. *J. Biol. Chem.* **254**, 6382–6386.

Kodama, T., and Yamada, K. (1979). An explanation of the shortening heat based on the enthalpy profile of the myosin ATPase reaction. *In* "Cross-Bridge Mechanism in Muscle Contraction" (H. Sugi and G. H. Pollack, Eds.), pp. 481–488. Univ. of Tokyo Press, Tokyo.

Kretsinger, R. H. (1980). Structure and evolution of calcium-modulated proteins. *CRC Crit. Rev. Biochem.* **8**, 119–174.

Kretzschmar, K. M. (1970). Energy production and chemical change during muscular contraction. Ph.D. Thesis, University of London.

Kretzschmar, K. M., and Wilkie, D. R. (1972). A new method for absolute heat measurement utilizing the Peltier effect. *J. Physiol. (London)* **224**, 18P–21P.

Kretzschmar, K. M., and Wilkie, D. R. (1975). The use of the Peltier effect for simple and accurate calibration of thermoelectric devices. *Proc. R. Soc. London Ser. B* **190**, 315–321.

Kuffler, S. W. (1952). Incomplete neuromuscular transmission in twitch system of frog's skeletal muscles. *Fed. Proc., Fed. Am. Soc. Exp. Biol.* **11**, 87.

Kuffler, S. W., and Vaughan Williams, E. M. (1953). Small-nerve junctional potentials. The distribution of small motor nerves to frog skeletal muscle, and the membrane characteristics of the fibres they innervate. *J. Physiol. (London)* **121**, 289–317.

Kushmerick, M. J. (1983). Energetics of muscle contraction. *Handb. Physiol. Section 10: Skeletal Muscle*, pp. 189–237. American Physiological Society.

Kushmerick, M. J., and Davies, R. E. (1969). The chemical energetics of muscle contraction. II. The chemistry, efficiency and power of maximally working sartorius muscles. *Proc. R. Soc. London Ser. B* **174**, 315–353.

Kushmerick, M. J., and Paul, R. J. (1976a). Aerobic recovery metabolism following a single isometric tetanus in frog sartorius at 0°C. *J. Physiol. (London)* **254**, 693–709.

Kushmerick, M. J., and Paul, R. J. (1976b). Relationship between initial chemical reactions and oxidative recovery metabolism for single isometric contractions of frog sartorius at 0°C. *J. Physiol. (London)* **254**, 711–727.

Kushmerick, M. J., and Paul, R. J. (1977). Chemical energetics in repeated contractions of frog sartorius muscle at 0°C. *J. Physiol. (London)* **267**, 249–260.

Kushmerick, M. J., Larson, R. E., and Davies, R. E. (1969). The chemical energetics of muscle contraction. 1. Activation heat, heat of shortening and ATP utilization for activation-relaxation processes. *Proc. R. Soc. London Ser. B* **174**, 293–313.

Lännergren, J. (1971). The effect of low-level activation on the mechanical properties of isolated frog muscle fibres. *J. Gen. Physiol.* **58**, 145–162.

Lännergren, J. (1975). Structure and function of twitch and slow fibres in amphibian skeletal muscle. *In* "Basic Mechanisms of Ocular Motility and Their Clinical Implications" (G. Lennerstrand and P. Bach-y-Rita, Eds.), pp. 63–84. Pergamon, Oxford.

Lännergren, J. (1978). The force–velocity relation of isolated twitch and slow muscle fibres of *Xenopus laevis*. *J. Physiol. (London)* **283**, 501–521.

Lännergren, J., Lindblom, P., and Johansson, B. (1982). Contractile properties of two varieties of twitch fibres in *Xenopus laevis*. *Acta Physiol. Scand.* **114**, 523–535.

Lebacq, J. (1980). Origin of the heat of shortening in striated muscle. Thése d'agregation Univ. Catholique de Louvain, Louvain.

LEVIN, A., and WYMAN, J. (1927). The viscous elastic properties of muscle. *Proc. R. Soc. London Ser. B* **101**, 218–243.

LEVY, R. M., UMAZUME, Y., and KUSHMERICK, M. J. (1976). Ca^{++} dependence of tension and ADP production in segments of chemically skinned muscle fibres. *Biochim. Biophys. Acta* **430**, 325–365.

LUFF, A. R. (1981). Dynamic properties of the inferior rectus, extensor digitorum longus, diaphragm and soleus muscles of the mouse. *J. Physiol. (London)* **313**, 161–171.

LUNDSGAARD, E. (1934). Phosphagen- und pyrophosphatumsatz in jodessigsaurevergifteten Muskel. *Biochem. Z.* **269**, 308–328.

LYMN, R. W. (1974a). Actin activation of the myosin ATPase: A kinetic analysis. *J. Theor. Biol.* **43**, 313–328.

LYMN, R. W. (1974b). Effect of modifier on three-step enzymic process: general rate equation. *J. Theor. Biol.* **43**, 305–312.

LYMN, R. W., and TAYLOR, E. W. (1970). Transient state phosphate production in the hydrolysis of nucleotide triphosphates by myosin. *Biochemistry* **9**, 2975–2983.

LYMN, R. W., and TAYLOR, E. W. (1971). Mechanism of adenosine triphosphate hydrolysis by actomyosin. *Biochemistry* **10**, 4617–4624.

MCCARTER, R. (1981). Studies of sarcomere length by optical diffraction. *In* "Cell and Muscle Motility" (R. M. Dowben and J. W. Shay, Eds.), pp. 35–62. Plenum, New York.

MCCROREY, H. L., GALE, H. H., and ALPERT, N. R. (1966). Mechanical properties of cat tenuissimus muscle. *Am. J. Physiol.* **210**, 114–120.

MCGLASHAN, M. L. (1978). "Chemical Thermodynamics." Academic Press, New York.

MAHLER, M. (1978a). Diffusion and consumption of oxygen in resting frog sartorius muscle. *J. Gen. Physiol.* **71**, 533–557.

MAHLER, M. (1978b). Kinetics of oxygen consumption after a single isometric tetanus of the frog sartorius muscle at 20°C. *J. Gen. Physiol.* **71**, 559–580.

MARÉCHAL, G. (1964). Phosphorylcreatine and ATP changes during shortening and lengthening of stimulated muscle. *Arch. Int. Physiol. Biochim.* **72**, 306–309.

MARÉCHAL, G., and BECKERS-BLEUKX, G. (1965). La phosphorylcréatine et les nucleotides adenyliques d'un muscle strié a la fin d'un étirement. *J. Physiol. (Paris)* **57**, 652–653.

MARÉCHAL, G., and MOMMAERTS, W. F. H. M. (1963). The metabolism of phosphocreatine during an isometric tetanus of frog sartorius muscle. *Biochim. Biophys. Acta* **70**, 53–67.

MARÉCHAL, G., and PLAGHKI, L. (1979). The deficit of the isometric tetanic tension redeveloped after a release of frog muscle at a constant velocity. *J. Gen. Physiol.* **73**, 453–467.

MARÉCHAL, G., GOFFART, M., and AUBERT, X. (1963). Nouvelles recherches sur les proprietes du muscle squelettique du paresseux (*Choloepus hoffman* Peters). *Arch. Int. Physiol. Biochim.* **71**, 236–240.

MARGOSSIAN, S. S., and LOWEY, S. (1973). Substructure of the myosin molecule. IV. Interaction of myosin and its subfragment with adenosine triphosphate and F-actin. *J. Mol. Biol.* **74**, 313–330.

MARSTON, S., and WEBER, A. (1975). The dissociation constant of the actin-heavy meromyosin subfragment 1 complex. *Biochemistry* **14**, 3868–3873.

MARSTON, S. A., and TAYLOR, E. W. (1980). Comparison of the myosin and actomyosin ATPase mechanism of four types of vertebrate muscle. *J. Mol. Biol.* **139**, 573–600.

MARSTON, S. B., RODGER, C. D., and TREGEAR, R. T. (1976). Changes in muscle crossbridges when β,γ-imido-ATP binds to myosin. *J. Mol. Biol.* **104**, 263–276.

MARSTON, S. B., TREGEAR, R. T., RODGER, C. D., and CLARK, M. L. (1979). Coupling between the enzymatic site of myosin and the mechanical output of muscle. *J. Mol. Biol.* **128**, 111–126.

MARTONOSI, A., and MALIK, M. N. (1972). Kinetics of formation and dissociation of H-meromyosin–ADP complexes. *Cold Spring Harbor Symp. Quant. Biol.* **37**, 184–185.

Mashima, H., Akazawa, K., Kushima, H., and Fujii, K. (1972). The force–load–velocity relation and the viscous-like force in the frog skeletal muscle. *Jpn. J. Physiol.* **22**, 103–120.

Matsubara, I., and Elliott, G. F. (1972). X-Ray diffraction studies on skinned single fibres of frog skeletal muscle. *J. Mol. Biol.* **72**, 657–669.

Matsubara, I., Goldman, Y. E., and Simmons, R. M. (1984). Changes in the lateral filament spacing of skinned muscle fibres when cross-bridges attach. *J. Mol. Biol.* **173**, 15–33.

Meyer, R. A., Kushmerick, M. J., and Brown, T. R. (1982). Application of ^{31}P-NMR spectroscopy to the study of striated muscle metabolism. *Am. J. Physiol.* **242**, C1–C11.

Moeschler, H. J., Schaer, J. J., and Cox, J. A. (1980). A thermodynamic analysis of the binding of calcium and magnesium ions to parvalbumin. *Eur. J. Biochem*, **111**, 73–78.

Moisescu, D. G., and Thieleczek, R. (1978). Calcium and strontium concentration changes within skinned muscle preparations following a change in the external bathing solution. *J. Physiol.* (*London*) **275**, 241–262.

Mommaerts, W. F. H. M., and Wallner, A. (1967). The break-down of adenosine triphosphate in the contraction cycle of the frog sartorius muscle. *J. Physiol.* (*London*) **193**, 343–357.

Mommaerts, W. F. H. M., and Schilling, M. O. (1964). The rapid-freezing method for the interruption of muscular contraction. *In* "Rapid Mixing and Sampling Techniques in Biochemistry" (B. Chance, R. H. Eisenhardt, Q. H. Gibson, and K. K. Lonberg-Holm, Eds.), pp. 239–254. Academic Press, New York.

Morel, J. E., Pinset-Härström, I., and Gingold, M. P. (1976). Muscular contraction and cytoplasmic streaming: a new general hypothesis. *J. Theor. Biol.* **62**, 17–51.

Morgan, D. L., and Julian, F. J. (1981a). Stiffness of frog single muscle fibers during tension transients and constant velocity shortening. *Biophys. J.* **33**, 225a.

Morgan, D. L., and Julian, F. J. (1981b). Tension, stiffness, unloaded shortening velocity and potentiation of frog muscle fibers at sarcomere lengths below optimum. *Biophys. J.* **33**, 225a.

Morgan, D. L., Mochon, S., and Julian, F. J. (1982). A quantitative model of intersarcomere dynamics during fixed-end contractions of single frog muscle fibers. *Biophys. J.* **39**, 189–196.

Moss, R. L. (1979). Sarcomere length-tension relations of frog skinned muscle fibres during calcium activation at short lengths. *J. Physiol.* (*London*) **292**, 177–192.

Mulhern, S. A., and Eisenberg, E. (1976). Further studies on interaction of actin with heavy meromysin and subfragment 1 in the presence of ATP. *Biochemistry* **15**, 5702–5708.

Mulieri, L. A., Luhr, G., Trefry, J., and Alpert, N. R. (1977). Metal film thermopiles for use with rabbit right ventricular papillary muscles. *Am. J. Physiol.* **233**, C146–C156.

Mulvany, M. J. (1975). Mechanical properties of frog skeletal muscles in iodoacetic acid rigor. *J. Physiol.* (*London*) **252**, 319–334.

Needham, D. M. (1971). "Machina Carnis." Cambridge Univ. Press, London and New York.

Noble, M. I. M., and Pollack, G. H. (1977). Molecular mechanisms of contraction. *Circ. Res.* **40**, 333–342.

Offer, G. (1974). The molecular basis of muscular contraction. *In* "Companion to Biochemistry" (A. T. Bull, J. R. Lagnado, J. O. Thomas, and K. F. Tipton, Eds.), pp. 623–671. Longmans, London.

Offer, G., and Elliott, A. (1978). Can a myosin molecule bind to two actin filaments? *Nature* (*London*) **271**, 325–329.

Ostwald, T. J., and MacLennan, D. H. (1974). Isolation of a high affinity calcium-binding protein from sarcoplasmic reticulum. *J. Biol. Chem.* **249**, 974–979.

Page, S. G., and Huxley, H. E. (1963). Filament lengths in striated muscle. *J. Cell Biol.* **19**, 369–390.

Paul, R. J. (1983). Physical and biochemical energy balance during an isometric tetanus and steady state recovery in frog sartorius at 0°C. *J. Gen. Physiol.* **81**, 337–354.

Peachey, L. D. (1965). The sarcoplasmic reticulum and transverse tubules of the frog's sartorius. *J. Cell Biol.* **25**(3Z), 209–231.

Podolin, R. A., and Ford, L. E. (1983). The influence of calcium on shortening velocity of skinned frog muscle cells. *J. Muscle Res. Cell Motil.* **4**, 263–282.

Podolsky, R. J. (1960). Kinetics of muscular contraction: the approach to the steady state. *Nature* (*London*) **188**, 666–668.

Podolsky, R. J. (1964). The maximum sarcomere length for contraction of isolated myofibrils. *J. Physiol.* (*London*) **170**, 110–123.

Podolsky, R. J., and Nolan, A. C. (1973). Muscle contraction transients, cross-bridge kinetics, and the Fenn effect. *Cold Spring Harbor Symp. Quant. Biol.* **37**, 661–668.

Podolsky, R. J., and Tawada, K. (1980). A rate limiting step in muscle contraction. *In* "Muscle Contraction: Its Regulatory Mechanisms" (S. Ebashi, K. Maruyama, and M. Endo, Eds.), pp. 65–78. Japan Scientific Societies Press, Tokyo.

Podolsky, R. J., and Teichholz, L. E. (1970). The relation between calcium and contraction kinetics in skinned muscle fibres. *J. Physiol.* (*London*) **211**, 19–35.

Potter, J. D., and Gergely, J. (1975). The calcium and magnesium binding sites on troponin and their role in the regulation of myofibrillar adenosine triphosphatase. *J. Biol. Chem.* **250**, 4628–4633.

Rall, J. A. (1978). Dependence of energy output on force generation during muscle contraction. *Am. J. Physiol.* **235**, C20–C24.

Rall, J. A. (1979). Effects on temperature on tension, tension-dependent heat, and activation heat in twitches of frog skeletal muscle. *J. Physiol.* (*London*) **291**, 265–275.

Rall, J. A. (1980a). Effects of deuterium oxide on the mechanics and energetics of skeletal muscle contraction. *Am. J. Physiol.* **239**, C105–C111.

Rall, J. A. (1980b). Effects of previous activity on the energetics of activation in frog skeletal muscle. *J. Gen. Physiol.*, **75**, 617–631.

Rall, J. A. (1982a). Energetics of Ca^{+2} cycling during skeletal muscle contraction. *Fed. Proc., Fed. Am. Soc. Exp. Biol.* **41**, 155–160.

Rall, J. A. (1982b). Sense and nonsense about the Fenn effect. *Am. J. Physiol.* **242**, H1–H6.

Rall, J. A., and Schottelius, B. A. (1973). Energetics of contraction in phasic and tonic skeletal muscles of the chicken. *J. Gen. Physiol.* **62**, 303–323.

Rall, J. A., Homsher, E., Wallner, A., and Mommaerts, W. F. H. M. (1976). A temporal dissociation of energy liberation and high energy phosphate splitting during shortening in frog skeletal muscle. *J. Gen. Physiol.* **68**, 13–27.

Ramsey, R. W., and Street, S. F. (1940). The isometric length–tension diagram of isolated skeletal muscle fibers of the frog. *J. Cell. Comp. Physiol.* **15**, 11–34.

Ranatunga, K. W. (1982). Temperature-dependence of shortening velocity and rate of isometric tension development in rat skeletal muscle. *J. Physiol.* (*London*) **329**, 465–483.

Rapoport, S. I. (1972). Mechanical properties of the sarcolemma and myoplasm in frog muscle as a function of sarcomere length. *J. Gen. Physiol.* **59**, 559–585.

Rayns, D. G., Simpson, F. O., and Bertaud, W. S. (1968). Surface features of striated muscle II guinea-pig skeletal muscle. *J. Cell Sci.* **3**, 475–482.

Reedy, M. K., Holmes, K. C., and Tregear, R. T. (1965). Induced changes in orientation of the cross bridges of glycerinated insect flight muscle. *Nature* (*London*) **207**, 1276–1280.

Reisler, E. (1980). Kinetic studies with synthetic myosin minifilaments show the equivalence of actomyosin and acto-HMM ATPases. *J. Biol. Chem.* **255**, 9541–9544.

Reisler, E., Smith, C., and Seegan G. (1980). Myosin minifilaments. *J. Mol. Biol.* **143**, 129–145.

Ricchiuti, N. V., and Mommaerts, W. F. H. M. (1965). Technique for myothermic measurements. *Physiologist* **8**, 259.

Ritchie, J. M. (1954). The relation between force and velocity of shortening in rat muscle. *J. Physiol.* (*London*) **123**, 633–639.

Ritchie, J. M., and Wilkie, D. R. (1958). The dynamics of muscular contraction. *J. Physiol.* (*London*) **143**, 104–113.

Roberts, D. V. (1977). "Enzyme Kinetics." Cambridge Univ. Press, London and New York.

Robertson, S. P., Johnson, J. D., and Potter, J. D. (1981). The time course of calcium exchange with calmodulin, troponin, parvalbumin and myosin in response to transient increases in calcium. *Biophys. J.* **34**, 559–569.

Ross, B. D., Radda, G. K., Gadian, D. G., Rocker, G., Esiri, M., and Falconer-Smith, J. (1981). Examination of a case of suspected McArdle's syndrome by ^{31}P nuclear magnetic resonance. *N. Engl. J. Med.* **304**, 1338–1343.

Rüdel, R., and Zite-Ferenczy, F. (1979a). Interpretation of light diffraction by cross-striated muscle as Bragg reflexion of light by the lattice of contractile proteins. *J. Physiol.* (*London*) **290**, 317–330.

Rüdel, R., and Zite-Ferenczy, F. (1979b). Do laser diffraction studies on striated muscle indicate stepwise sarcomere shortening? *Nature* (*London*) **278**, 573–575.

Sandberg, J. A., and Carlson, F. D. (1966). The length dependence of phosphorylcreatine hydrolysis during an isometric tetanus. *Biochem. Z.* **345**, 212–231.

Sandow, A. (1965). Excitation–contraction coupling in skeletal muscle. *Pharmacol. Rev.* **17**, 265–320.

Sartorelli, L., Fromm, H. J., Benson, R. W., and Boyer, P. D. (1966). Direct and ^{18}O-exchange measurements relevant to possible activated or phosphorylated states of myosin. *Biochemistry* **5**, 2877–2884.

Scopes, R. K. (1973). Studies with a reconstituted muscle glycolytic system. The rate and extent of creatine phosphorylation by anaerobic glycolysis. *Biochem. J.* **134**, 197–208.

Scopes, R. K. (1974). Studies with a reconstituted glycolytic system. The anaerobic glycolytic response to simulated tetanic contraction. *Biochem. J.* **138**, 119–123.

Seraydarian, K., Mommaerts, W. F. H. M., and Wallner, A. (1962). The amount and compartmentation of adenosine diphosphate in muscle. *Biochim. Biophys. Acta* **65**, 443–460.

Simmons, R. M., and Jewell, B. R. (1974). Mechanics and models of muscular contraction. *Recent Adv. Physiol.* No. 9, pp. 87–114.

Sleep, J. A. (1981). Single turnovers of adenosine 5′-triphosphate by myofibrils and actomyosin subfragment 1. *Biochemistry* **20**, 5043–5051.

Sleep, J. A., and Boyer, P. D. (1978). Effect of actin concentration on the intermediate oxygen exchange of myosin. *Biochemistry* **17**, 5417–5422.

Sleep, J. A., and Hutton, R. L. (1978). Actin mediated release of ATP from myosin ATP-complex. *Biochemistry* **17**, 5423–5430.

Sleep, J. A., and Smith, S. J. (1981). Actomyosin ATPase and muscle contraction. *Curr. Top. Bioenerg.* **11**, 239–286.

Sleep, J. A., and Taylor, E. W. (1976). Intermediate states of actomyosin adenosine triphosphatase. *Biochemistry* **15**, 5813–5817.

Sleep, J. A., Hackney, D. D., and Boyer, P. D. (1978). Characterization of phosphate oxygen exchange reactions catalyzed by myosin through measurement of the distribution of ^{18}O-labeled species. *J. Biol. Chem.* **253**, 5235–5238.

Smith, D. S. (1966). The organization and function of the sarcoplasmic reticulum and T-system of muscle cells. *Prog. Biophys. Mol. Biol.* **16**, 107–142.

Smith, I. C. H. (1972a). Energetics of activation in frog and toad muscle. *J. Physiol.* (*London*) **220**, 583–599.

Smith, I. C. H. (1972b). The activation heat of striated muscle. Ph.D. thesis, University of London.

Smith, I. C. H. (1973). Analysis for instrument lag and for smoothing. *J. Phys.* **6**, 673.

SMITH, R. S., and OVALLE, W. K., Jr. (1973). Varieties of fast and slow extrafusal muscle fibres in amphibian hind limb muscles. *J. Anat.* **116**, 1–24.

SMITH, S. J., and WOLEDGE, R. C. (1982). Thermodynamic analysis of calcium binding to frog parvalbumin. *J. Muscle Res. Cell Motil.* **3**, 507.

SOLANDT, D. Y. (1936). The effect of potassium on the excitability and resting metabolism of frog's muscle. *J. Physiol.* (*London*) **86**, 162–170.

SOMLYO, A. V., GONZALEZ-SERRATOS, H., SHUMAN, H., MCCLELLAN, G., and SOMLYO, A. P. (1981). Calcium release and ionic changes in the sarcoplasmic reticulum of tetanized muscle: An Electron-Probe Study. *J. Cell Biol.* **90**, 577–594.

SPRONCK, A. C. (1965). Evolution temporelle de l'hydrolyse de la phosphocréatine et de la synthèse d'hexose diphosphate pendant et après cinq secousses simples, à 0°C, chez le sartorius de *Rana temporaria* intoxiqué par l'acide mono iodoacétique. *Arch. Int. Physiol. Biochem.* **73**, 241–259.

SPRONCK, A. C. (1970). Evolution temporelle de l'hydrolyse et de la resynthese de la phosphorylcréatine pendant et après la contraction isometrique du sartorius de *Rana temporaria* L. Ph.D. thesis, University of Liege.

SPURWAY, N. C. (1982). Histochemistry of frog myofibrillar ATPase *IRCS Med. Sci. Physiol.* **10**, 1042–1043.

SPURWAY, N. C. (1984). Quantitative histochemistry of frog skeletal muscles. *J. Physiol.* (*London*) **346**, 62P.

SQUIRE, J. (1981a). "The Structural Basis of Muscular Contraction." Plenum, New York.

SQUIRE, J. (1981b). Muscle regulation: A decade of the steric blocking model. *Nature* (*London*) **291**, 614–615.

STEIN, L. A., SCHWARZ, R. P. JR., CHOCK, P. B., and EISENBERG, E. (1979). Mechanism of actomyosin adenosine triphosphatase. Evidence that adenosine 5′-triphosphate hydrolysis can occur without dissociation of the actomyosin complex. *Biochemistry* **18**, 3895–3909.

STEIN, L. A., CHOCK, P. B., and EISENBERG, E. (1981). Mechanism of actomyosin ATPase: Effect of actin on the ATP hydrolysis step. *Proc. Natl. Acad. Sci. U.S.A.* **78**, 1346–1350.

STEPHENSON, D. G., and WILLIAMS, D. A. (1982). Effects of sarcomere length on the force–pCa relation in fast- and slow-twitch skinned muscle fibres from the rat. *J. Physiol.* (*London*) **333**, 637–653.

STREET, S. F., and RAMSEY, R. W. (1965). Sarcolemma: Transmitter of active tension in frog skeletal muscle. *Science* **149**, 1379–1380.

SUGI, H. (1972). Tension changes during and after stretch in frog muscle fibres. *J. Physiol.* (*London*) **225**, 237–253.

SZABO, Z. G. (1969). Kinetic characterization of complex reaction systems. *In* "Comprehensive Chemical Kinetics" (C. H. Beauford and C. F. H. Tipper, Eds.), Vol. II, pp. 1–80. Elsevier, Amsterdam.

TAYLOR, E. W. (1977). Transient phase of adenosine triphosphate hydrolysis of myosin, heavy meromyosin, and subfragment 1. *Biochemistry* **16**, 732–740.

TAYLOR, E. W. (1979). Mechanism of actomyosin ATPase and the problem of muscular contraction. *Crit. Rev. Biochem.* **6**, 103–164.

TAYLOR, S. R. (1974). Decreased activation in skeletal muscle fibres at short muscle length. *Ciba Found. Symp.* **24**, 93–115.

TER KEURS, H. E. D. J., LUFF, A. R., and LUFF, S. E. (1981). The relationship of force to sarcomere length and filament lengths of rat extensor digitorum longus muscle. *J. Physiol.* (*London*) **317**, 24P.

TER KEURS, H. E. D. J., LUFF, A. R., and LUFF, S. E. (1984). Force–sarcomere length relation and filament length in rat extensor digitorum muscle. *Adv. Exp. Med. Biol.* **170**, 511–522.

THOMAS, D. D., and COOKE, R. (1980). Orientation of spin-labeled myosin heads in glycerinated muscle fibers. *Biophys. J.* **32**, 891–906.

Thomas, D. D., Ishiwata, S., Seidel, J. C., and Gergely, J. (1980). Submillisecond rotational dynamics of spin-labeled myosin heads in myofibrils. *Biophys. J.* **32**, 873–890.

Tirosh, R., Liron, N., and Oplatka, A. (1979). A hydrodynamic mechanism for muscular contraction. *In* "Cross-bridge Mechanism in Muscle Contraction" (H. Sugi and G. H. Pollack, Eds.), pp. 593–610. Univ. of Tokyo Press, Tokyo.

Tonomura, Y., and Kitagawa, S. (1960). The initial phase of actomyosin–adenosine triphosphatase. II. Factors influencing activity. *Biochim. Biophys. Acta* **40**, 135–140.

Trentham, D. R., Bardsley, R. G., Eccleston, J. F., and Weeds, A. G. (1972). Elementary processes of the magnesium ion-dependent adenosine triphosphatase activity of heavy meromyosin. *Biochem. J.* **126**, 635–644.

Trentham, D. R., Eccleston, J. F., and Bagshaw, C. R. (1976). Kinetic analysis of ATPase mechanisms. *Q. Rev. Biophys.* **9**, 217–281.

Ullrick, W. C. (1967). A theory of contraction for striated muscle. *J. Theor. Biol.* **15**, 53–69.

Wagner, P. D., and Weeds, A. G. (1979). Determination of the association of myosin subfragment 1 with actin in the presence of ATP. *Biochemistry* **18**, 2260–2266.

Walsh, T. H., and Woledge, R. C. (1970). Heat production and chemical change in tortoise muscle. *J. Physiol.* (*London*) **206**, 457–469.

Wearst, R. C., ed. (1976–1977). "CRC Handbook of Chemistry and Physics," 57th Ed. CRC Press, Cleveland, Ohio.

Weber, A., and Hasselbach, W. (1954). Die Erhohung der rate der ATP-spaltung furch myusin- und aktomyosin-gele fur Geginn der Spaltung. *Biochim. Biophys. Acta* **15**, 237–245.

Weber, A., and Murray, J. M. (1973). Molecular control mechanisms in muscle contraction. *Physiol. Rev.* **53**, 612–673.

Wells, J. B. (1965). Comparison of mechanical properties between slow and fast mammalian muscles. *J. Physiol.* (*London*) **178**, 252–269.

Wendt, I. R., and Gibbs, C. L. (1973). Energy production of rat extensor digitorum longus muscle. *Am. J. Physiol.* **224**, 1081–1086.

Werber, M. M., Szent-Gyorgyi, A. G., and Fasman, G. D. (1972). Fluorescence studies on heavy meromyosin-substrate interaction. *Biochemistry* **11**, 2872–2883.

White, H., Maggs, A., and Trinick, J. (1980). Determination of the percentage of free actin subunits in rigor blowfly myofibrils by fluorescent microscopy. *Fed. Proc., Fed. Am. Soc. Exp. Biol.* **39**, 1962.

White, H. D. (1977). Magnesium-ADP binding to acto-myosin-S1 and acto-heavymeromyosin. *Biophys. J.* **17**, 40a.

White, H. D., and Taylor, E. W. (1976). Energetics and mechanism of actomyosin adenosine triphosphatase. *Biochemistry* **15**, 5818–5826.

Wilkie, D. R. (1960). Thermodynamics and interpretations of biological heat measurements. *Prog. Biophys. Biophys. Chem.* **10**, 259–298.

Wilkie, D. R. (1968). Heat, work and phosphorylcreatine break-down in muscle. *J. Physiol.* (*London*) **195**, 157–183.

Wilkie, D. R. (1974). The efficiency of muscular contraction. *J. Mechanochem. Cell Motil.* **2**, 257–267.

Wilkie, D. R. (1975). Muscle as a thermodynamic machine. *Ciba Found. Symp.* **31**, 327–355.

Winegrad, S. (1968). Intracellular calcium movements of frog skeletal muscle during recovery from tetanus. *J. Gen. Physiol.* **51**, 65–83.

Wohlish, E., and Clamann, H. G. (1931). Quantitative Untersuchungen zum Problem der thermoelastichen Eigenschaften des Skelettmuskels, VIII. Mitteilung ueber tierische Gewebe mit Faserstruktur. *Z. Biol.* **91**, 399–438.

Woledge, R. C. (1961). The thermoelastic effect of change of tension in active muscle. *J. Physiol.* (*London*) **155**, 187–208.

WOLEDGE, R. C. (1963). Heat production and energy liberation in the early part of a muscular contraction. *J. Physiol. (London)* **166**, 211–224.

WOLEDGE, R. C. (1968). The energetics of tortoise muscle. *J. Physiol. (London)* **197**, 685–707.

WOLEDGE, R. C. (1971). Heat production and chemical change in muscle. *Prog. Biophys. Mol. Biol.* **22**, 37–74.

WOLEDGE, R. C. (1973). *In vitro* calorimetric studies relating to the interpretation of muscle heat experiments. *Cold Spring Harbor Symp. Quant. Biol.* **37**, 629–634.

WOLEDGE, R. C. (1982). Is labile heat characteristic of muscle with a high parvalbumin content? Observations on the retractor capitis muscle of the terrapin *Pseudemys elegans scripta*. *J. Physiol. (London)* **324**, 21P.

WOLEDGE, R. C., and REILLY, P. (1981). Isotachophoretic analysis of metabolites in muscle extracts. *In* "Biochemical and Biological Applications of Isotachophoresis" (A. Adam and C. Scots, Eds.), Elsevier, Amsterdam.

WRAY, J. S., and HOLMES, K. C. (1981). X-ray diffraction studies of muscle. *Annu. Rev. Physiol.* **43**, 553–565.

YAMADA, K. (1970). The increase in the rate of heat production of frog's skeletal muscle caused by hypertonic solutions. *J. Physiol. (London)* **208**, 49–64.

YAMADA, K., and KOMETANI, K. (1984). Dependence of the shortening heat on sarcomere length in fibre bundles from frog semitendinosus muscles. *Adv. Exp. Med. Biol.* **170**, 853–862.

YOUNT, R. G., OJALA, D., and BABCOCK, D. (1971). Interaction of P-N-P and P-C-P analogs of adenosine triphosphate with heavy meromyosin, myosin, and actomyosin. *Biochemistry* **10**, 2490–2496.

YU, L. C., DOWBEN, R. M., and KORNACKER, K. (1970). The molecular mechanism of force generation in striated muscle. *Proc. Natl. Acad. Sci. U.S.A.* **66**, 1199–1205.

Author Index

C

D

E

F

G

I

N

O

S

Z

Subject Index

Q

R

T

U

V

W

X

Y

Z

MONOGRAPHS OF THE PHYSIOLOGICAL SOCIETY

Published by EDWARD ARNOLD

1 *H. Barcroft and H. J. C. Swan*
Sympathetic Control of Human Blood Vessels, 1953*
2 *A. C. Burton and O. G. Edholm*
Man in a Cold Environment, 1955*
3 *G. W. Harris*
Neural Control of the Pituitary Gland, 1955*
4 *A. H. James*
Physiology of Gastric Digestion, 1957*
5 *B. Delisle Burns*
The Mammalian Cerebral Cortex, 1958*
6 *G. S. Brindley*
Physiology of the Retina and Visual Pathway, 1960 (2nd edition, 1970)
7 *D. A. McDonald*
Blood Flow in Arteries, 1960*
8 *A. S. V. Burgen and N. G. Emmelin*
Physiology of the Salivary Glands, 1961
9 *Audrey U. Smith*
Biological Effects of Freezing and Supercooling, 1961*
10 *W. J. O'Connor*
Renal Function, 1962*
11 *R. A. Gregory*
Secretory Mechanisms of the Gastro-Intestinal Tract, 1962*
12 *C. A. Keele and Desiree Armstrong*
Substances Producing Pain and Itch, 1964*
13 *R. Whittam*
Transport and Diffusion in Red Blood Cells, 1964
14 *J. Grayson and D. Mendel*
Physiology of the Splanchnic Circulation, 1965*
15 *B. T. Donovan and J. J. van der Werff ten Bosch*
Physiology of Puberty, 1965*
16 *I. de Burgh Daly and Catherine Hebb*
Pulmonary and Bronchial Vascular Systems, 1966*
17 *I. C. Whitfield*
The Auditory Pathway, 1967*
18 *L. E. Mount*
The Climatic Physiology of the Pig, 1968*
19 *J. I. Hubbard, R. Llinás and D. Quastel*
Electrophysiological Analysis of Synaptic Transmission, 1969*
20 *S. E. Dicker*
Mechanisms of Urine Concentration and Dilution in Mammals, 1970*
21 *G. Kahlson and Elsa Rosengren*
Biogenesis and Physiology of Histamine, 1971*
22 *A. T. Cowie and J. S. Tindal*
The Physiology of Lactation, 1971*
23 *P. B. C. Matthews*
Mammalian Muscle Receptors and their Central Actions, 1972
24 *C. R. House*
Water Transport in Cells and Tissues, 1974
25 *P. P. Newman*
Visceral Afferent Functions of the Nervous System, 1974

Published by CAMBRIDGE UNIVERSITY PRESS

28 *M. J. Purves*
The Physiology of the Cerebral Circulation, 1972

29 *D. McK. Kerslake*
The Stress of Hot Environments, 1972

30 *M. R. Bennett*
Autonomic Neuromuscular Transmission, 1972

31 *A. G. Macdonald*
Physiological Aspects of Deep Sea Biology, 1975

32 *M. Peaker and J. L. Linzell*
Salt Glands in Birds and Reptiles, 1975

33 *J. A. Barrowman*
Physiology of the Gastro-intestinal Lymphatic System, 1978

35 *J. T. Fitzsimons*
The Physiology of Thirst and Sodium Appetite, 1979

39 *R. J. Linden and C. T. Kappagoda*
Atrial Receptors, 1982

40 *R. Elsner and B. Gooden*
Diving and Asphyxia, 1983

Published by ACADEMIC PRESS

34 *C. G. Phillips and R. Porter*
Corticospinal Neurones: Their Role in Movement, 1977

36 *O. H. Peterson*
The Electrophysiology of Gland Cells, 1980

37 *W. R. Keatinge and M. Clare Harman*
Local Mechanisms Controlling Blood Vessels, 1980

38 *H. Hensel*
Thermoreception and Temperature Regulation, 1981

41 *R. C. Woledge, N. Curtin, and E. Homsher*
Energetic Aspects of Muscle Contraction, 1985

42 *C. Bernard*
Memoir on the Pancreas (*translation from the French by J. Henderson*), 1985

Volumes marked * are now out of print